ADVANCES IN ENVIRONMENTAL ENGINEERING RESEARCH IN POLAND

Advances in Environmental Engineering Research in Poland

Edited by:

Małgorzata Pawłowska & Lucjan Pawłowski
Lublin University of Technology, Lublin, Poland

Routledge
Taylor & Francis Group

Routledge is an imprint of the Taylor & Francis Group, an informa business

Typeset by MPS Limited, Chennai, India

Library of Congress Cataloging-in-Publication Data

Applied for

Published by: Routledge
Schipholweg 107C, 2316 XC Leiden, The Netherlands
e-mail: Pub.NL@taylorandfrancis.com
www.routledge.com – www.taylorandfrancis.com

ISBN: 978-0-367-77494-3 (hbk)
ISBN: 978-1-032-05593-0 (pbk)
ISBN: 978-1-003-17166-9 (eBook)
DOI: 10.1201/9781003171669

Table of contents

Editorial

The economic and social development is inextricably connected with the changes occurring in the natural environment. Constant increase of production and standards of living requires acquisition of new mineral resources, metal ores, greater energy and water consumption as well as occupying new areas of land for plant cultivation and animal husbandry, which is evident from the statistics related to the recent years. For example, global extraction of fossil fuels and metal ores increased by a yearly average of 1.9% and 2.7%, between 1970 and 2017, respectively (IRP, 2019), global energy consumption was growing by an average of 2%/year over the period of 2000-2018), although a slowdown occurred in 2019 in relation to the previous year (it reached only 0.6%/year). The year 2020 will be unique in terms of energy consumption and demand for certain raw materials due to the coronavirus pandemic. International Energy Agency estimates that slowing down in the economic activity will cause a decrease in global energy consumption in 2020 by about 6%. Undoubtedly, this is good news for the environment. It has already been observed that due to a lower consumption in fossil fuels, the environmental pollution in Italy, Spain or USA was reduced by up to 30% (Muhammad et al. 2020). Additionally, daily global emission of CO_2 in April 2020 dropped by an average of 17% compared with the mean level noted in 2019, and the NO_2 emission in USA declined by 25.5% compared to the previous years (Le Quéré et al. 2020; Berman & Ebisu, 2020).

However, it can be expected that after a temporary standstill of economy, an intensive development will occur, accompanied with increased emissions. There is hope that the development of the renewable energy market will largely meet the increased primary energy demand, and thus the amount of emitted pollutants will not revert to the pre-pandemic state.

While using valuable materials from the environment, people produce gaseous pollutants, solid wastes and wastewater. It is estimated that globally, municipal wastes alone are produced in the amount of 2.01 billion tonnes per year, i.e. 0.74 kilogram per person per day, which will grow to 3.40 billion tonnes by 2050 (World Bank data, 2018). In turn, there is growing awareness that in long-term such process will lead to a disaster. Thus, actions aimed at improving the condition of the environment and preventing its further degradation are taken both internationally and in individual countries. Today, people are aware that the growing requirements in terms of economic standards of living must be accompanied by a technological development in the area of environmental engineering, including the actions taken to preserve the natural environment in equilibrium or restore it if has degraded.

This monograph presents selected issues connected with mitigating the negative impact on the environment in terms of several key areas of environmental engineering, namely acquisition of energy from renewable sources, management of sewage sludge accounting for phosphorus recovery, removal of pharmaceuticals from wastewater, as well as environmental aspects of producing and transporting heat to recipients.

Małgorzata Pawłowska
Lucjan Pawłowski

REFERENCES

Berman J.D., Ebisu K. 2020, Changes in U.S. air pollution during the COVID-19 pandemic. *Science of The Total Environment* 739: 139864.

Le Quéré, C., Jackson C.B., Jones M.W. et al. 2020. Temporary reduction in daily global CO_2 emissions during the COVID-19 forced confinement, *Nature Climate Change* 10: 647–653.

ENERDATA, Global Energy Statistical Yearbook, 2020, https://yearbook.enerdata.net/total-energy/world-consumption-statistics.html

IEA International Energy Agency, https://www.iea.org/data-and-statistics/charts/change-in-global-primary-energy-demand-1900-to-2020e

IRP GLOBAL RESOURCES OUTLOOK, 2019 https://www.resourcepanel.org/sites/default/files/documents/document/media/unep_252_global_resource_outlook_2019_web.pdf

Muhammad S., Long X., Salman M. 2020. COVID-19 pandemic and environmental pollution: A blessing in disguise? *Science of The Total Environment* 728: 138820.

World Bank data, 2018 https://datatopics.worldbank.org/what-a-waste/trends_in_solid_waste_management.html

About the editors

Małgorzata Pawłowska Ph.D., Sc.D. is a researcher and lecturer working at the Faculty of Environmental Engineering of Lublin University of Technology. She received her M.Sc. of the philosophy of nature and the protection of the environment at the Catholic University of Lublin in 1993. In 1999, she received Ph.D. in agrophysics at the Institute of Agrophysics of the Polish Academy of Sciences, and in 2010 she obtained a postdoctoral degree in the technical sciences in the field of environmental engineering at the Wrocław University of Technology. In 2018, she was awarded the title of professor of technical sciences.

The scientific interests of prof. Małgorzata Pawłowska focus mainly on the issues related to the reduction the greenhouse gases concentrations in the atmosphere, energy recovery of organic waste, and the possibility of using the waste from the energy sector in the reclamation of degraded land. A measurable outcome of her research is the authorship or co-authorship of 105 papers, including 40 articles in scientific journals, 4 monographs, 24 chapters in monographs, co-editor of 5 monographs, co-authorship of 15 patents and dozens of patent applications. Prof. Małgorzata Pawłowska has participated in the implementation of 9 research projects concerning, first of all, the prevention of pollutant emissions from landfills and the implementation of sustainable waste management.

In the years 2013-2019 she was the head of the Department of Alternative Fuels Engineering at the Institute of Renewable Energy Sources Engineering. Currently, she heads the Department of Biomass and Waste Conversion into Biofuels.

Lucjan Pawłowski, Ph.D., Sc.D., was born In Poland, 1946. Director of the Institute of Environmental Protection Engineering of the Lublin University of Technology, Member of the European Academy of Science and Arts, Member of the Polish Academy of Science, Deputy President of the Engineering Science Division of the Polish Academy of Science, honorary professor of China Academy of Science. He received his Ph. D. in 1976, and D. Sc. (habilitation) in 1980, both at the Wrocław University of Technology. He started research on the application of ion exchange for water and wastewater treatment. As a result, together with B. Bolto from CSIRO Australia, he has published a book "Wastewater Treatment by Ion Exchange" in which they summarized their own results and experience of the ion exchange area. In 1980 L. Pawłowski was elected President of International Committee "Chemistry for Protection of the Environment". He was Chairman of the Environmental Chemistry Division of the Polish Chemical Society from 1980-1984. In 1994 he was elected the Deputy President of the Polish Chemical Society and in the same year, the Deputy President of the Presidium Polish Academy of Science Committee "Men and Biosphere". In 1999 he was elected President of the Committee "Environmental Engineering" of the Polish Academy of Science. In 1991 he was elected the Deputy Reactor of the Lublin University of Technology, and held this post for two terms (1991-1996). He has published 19 books, over 128 papers, and authored 98 patents, and is a member of the editorial board of numerous international and national scientific and technical journals.

Renewable sources of energy

CHAPTER 1

Sustainable development and renewable sources of energy

A. Pawłowski
Faculty of Environmental Engineering, Lublin University of Technology, Poland

ABSTRACT: Sustainable development is now the core idea in the UN's and the EU's policies. It assumes changes in human behaviour in relation to other people and the environment and it calls for ending our destructive attitude to the nature and beginning of a real protection. If this change really happens, we will witness Sustainable Development Revolution for the sake of the present and future generations.Perhaps the biggest challenge relates to the energy sector. Our civilization cannot sustain itself without access to electricity and the demand for energy is still growing: it is assumed, that it will double to the year 2040, in comparison with the year 2010. The biggest challenge is, that there are different ways of electricity production with different environmental consequences. Now we are witnessing a shift from coal technologies to sustainable renewable sources of energy. Although they are not environmentally neutral, still are much better than traditional coal power plants. The paper discusses chosen problems connected with the renewable energy sector, showing positive and negative factors, that are influencing the present situation.

1.1 INTRODUCTION

Sustainable development idea was introduced by the United Nations in a report "Our Common Future" in 1987. From 1992, when the first Earth Summit in Rio de Janeiro was held, it is a central idea of international policy.

Sustainable development is a complex answer to the problems of the modern world, is development that meets the needs of the present without compromising the abilities of future generations to meet their own needs (WCED 1987). Such development has 3 pillars: ecological, social and economic. In 2015 the discussion was extended, when the United Nations introduced 17 Sustainable Development Goals. They refer to basic human needs like, access to clean water, reduction of poverty or fighting climate change, as well as to the good quality infrastructure and general conditions of human life.

The use and exploitation of the natural environment leads to a number of negative environmental impacts, both direct and indirect, reversible and irreversible, and of different scale and extent: from the impacts of local events, like smog, up to far-reaching global problems, like climate change (Ovchynnikova 2020).

Also on social level there are many challenges (Rydzewski 2019). Wars and terrorist attacks still do happen. There are very deep inequalities between nations and different social groups. There are even problems with assuring fulfilment of basic human needs, like access to water. Lack of such access is touching about 1 billion people, mainly in Africa. It is estimated that infected water is the cause of 80% of the diseases in developing countries and kills about 3 million people every year (Water Safe…, 2004).

From economic point of view, it is worth to notice, that technologically we can do almost everything now, but we don't have money for everything, so we must choose. Also, our help for poor countries is not sufficient. Despite the many UN programs intended to bridge the gap between these two groups of countries, the gap is still growing (Greig et al. 2007).

So, problems do exist, however they were recognized and let's hope that international cooperation in introduction of the sustainable development will finally be much more successful (Sztumski 2019).

DOI 10.1201/9781003171669-1

Table 1.1. Key stages in mankind's development (Pawlowski 2011).

Name of stage	Time period referred to
Hunter-gatherer era	Upper Palaeolithic
The Agricultural Revolution	Starting c. 9000 years ago in Asia, some 4000 years later in Europe.
The Scientific Revolution	1543 – symbolic beginning marked by the publication of Copernicus's "On the Revolutions of the Heavenly Spheres".
The Industrial Revolution	1687 – development, with the appearance of Newton's "Principia Mathematica". 1769 – symbolic milestone: Watt's improvements to the steam engine. Further stage (1860–1914) starting with the use of oil (in the internal combustion engine) and electricity.
The Sustainable Development Revolution	Watershed dates: 1969 – U'Thant's Report. 1987 – definition of sustainable development adopted by the UN. 1992 – UN Earth Summit in Rio de Janeiro. 2002 – UN Earth Summit in Johannesburg. 2002 – UN Earth Summit in Rio de Janeiro. 2015 – introduction of the Sustainable Development Goals.

I think that the current debate on sustainable development – if successful – does deserve to be called a revolution, like there were agricultural, scientific and industrial revolutions in the past – see Table 1.1 in its own right. While it is true that this kind of development has not been introduced yet, many recent international initiatives are heading in that direction. Man's current influence on the biosphere is without doubt global, and hence it requires a worldwide sustainable response. The idea of sustainable development emphasises the necessity for basic change in mankind's behaviour in relation to the environment as a toll to stop further destruction and thus save the planet. It is trying to cure the causes not only the consequences of the present problems.

1.2 RENEWABLES AND SUSTAINABILITY

Energy is not highlighted as basic human need in classical approaches, like Abraham Maslow's model. He pointed out at five groups of needs, forming a hierarchical arrangement, which means that higher-order needs can only be met after lower-order needs have also been met. The division into these groups, starting from the most elementary ones, is as follows (Maslow 1954):

- physiological needs (e.g. the need to eat, to drink, to avoid pain),
- safety needs (e.g. the need for care, support, peace),
- needs of love and belonging (e.g. the need for relationship, the need to be loved),
- needs for prestige and esteem (e.g. the need to confirm one's own self-respect),
- needs for self-realisation (the need to have goals, strive to develop one's own capabilities).

Apart from the above hierarchy, Maslow also pointed at cognitive needs (need for knowledge, understanding) and aesthetic needs (need for harmony and beauty).

The present human civilization however cannot sustain itself without energy. If we don't perceive it as a basic need, it is so, because energy is almost everywhere. As for now, almost 90% of people has access to electricity (UNDP 2017). However, even few days without electricity could be fatal.

I am convinced, that renewable energy market is one of the main pillars of Sustainable Development Revolution. However, there are different ways of electricity production with different environmental consequences, and among energy carriers fossil fuels still count.

As for now still the most popular energy carrier is coal. It is perceived as the worst, since its combustion is responsible for large emissions of carbon dioxide, which is the most important gas contributing to the global warming. The enhanced greenhouse effect is perhaps the biggest threat to the whole biosphere. Excessive emissions of CO_2 and other greenhouse gases (methane, freons, ozone and nitrogen oxides) may melt the polar caps, cause flooding of the vast coastal zones of seas and oceans, or increase the area affected by climate anomalies. Other problems include a decline in agricultural production (due to the increased population of the wintering pests) as well as certain diseases (fungal, bacterial, and viral), and seed germination problems caused by the rising temperature. Of course, with present technology it is possible to create – for example – seeds that can resist extreme weather conditions, but they are not for free. Such seeds are patented and controlled by the global companies like Monsanto and unavailable for poor communities (Klein 2014). But first of all, the total cost of growing number of serious weather anomalies connected with global climate change is already so high, that it is perceived as one of the factors, that may even destroy the global economy.

The list of countries with the highest emissions of carbon dioxide are presented in Table 1.2. Unfortunately, in case of most of these countries the emissions are still growing.

So, there are two important reasons, why energy issues are at the core of Sustainable Development Revolution.

First of all, we need energy in our everyday life, and for most of people it is impossible to imagine the World without electricity.

Secondly, because of the huge pollution and destruction of the atmosphere from burning of the fossil fuels, we need the shift from fossil fuels to renewable and clean sources of energy, to save our planet.

The already mentioned 17 Sustainable Development Goals are supporting renewables, which is directly indicated as Goal no 7: Affordable and Clean Energy. It is not only about better energy efficiency but also about using renewables as much as possible. Many other goals are interconnected with this, like:

- Goal no 9: Industry, Innovation and Infrastructure – production of renewable installations is an industrial issue, to make new devices better we need innovations and of course also development of infrastructure is important: building of new renewable plants, connections between them and connection to the electrical network.
- Goal no 8: Decent work and Economic Growth – Clean Renewable Energy is a huge part of the market, which means it is giving a lot of work for millions of people. According to the UNDP, the renewable energy sector in 2017 gave employment to as much as 10.3 million people.

Also, in the case of European Union renewables are in favour. First program was called 3 x 20 to the year 2020, and its basic principles were as follows (EC 2009):

- 20 % better energy efficiency,
- 20 % less of emission of the greenhouse gases,
- 20 % of energy of the member states from renewables,

Table 1.2. Top 10 countries with the highest emission of carbon dioxide in 2017 (Wikipedia 2020).

No	Country	Emission of CO_2 in Mt per year	Trend in emission level
1.	China	10,877.218	▲
2.	USA	5,107.393	▼
3.	European Union	3,548.345	▼
4.	India	2,454.774	▲
5.	Russia	1,764.866	▼
6.	Japan	1,320.776	▲
7.	Germany	796.529	▼
8.	South Korea	673.324	▲
9.	Iran	671.450	▲
10.	Saudi Arabia	638.762	▲

Next new horizon was established, covering the time to the year 2030, with new goals (EC 2018):

- further improvement in energy efficiency,
- further reduction of the greenhouse gases emissions (40% to the year 2030, in relation to the level of emissions from 1990),
- 27% of energy of the member states from renewables.

Even more may be achieved.

In case of the European Union, according to the report from German Space Agency, till 2050 as much as 96% of energy may be produced from renewable sources (Teske 2012).

In 2011 Energy Research Institute of University of Melbourne, together with Beyond Zero Emissions organization, announced, during the next decade Australia may have 100% electricity demand covered by renewables, 60% supposed to be delivered by the solar power and 40% by wind power (Klein 2014; Wright & Hearps 2011).

In 2013 American National Oceanic and Atmospheric Administration presented interesting research, according to which by 2030 as much as 70% of electricity in the USA could be delivered by renewables (mostly from wind and sun) and it is going to be profitable (Klein 2014; MacDonald & Clack 2014).

It is also wort to mention, that even now there are countries that almost in 100% are meeting their energy demands using renewables. Among them is Norway, the country, where near all the electricity is from hydropower (Salay 1997). Other countries may follow this path. Let us mention Germany, where by the year 2050 as much as 100% of electricity is going to be produced from renewables (von Weizsäcker & Wijkman 2018).

What is the present situation of renewable energy sources market?

Renewables account for 29,35 % installed of World's electricity capacity (IEA 2020). Table 1.3 presents share of different renewables in the global market.

As for now, in most regions 4 sources of renewable energy dominate: hydropower, wind energy, bioenergy and solar energy.

In the case of biomass and biogas burning, as well as while using geothermal energy sources, emissions to the environment are generated (mainly NO_2 and SO_2 for biomass and biogas, and $CO_2$and hydrogen sulphide for geothermal power), but their level is much lower than in the case of fossil fuels (Boyle 1996; Bułkowska et al. 2016; Cao & Pawłowski 2011).

For other renewable energy sources (RES), emissions are close to zero, although production of the equipment used to generate RES means using metals and technologies that are not environmentally neutral (Velkin & Shcheklein 2017).

Table 1.4 presents growth in renewable energy production in the last 3 years.

So, the most popular worldwide renewable source of energy is hydropower with 4 049 TWh of electricity produced annually and 69% share in the market. However, the biggest annual growth for renewable source of energy is for solar and wind energy – respectively 16 and 12%.

There are many differences between different regions of the World – see Table 1.5.

Apart of hydropower, in Africa also wind energy and solar energy are popular, in Asia all renewables are in use, even generally less popular geothermal energy and in Europe and North America wind, solar and bio-energy. It is worth to take a closer look at the situation in particular countries of the European Union, due to ambitious energy policy of the Community – see Table 1.6.

Table 1.3. Renewable electricity generation by energy use (IRENA 2018).

No	Source of energy	Electricity generated [TWh]	Share in the market [%]
1.	Hydropower	4 049	69%
2.	Wind energy	958	16%
3.	Bioenergy*	467	8%
4.	Solar energy	83	6%
5.	Geothermal energy	1	1%

*Bioenergy includes i.a. solid biofuels (like biomass), liquid biofuels and biogas

Table 1.4. Annual growth for renewable electricity generation by source, 2018–2020 (EIA 2020).

	Year		
Source	2018	2019	2020
Hydrpower	4%	2%	1%
Wind	11%	12%	12%
Bionergy	7%	8%	3%
Solar	33%	22%	16%
Other	4%	7%	3%

Table 1.5. Renewable energy generation by region in 2019 in GWh (IRENA 2020).

Region	Hydropower	Wind energy	Solar energy	Bioenergy	Geothermal
Africa	134 043	14 117	8 677	3 233	5 005
Asia	1 772 409	439 991	293 102	164 901	26 333
Central America	29 160	5 838	2 625	6 066	3 969
Eurasia	268 215	20 195	8 322	2 884	7 857
Europe	613 125	383 587	131 753	191 140	12 668
Middle East	18 721	1 337	6 163	384	-
North America	731 681	320 560	90 349	81 174	24 148
Oceania	44 307	17 285	10 262	4 204	8 215
South America	655 424	60 003	10 781	68 565	214

Table 1.6. Share of energy from renewable sources in the EU – % of gross final energy consumption (Eurostat 2018).

Country	2005	2015	2018	2020 target
EU-27	10.2	17.9	18.9	20
Austria	24.4	33.5	33.4	34
Belgium	2.3	8.0	9.4	13
Bulgaria	9.2	18.3	20.5	16
Croatia	23.7	29.0	28.0	20
Cyprus	3.1	9.9	13.9	13
Czech Republic	7.1	15.1	15.1	13
Denmark	16.0	30.9	36.1	30
Estonia	17.4	28.2	30.0	25
Finland	28.8	39.3	41.2	38
France	9.6	15.0	16.6	23
Germany	7.2	14.9	16.5	18
Greece	7.3	15.7	18.0	18
Hungary	6.9	14.5	12.5	13
Italy	7.5	17.5	17.8	17
Ireland	2.8	9.1	11.1	16
Latvia	32.3	37.5	40.3	40
Lithuania	16.8	25.8	24.4	23
Luxembourg	1.4	5.0	9.1	11
Malta	0.1	5.1	8.0	10
Netherlands	2.5	5.7	7.4	14
Poland	6.9	11.7	11.3	15
Portugal	19.5	30.5	30.3	31
Romania	17.6	24.8	23.9	24
Slovakia	6.4	12.9	11.9	14
Slovenia	16.0	21.9	21.1	25
Spain	8.4	16.2	17.4	20
Sweden	40.7	53.0	54.6	49

The situation turns out to be very diverse (Stoenoiu 2018) and there are different targets to be achieved for particular countries. The best situation is in Sweden (54.6% energy from renewables), Finland (41.2% energy from renewables) and Latvia (40.3% energy from renewables). The lowest amounts energy from renewables occur in The Netherlands (7.4%), Malta (8%) and Luxembourg (9.1%). It is worth to notice, that most of the European countries meet the 2020 target, however there are exceptions. It looks like the goals will not be met in Belgium, France, Germany, Greece, Ireland, Luxembourg, Malta, The Netherlands, Poland, Slovakia, Slovenia and Spain. Fortunately, the deficiencies will be quite small, around 1% on average.

To sum up, renewable has the future. As for now, from the world's perspective. Let's take a closer look at the challenges connected with using of such sources of energy.

1.3 HYDROPOWER

The main task of hydropower plants is to produce electricity. There is no need to produce steam as in coal-fired and nuclear power plants – hydropower itself drives the turbine that supplies energy to the power generator (Pawłowska 2011).

From the technical point of view, there are four basic types of hydropower plants (Kucowski et al. 1993):

- run-of-the-river plants (without a dam),
- conventional plants (with a dam and reservoir),
- pumped storage plants (reservoir is located above the plant; when the demand for energy is lower, for instance at night, the plant will pump water back to the reservoir to replenish the storage of water),
- tidal power plants: using the power of tides to drive the turbine in both directions.

Also, the OTEC plants (ocean thermal energy conversion plants) are an interesting solution. They are located in oceans near the equator, in the tropics, in Hawaii, Japan, or Indonesia. In this zone, the temperature of water is about 30°C at the surface, but only 7°C at the depth of 300–500 m. This difference is used as a source of heat converted to electricity.

10 biggest hydropower stations are presented in Table 1.7.

The largest hydropower plant in the world is in China – the Three Gorges Dam on the Yangtze River. The plant was opened on 24 June 2003, but the full capacity of its storage reservoir was achieved in October 2010 (39.3 billion cubic meters, the reservoir stretches 700 km along the Mudong River). In 2008 it generated 80.8 TWh, and together with Gezhouba Dam the complex generated 97.9 TWh (Wikipedia 2018).

Table 1.7. The biggest hydropower plants in the World (Irena 2020; Wikipedia 2020).

No	Name/Country	River	Installed capacity [MW]	Annual production [TWh]	Year of completion
1.	Three Gorges Dam, China	Yangtze	22 500	98.8	2012
2.	Itaipu Dam, Brazil/ Paraguay	Parana	14 000	103	1991, 2003
3.	Xiluodu Dam, China	Jinsha	13 860	55.2	2014
4.	Belo Monte Dam, Brazil	Xingu	11 233	39.5	2019
5.	Guri Dam, Venezuela	Caroni	10 235	53.41	1978, 1986
6.	Tucurui Dam, Brazil	Tocantins	8 370	41.43	1984, 2007
7.	Grand Coulee Dam, USA	Columbia	6 809	20	1950, 1991
8.	Xiangjiaba Dam, China	Jinsha	6 448	30.7	2014
9.	Longtan Dam, China	Hongshui	6 426	18.7	2009
10.	Sayano-Shushenskaya, Russia	Yenisei	6 400	26.8	1989, 2014

In the Top 10 we have 3 other dams from China. Taking into account the full list of biggest hydropower plants in the World (71 facilities with installed capacity above 2000 MW), 31% of them are in China.

Second World's biggest hydropower plant is the Itaipu dam on the Parana river on the Brazil – Paraguay border, which capacity is 103 TWh (Power Technology 2020).

In many other countries there are no such powerful rivers, but even small hydropower plants (below 10 or 5 MW) offer a number of benefits such as the following:

- electricity is produced without causing any pollution,
- protection by dams against flooding is provided,
- it is stable source of energy in comparison with some other renewable sources of energy, like wind or sun.

There are also a number of environmental impacts that give rise to serious reservations (Gicquel & Gicquel 2013):

- change of the local environmental conditions (especially hydrological), the landscape, or even the climate, which is an obstacle in the case of projects located near natural areas, like national parks, protected by law,
- obstacles to the natural migrations of fish,
- necessary relocation of inhabitants of human settlements that will be covered with water (in the case of the Three Gorges Dam on the Yangtze River, as many as 1.4 million people had to be relocated),
- problems keeping the optimum purity of water in storage reservoirs.

Therefore, the proper location of the dam is extremely important. There is another problem however: in case of extreme weather conditions and floods, as the consequence of the climate change, the amount of water in the river may be so big, that the dam may collapse. In case of the mentioned Three Gorges Dam it almost happen in 2020 (Trivedi 2020).

1.4 WIND ENERGY

Wind turbines are sometimes installed as standalone units to produce electricity for a single household. However, they are usually grouped in wind farms including tens or even hundreds of turbines. It means that they require large areas and by this are changing the landscape (Pawłowska 2011).

World's largest onshore wind power stations are presented in Table 1.8. Most of them are new and built in China. Gansu Wind Farm is especially impressive, since the peak capacity of this plant supposed to be 20 000 MW, so more than the total capacity of all other wind power plants from Top 10.

Table 1.8. World's largest onshore wind power stations (Power Technology 2020; Wikipedia 2020).

No.	Name/Country	Current Capacity [MW]	Year of Completion
1.	Gansu Wind Farm (Jiuquan Wind Power Base), China	6 800*	2020
2.	Zhang Jiakou, China	3000	2020
3.	Urat Zhongqi, Bayannur City, China	2100	2020
4.	Hami Wind Farm, China	2000	2013
5.	Damao Qi, Baotou City, China	1600	2013
	Jaisalmer Wind Park, India	1600	2020
7.	Alta (Oak Creek-Mojave), USA	1548	2011
8.	Muppandal Wind farm, India	1500	2001
9.	Hongshagang, Town, Minqin County, China	1000	2013
10.	Kailu, Tongliao, China	1000	2013

*Further 13 2000 MW under construction

From technical point of view there are at least two basic kinds of wind turbines.

- turbines with a horizontal axis of rotation,
- turbines with a vertical axis of rotation.

One of the most important challenges connected with wind energy is the height of the turbine. Wind speed increases with the height, so the higher is the turbine the more energy may be produced. However, building higher turbines is much more expensive. But there may be a solution to this problem, which is called BAT – Altaeros Buoyan Airborne Turbine. In this case, the turbine is installed inside of the balloon, which can fly to around 300 m above the ground level (Deodhar et al. 2015).

Another important challenge is posed by the high variability of wind conditions – both during the day/night and throughout the seasons. As a result, variations in the actual capacity of the wind farm cannot be avoided.

This problem may be eliminated by using more effective methods for the storage of electricity produced in periods when winds are stronger. Interesting solution is to generate hydrogen to carry energy. This 'carrier' could be used when winds are very weak or when there are is no wind at all.

And again, as it was in case of the solar plants, production of wind energy is not connected with any pollution of the environment, however to produce such turbines we need many rare metals, like neodymium. Mining them is responsible for quite big pollution (Helder 2019), and this problem is very rarely considered.

1.5 BIOENERGY

In case of bioenergy is it energy derived from biological sources and processes. Two most popular sources of that energy are biomass and biofuels.

1.5.1 *Biomass*

Biomass materials include straw, peat, wood waste (from agriculture, forestry, or paper industry), as well as timber from special plantations (Bułkowska et al. 2016; Pawłowska 2011). In Europe, Osier Willow (*salix viminalis*) is a popular source of biomass. Other sources include poplar (*populus*), Black Locust (*robinia pseudacacia*) and rose (*rosa multiflora*).

The characteristic features of these species include (Mackow et al. 1993):

- fast growth, even 10 times faster than the natural growth of biomass in an ordinary forest,
- high calorific value: two tonnes of dry timber equal one tonne of coal,
- no special requirements as regards soil and climate conditions – fallow land, land set aside, or even polluted or flooded land can be used,
- no special requirements as to temperature and climate – resistant to low temperatures,
- high resistance to pests and diseases,
- extended use of plantation (in the case of willows, crops are harvested every 2–3 years and the plantation may be used for up to 25 years).

The list of the biggest biomass power stations is presented in Table 1.9.

Although there are no installations from China in Top 10, this is the country in which in 2019 half of the World's new capacity was installed – 3300 MW.

However, there are some problems with biomass (Pawłowska et al. 2017):

- The dynamic development of agricultural crops for biofuels in the European Union countries caused in the years 2007–2010 a dramatic 2.5-fold increase in the food prices index (according to the United Nations Food and Agriculture Organization).
- In addition, the use of biofuels for transport caused, that especially in developing countries, the tropical forests are vanishing more rapidly than before and replaced with biofuel plants. It has negative effect on the climate, since forests play important role in stabilization of the climate.

Table 1.9. World's biggest biomass power stations (Irena 2020; Wikipedia 2020).

No	Name/Country	Capacity [MW]	Year of Completion
1.	Drax, North Yorkshire, UK	2 595	1986
2.	Alholmens Kraft, Alholmen, Finland	265	2001
3.	Maasvlakte, Rotterdam, The Netherlands	220	2013
4.	Połaniec, Poland	205	2012
	Atikokan Generating Station, Ontario, Canada	205	2003
6.	Rodenhuize, Belgium	180	2010
7.	Kymijärvi, Finland	160	2012
8.	Ashdown Paper Mill, USA	157	1968
9.	Wisapower, Finland	150	2004
10.	Vaasa, Finland	140	1982

What's more, converting natural ecosystems to produce energy plants in itself is responsible for huge emissions of CO_2, even 400 times more than using such source of energy will save a year!

- The problem is also with transporting biomass over longer distances, even between the continents. In case of such practices energy efficiency is dramatically falling!

So, what about the future of biomass? The problems do exist, but we are aware of them and all of them are possible to overcome. The most important issue is that biomass is everywhere. And if we compare the available potential with the present use, we will find out that only 2/5 of biomass resources are in use (Vlosky & Smithhart 2011), so this source of energy has the future.

1.5.2 *Biogas*

Biogas is produced as a result of fermentation in municipal landfills, in wastewater treatment plants, or in purpose-built biogas plants (Bułkowska et al. 2016; Pawłowska 1999).

Biogas plants installations include digester and gas holder (in a chamber, if need in separate chambers). The gas is produced in a digester in a process of anaerobic digestion (Kumar 2017).

The anaerobic digestion process takes place in the digester and the produced biogas will be collected in the gas holder (Kumar 2017). Biogas consists mainly of methane and CO_2. Methane is particularly important, as it can produce both electricity and heat when burned (Pawłowski 2011).

Biogas burning generates a small quantity of pollution in the form of an additional emission of CO_2, but it reduces the emission of methane – a significant greenhouse gas, which heat absorption coefficient is much higher than in case of CO_2.

Important advantage of using biogas is that the installation is not expensive.

In the USA in 2019 biogas from landfills, wastewater treatment and animal waste plants all together made possible to generate 11.7 billion KWh of electricity (EIA 2020). Carbon Cycle Energy (C2e), power plant in Colorado has the capacity of 290 000 MWh alone (Kumar 2019).

In China the capacity of biogas plants is 350 MW (Scarlet et al. 2018).

However, the world leader in production of electricity from biogas is the European Union, where 10GW of energy is generated from this source annually, 2/3 of the total world production (Scarlet et al. 2018). The most of the installations are in Germany, Italy, France and Czech Republic.

The World's biggest single biogas plant is in Korsko in Denmark, where 1 million tonnes of food waste and agricultural products are processed annually, which translates into 45.4 MW of electricity (Kumar 2017).

It is also worth to mention the foundation of European Biogas Association (EBA) in 2009 in Brussels. It is non-profit organization that is promoting biogas production and use in Europe. In 2019 they had 100 members from 28 countries (EBA 2019).

Even if amounts of energy produced from bioenergy are still smaller, than in case of water or wind, this market is developing dynamically and has huge potential for the future.

1.6 SOLAR ENERGY

Solar energy may be used for production of electricity or heat. Every year 7,500 times more of solar energy reaches the Earth (86,000 TW), in relation to the primary energy consumed by the entire human civilization!

There are two basic methods (Fanhi & Fanchi Ch 2013; Kucowski et al. 1993):

- Heliothermic method – conversion of light energy into heat used to drive the turbine and power generating unit. Solar radiation is absorbed by the solar collectors. Collectors may be either flat-plate (heating up to about 90°C) or concentrating (parabolic, trough or dish – reaching 750°C in the case of dish-type collectors).
- Helioelectric method – direct conversion of solar energy into electricity. From the technical point of view, this method is based on photocells (made of silica, gallium arsenide, or cadmium sulphide).

Solar collectors should be used in areas with rather sunny weather throughout the year.

Photocells offer more flexibility – they can convert both direct radiation and dispersed energy (i.e. they can work even on cloudy days).

In most European countries insolation is rather moderate and subject to high variations. Therefore, solar energy is used as a secondary source of energy in water heating systems or in heating systems in buildings. Space-based solar power plants (SBSP) would solve this problem. However, the transmission of electricity back to Earth is a technical barrier.

In another interesting project, the European Union is planning on building of a 400-billion-Euro solar power plant on the Sahara desert, which could cover about 20% of the EU's energy demand (Matlack 2010; Meinhold 2009).

The World's biggest PV stations are presented in Table 1.10.

The situation is very dynamic, since 7 out of 10 presented installations were built in last two years. Till 2019 the biggest PV plant was on the Tenger desert in Zhongwei, Ningxia, China (Power Technology 2018). It was opened in 2016. Four years later it is only the 5th biggest installation.

Now four out of ten biggest PV power stations in the World are in India. The biggest of them (number one on the list) is Bhadla Solar Park with capacity of 2245 MW. It was built in Bhadla village in Jodhpur district of Rajasthan. It covers an area of 5,783 ha. This region is famous of the possible best solar irradiation conditions: 5.72KWh/m^2/day (NS Energy 2020)

Even more ambitious project is under construction in the United Arab Emirates. It is called Mohammed bin Rashid Al Maktoum Solar Park, after Vice President and Prime Minister of this country. Now it has the capacity of 1547 MW, which makes it 7th biggest PV plant in the world. However, the total capacity of the entire project is planned to reach 5,000 MW in 2030 (Power Technology 2018).

Table 1.10. World's largest PV power stations (Irena 2020, Wikipedia 2020).

No	Name/Country	Capacity [MW]	Year of completion
1.	Bhadla Solar Park, India	2245	2020
2.	Huanghe Hydropower Golmud Solar Park, China	2200	2020
3.	Pavagada Solar Park, India	2050	2019
4.	Benban Solar Park, Egypt	1650	2019
5.	Tengger Desert Solar Park, China	1547	2016
6.	Noor Abu Dhabi, United Arab Emirates	1177	2019
7.	Mohammed bin Rashid Al Maktoum Solar Park, United Arab Emirates	1013*	2020
8.	Kurnool Ultra Mega Solar Park, India	1000	2017
9.	Datong Solar Power Top Runner Base, China	1000**	2016
10.	NP Kunta, India	900***	2020

*will be 5000 MW, **will be 3000 MW, ***will be 1500 MW

Table 1.11. World's largest geothermal power plants (Statista 2020).

No	Name/Country	Capacity [MW]	Year of completion
1.	The Geysers, USA	1 520	1960
2.	Cerro Prieto, Mexico	820	1973
3.	Lardello, Italy	769	1913
4.	Olkaria, Kenya	727	1981, 2015
5.	Imperial Valley, USA	503	1982
6.	Sarulla, Indonesia	330	2017
	Tiwi, New Zealand	330	1972
8.	Hellisheidi, Iceland	302	2006
9.	Coso, USA	270	1987
10.	Darjat, Indonesia	255	1988

In case of the solar plants of course there is no emission of any pollution, however producing solar collectors or photocells is polluting the environment and they are hazardous waste after dismantling. This side of this technology will need further attention in the future.

1.7 GEOTHERMAL ENERGY

Geothermal projects use the natural heat of our planet – mostly in underground waters, but also energy stored in the Earth – in underground waters, water vapour and hot rocks (Cholewa & Siuta-Olcha 2010). Liquid magma is the main source of this heat. Natural decomposition of radioactive elements is an additional factor.

The technological point is to create an artificial cycle: boreholes are drilled to collect hot water, this water is used, and then pumped back to the Earth for re-heating.

The biggest geothermal plants are presented in Table 1.11.

The largest geothermal power producer is located in the USA – The Geysers (north of San Francisco) with 15 plants and a capacity of 1520 MW, while Iceland (World Atlas 2019) has the highest share of geothermal energy in a country's total energy balance – 87% of heat and 24% of electricity (near all the rest of electricity, 75,4%, is from hydropower). In Top 10 we can find also Lardello geothermal plant from Italy, which is the oldest such facility in the World, build in 1913 (Unwin 2019).

Geothermal waters are gaining popularity as the source of energy used to produce both electricity and heat. The use of this RES depends to a large extent on the capacity of the source and temperature of water. Electricity can be produced at a temperature of 150–200°C (Pawłowska 2011). In many countries the temperature is lower, so only production of heat is possible. Still it counts.

1.8 CONCLUSIONS

The sustainable development concept anticipates major civilisational change on the ecological, social and economic levels. The tremendous scope of these changes is leading to a new order as revolutionary as the aforementioned ones representing breakthroughs in human history and conventionally dubbed Revolutions.

The objective assessment of the present development of humankind is made difficult by the fact that many negative processes are still ongoing. The sector that mostly pollutes the environment is energy sector, especially connected with power plants burning coal. But a change is coming, Sustainable Development Revolution has begun.

Every year the amount of energy produced from coal is shrinking, and the amounts of energy produced from renewable sources of energy is growing. Strong political support for renewables from the UN and the EU means, that sooner or later they will become the main source of energy on the Earth. There is no doubt, that renewable energy market is one of the main pillars of sustainability. We must be aware however, that not in all aspects renewables are environmentally neutral. For most of the renewables emissions are close to zero, but manufacturing the installations that can capture such energy is polluting the environment. But technological progress enables to produce devices, which are not only more efficient, but also better for the environment.

We must remember however, that for the real introduction of sustainable development, even the best technology is not enough. Sustainable Development Revolution requires not only care about the environment, but also about other people, so i.a. reducing inequality and poverty. It is possible to achieve only when we all will cooperate together: scientist from different countries, and different specialties with different societies, but also ordinary people. For the better common future.

REFERENCES

Boyle, G. 1996. *Renewable Energy: Power for Sustainable Future.* Oxford: The Open University and Oxford University Press.

Bułkowska, K., Gusiatyn, Z.M., Klimiuk, E., Pawłowski, A., Pokój, T. 2016. *Biomass for Biofuels*. Boca Raton, London, New York, Leiden: CRC Press, Taylor & Francis Group, A Balkema Book.

Cao, Y., Pawłowski, A. 2013. Biomass as an Answer to Sustainable Energy: opportunity versus Challenge, in *Environment Protection Engineering* 39 (1): 153–161.

Cholewa, T., Siuta-Olcha, A. 2010. *Energetyka – dziś i jutro.* Monografie Komitetu Inżynierii Środowiska PAN, vol. 67, KIŚ: Lublin.

Deodhar, N., Vermillion, C., Tkacik P. 2015. A case study in experimentally-infused plant and controller optimization for airborne wind energy systems. *IEEE Xplore:* July 30.

EBA, 2019. *European Biogas Association.* http://european-biogas.eu/ [1.06.2019].

EC (European Comission), 2009. *2020 climate & energy package.* Brussels.

EC (European Comission), 2018. 2030 *climate & energy framework*. Brussels.

EIA. 2020. *Landfill gas and biogas.* https://www.eia.gov/energyexplained/biomass/landfill-gas-and-biogas.php [1.11.2020].

Eubia. 2020. *Biogas.* https://www.eubia.org/cms/wiki-biomass/bioenergy-drivers/biogas/ [1.11.2020].

Eurostat. 2018. *Share of energy from renewable sources in the EU – % of gross final energy consumption.* https://ec.europa.eu/eurostat/statistics-explained/index.php?title=File:Share of energy_from_renewable_sources,_2004-2018_(%25_of_gross_final_energy_consumption).png [1.11.2020].

Fanchi, J.R., Fanchi C.J. 2013. *Energy in the 21st century.* World Scientific: Singapore.

IEA. 2005, 2019. *International Energy Annual.* Washington D.C.

IEA. 2020. *Global energy review 2020 – renewables.* https://www.iea.org/reports/global-energy-review-2020/renewables [1.11.2020].

IEA. 2020. *Annual growth for renewable electricity generation by source, 2018–2020.* https://www.iea.org/data-and-statistics/charts/annual-growth-for-renewable-electricity-generation-by-source-2018–2020 [1.11.2020].

Irena. 2020. *Irena Renewable Energy Statistics 2020.* https://www.irena.org/-/media/Files/IRENA/Agency/Publication/2020/Jul/IRENA_Renewable_Energy_Statistics_2020.pdf [1.11.2020].

Gicquel, R., Gicquel M. 2012. *Introduction to global energy issues.* CRC Press, Taylor & Francis Group, A Balkema Book: Boca Raton, London, New York, Leiden.

Greig, A., Hulme, D. & Turner, M. 2007. *Challenging Global Inequality. Development Theory and Practice in the 21st Century.* Basingstoke Hants: Palgrave Macimillan.

Klein, N., 2014. *This changes everything.* Simon & Schuster: New York.

Kumar, P. 2017. *Largest Biogas Plants.* https://www.nsenergybusiness.com/news/newslargest-biogas-plants-061017-5943061/# [1.11.2020].

Kucowski, J., Laudyn, D. & Przekwas, M. 1993. *Energetyka a ochrona srodowiska.* Warsaw: Wydawnictwo Naukowo-Technczne.

MacDonald, A., Clack Ch., 2014. *Low Cost and Low Carbon Emission Wind and Solar Energy Systems Are Feasible for Large Geographic Domains*. Earth System Research Laboratory. National Oceanic and Atmospheric Administration. Silver Spring, Maryland.

Maćkow, J., Paczosa, A. & Skirmuntt, G. 2004. *Eko-generacja przyszłosci.* Katowice, Warsaw: WNS.

Maslow, S. H. 1954. *Motivation and personality.* Harper & Brothers: New York.

Matlack, C. 2011. *A Consortium Wants to Invest $ 560 Billion in Sahara Solar Panels*. http://www.businessweek.com/magazine/content/10_38/b4195012469892.htm [1.03.2011].

Meinhold, B. 2011. *World's Largest Solar Project Planned for Saharan Desert.* http://inhabitat.com/worlds-largest-solar-project-sahara-desert/ [1.03.2011].

NS Energy. 2020. *Bhadla Solar Park Rajasthan.* https://www.nsenergybusiness.com/projects/bhadla-solar-park-rajasthan/ [1.11.2020].

Ovchynnikova O. 2020. Risk and Uncertainty in Sustainable Development: Undertaking Politics of the Climate Change in the United States. *Problemy Ekorozwoju/ Problems of Sustainable Development* 15(1): 229–235.

Pawłowski, A., 2011. *Sustainable Development as a Civilizational Revolution. Multidimensional Approach to the Challenges of the 21st century.* Boca Raton, Londyn, Nowy Jork, Leiden: CRC Press, Taylor & Francis Group.

Pawłowski A., Pawłowska M., Pawłowski L., 2017. Mitigation of Greenhouse Gases Emissions by Management of Terrestrial Ecosystem. *Ecological Chemistry and Engineering S* 24(2): 213–222.

Power Technology, 2018, *The world's biggest solar power plants.* https://www.power-technology.com/features/the-worlds-biggest-solar-power-plants/ [1.06.2019].

Power Technology. 2019. *Top 10 biggest wind farms.* https://www.power-technology.com/features/feature-biggest-wind-farms-in-the-world-texas/ [1.11.2020].

Power Technology. 2020. *The Itaipu Hydroelectric Dam Project, Brazil.* https://www.power-technology.com/projects/itaipu-hydroelectric/ [1.11.2020].

Rydzewski, P. 2019. Social Dimensions of Sustainable Development in International Public Opinion. *Problemy Ekorozwoju/ Problems of Sustainable Development* 14(1): 53–62.

Salay, J. 1997. Energy, *From Fossil Fuels to Sustainable Energy Resources.* Uppsala: The Baltic University Programme.

Scarlet, N., Dallemand, J.-F., Fahl, F. 2018. Biogas: Developments and perspectives in Europe. *Renewable Energy* 129(A): 457–472.

Statista. 2020. *Ranking of largest geothermal plants worldwide as of March 2020.* https://www.statista.com/statistics/525206/geothermal-complexes-worldwide-by-size/ [1.11.2020].

Stoenoiu, C.E. 2018. Pattern of Energy Productivity and Gross Domestic Product Among European Countries. *Problemy Ekorozwoju/ Problems of Sustainable Development 13*(2): 113–123.

Sztumski, W. 2019. For Further Social Development, Peaceful, Safe and Useful for People. *Problemy Ekorozwoju/ Problems of Sustainable Development* 14(2): 25–32.

Teske, S. 2012. *Energy Revolution. A Sustainable EU27 Energy Outlook.* Greenpeace International and the European Renewable Energy Council. Brussells.

Trivedi, A. 2020. The big China disaster that you're missing. *The Japan Times* 09.01. https://www.japantimes.co.jp/opinion/2020/09/01/commentary/world-commentary/big-china-disaster/ [1.11.2020].

UNDP. 2017. *Sustainable Development Goals: Goal 7: Affordable and clean energy.* https://www.undp.org/content/undp/en/home/sustainable-development-goals.goal-7-affordable-and-clean-energy.html [1.11.2020].

Unwin, J. 2019. *The oldest geothermal plant in the world.* https://www.power-technology.com/features/oldest-geothermal-plant-larderello/Unwin [1.11.2020].

Velkin, V.I., Shcheklein, S.E., 2017. Influence of RES Integrated Systems on Energy Supply Improvements and Risks. *Problemy Ekorozwoju/ Problems of Sustainable Development* 12(1): 123–129.

Vlosky, R., Smithhart, R., 2011. A Brief Global Perspective on Biomass for Bioenergy and Biofuels. *Journal of Tropical Forestry and Environment* 1(1): 1–13.

von Weizsäcker, E.U., Wijkman A. 2018. *Come on! Capitalism, short-termism, population and the destruction of the planet – A Report to the Club of Rome.* Springer: New York.

Water, Safe, Strong and Sustainable. Vision on European Water Supply and Sanitation in 2030. 2004. Brussels: Water Supply and Sanitation Platform.

WCED. 1987. *Our Common Future. The Report of the World Commission on Environment and Development.* New York: Oxford University Press.

Wikipedia. 2020, https://en.wikipedia.org/ [1.11.2020].

World Atlas, 2019, *Largest geothermal power plants in the World.* https://www.worldatlas.com/articles/largest-geothermal-power-plants-in-the-world.html [1.06.2019].

Wright M., Hearps P., 2011. *Zero Carbon Australia 2020: Stationary Energy Sector Report.* University of Melbourne, Melbourne, Australia.

CHAPTER 2

Assessment of the contemporary conditions for the renewable energy development in Poland

D. Gryglik
Faculty of Civil Engineering, Architecture and Environmental Engineering, Lodz University of Technology, Lodz, Poland

ABSTRACT: The development of civilization depends on a stable access to energy carriers. As a consequence of the exponentially growing population in less economically developed countries and the increasingly consuming lifestyle of people in more economically developed ones, the demand for energy in the world is still increasing. Considering the consumption of the currently known fossil fuel resources and the destructive impact of their combustion on the natural environment, we strive towards generating energy by other methods. Renewable sources are currently booming around the world. The article analyzed the conditions for the development of renewable energy sources in Poland and worldwide. Particular attention was paid to the current situation in Poland. Economic, legal, social and environmental issues under local conditions were taken into account. The specificity of the Polish energy mix was indicated. The unique difficulties and ambiguity of changing Poland's energy mix against other European countries were emphasized.

2.1 INTRODUCTION

Global environmental changes and the resulting consequences are one of the main threats facing humanity today. This is the price of economic development and the exponentially increasing world population, which was confirmed by the report entitled "Healthy planet, healthy people" [Global Environment Outlook 2019) prepared on the initiative of the United Nations Environment Program (UNEP) and published during the United Nations Environment Assembly (UNEA-4), which took place on March 11–15, 2019 in the capital of Kenya - Nairobi. The report was prepared as a result of extensive consultation and cooperation of 250 people (scientists, experts and politicians) from over 70 countries. They highlighted the fact that while nearly a billion people still do not have free access to electricity, the global energy consumption has more than doubled since 1990, and it is expected that by 2040, the global energy demand will increase by up to 30%. The report emphasizes that in order to achieve Sustainable Development Goals (SDGs) and achieve other internationally agreed environmental goals, the changes in the approach to previous habits are necessary and we have to work out the new ones. Transformation should cover entire social systems and structures, cultural values and norms, as well as sectors of economic activity, including industry, agriculture, construction, transport and energy.

Presently, the renewable energy sources (RES), as well as energy-saving green buildings and low-carbon transport should become the norm. However, many countries around the world still face the dilemma of achieving clean energy goals and reducing environmental degradation, or combusting fossil fuels to support the economic development.

Currently, the conventional energy resources are of key importance to ensure a high standard of living and maintain stable economic development in the society of most developed countries.

They guarantee that the energy and heat supply needs are met and constitute the basis for the functioning of the economy and technological development.

DOI 10.1201/9781003171669-2

On the other hand, there is a growing awareness that the production of energy by burning conventional and consumable resources is at the top of the list of the most serious threats to the environment, and thus the safe life of our and subsequent generations.

2.2 WHY RENEWABLE ENERGY RESOURCES?

Bearing in mind the vision of consumption of the currently known fossil fuel resources and the destructive impact of their combustion on the natural environment, we strive towards generating energy by other methods. Renewable resources are currently booming around the world. The universal advantages of using renewable energy in every place can be summarized in following points:

1. In the foreseeable time, the economically profitable coal, oil and gas resources will be depleted due to the wasteful exploitation of natural reserves and the ever-growing population of people.
2. The implementation of renewable energy technology and thus withdrawal from the burning of fossil fuels will reduce the environmental burden of harmful gas and dust emissions, which gives the opportunity to reduce the greenhouse effect and the incidence of the smog phenomenon resulting from the low-stack emission.
3. Many power plants based on renewable sources (except for large water-power station) are sources of low power, which means that they can be power supply source on a local scale, which in turn forces decentralization of sources and increases the country's energy security.
4. The production of energy from local renewable sources allows countries to become independent of energy imports, and thus reduces the risk of international conflicts.

2.2.1 *Fossil fuel resources are not infinite*

The large-scale use of conventional energy sources began in the nineteenth century and lasted in an unlimited and inefficient manner for many decades. The idea of systematic protection of mineral deposits as a component of the natural environment, the exploitation of which causes an irreversible reduction of its resources, was born at the beginning of the 20th century. The noticeable depletion of known mineral deposits was first time highlighted at the Conference of Governors convened by President T. Roosevelt in 1908 (Proceedings of a Conference of Governors 1908). Then, in 1956, an American geologist Marion King Hubbert suggested the theory that for any given geographical area, from an individual oil-producing region to the planet as a whole, the rate of petroleum production tends to follow a bell-shaped curve (Hubbert 1956). At the be-ginning of the curve (before the peak) the production speed increases due to the speed of detection and the addition of infrastructure; however, at some point it reaches its maximum and then decreases rapidly due to the depletion of resources. In 1972, further alarming speculation was provided by the Club of Rome members report entitled "The Limits to Growth", which simulated the dependence of the exponential economic growth and population under the conditions of limited supply of resources (Meadows et al. 1972). Despite the increasingly frequent reports of the need to protect mineral deposits, society and decision makers did not acknowledge this for a long time. Nowadays, the awareness of limited fossil fuel resources and of the need to develop other methods to meet the global energy demand is deeply rooted in human minds.

Nevertheless, the fossil deposits are used not only for energy production. They are also sources of necessary raw materials, the basis for the development of industrial activity and economic activation. Therefore, meeting the demand for mineral resources should be supported by the development of technology in the direction of saving resources, higher efficiency of equipment, the use of secondary raw materials, recycling and searching for substitutes.

2.2.2 *The implementation of renewable energy technology will reduce the emission of harmful substances into the atmosphere*

It is widely known that the exploitation of coal, which is the basic raw material for the production of electricity and heat, is associated with a negative impact on the natural environment.

Many studies confirm that the main reason for the increasing greenhouse gas emissions to the atmosphere is the steady increase in GDP and growth of population, and that the development of alternative energy sources has a positive effect on reducing emissions (de Souza Mendonça et al. 2020).

The first formal document confirming the international awareness of the harmfulness of emissions greenhouse gases (GHGs) to the atmosphere was the United Nations Framework Convention on Climate Change (UNFCCC) setting out the assumptions for international cooperation on reducing the greenhouse gas emissions responsible for global warming. The convention was signed during the United Nations Conference on Environment and Development held in 1992 in Rio de Janeiro, but it entered into force only after 2 years in 1994. Since then, the so-called Conferences of the Parties (COP) have been held every year in December, during which the implementation of UNFCCC provisions by signatories and convention-related legal instruments is reviewed. The first of them – COP-1 took place in 1995 in Berlin, and the last (COP-25) took place in 2019 in Madrid.

Although the environmental protection against emissions has been at the forefront of the today's problems in most economies around the world, many disputes and lively debates between representatives of many countries of the world during the COPs reflect the complexity of the problem well. Depending on different starting points (e.g. domestic fossil fuel resources, countries' renewable energy potential and economic conditions as an ability to further increase it) major controversies are generated over the closer and further goals of energetics strategy.

Nevertheless, in the last climate and energy package, the EU pledged to cut the greenhouse gas emissions by 2030, by at least 40% percent compared to the 1990 levels, and then unveiled even a plan to reach net zero by 2050.

2.2.3 *Diversification of energy sources and political independence*

Crude oil is still the dominant raw material in the structure of primary energy in the world. The geographical distribution of the oil reserves documented so far is clearly focused in only several places. About 60% of crude oil reserves are located in the Middle Eastern countries and ten countries with the largest oil reserves have about 81.5% of global resources. This is an unfavorable situation for the economic and political reasons, all the more so as some of these regions are economically and politically unstable. Indeed, the tendency to increase the fossil primary energy carriers imports can be seen in all economic giants of the world with China and the United States of America at the forefront. The same situation takes also place in Europe. As history has shown many times, the resource monopolists can easily use the energy resources for political games. The history of the 1973 oil crisis and the Arab embargo on oil supplies and then, the gas conflict in Ukraine (2009) clearly shows the problem of the world's dependence on energy import.

Therefore, the importance of renewable energy sources should be appreciated. If they become a significant pillar of local energy in individual countries, they will eliminate there the need to import some carriers and increase their energy independence index.

2.3 RENEWABLE ENERGY DEVELOPMENT IN THE WORLD

As a result of the very high demand for fossil energy resources that developed after the Second World War, some of their deposits were depleted, especially in the second half of the 20th century. Today, the hard coal resources in Western Europe are almost exhausted. According to Eurostat, in 2017 hard coal was mined only in five of the 28 European Union countries. The highest share in the total EU output was recorded in Poland (83.0%) (Eurostat).

For its own consumption, the EU also needs energy which is imported from third countries. In 2018, the main imported energy product was petroleum products, accounting for almost two thirds of energy imports into the EU, followed by gas (24 %) and solid fossil fuels (8 %). Annually, more than half (55.1%) of the gross energy available in the EU-28 is imported (Eurostat). The dependence of the European Union (EU) on the energy imports, in particular crude oil and natural gas, raises concerns in relation to the energy policy regarding the security of energy supply. That is why many countries are taking steps to completely abandon the use of hard coal for the energy purposes. The United Kingdom, the birthplace of the industrial revolution, intends to phase out the coal energy by 2025, and France announced the closure of its coal power plants even earlier, by 2022.

Belgium has closed its last coal-fired power station in 2016, and Sweden (shortly after Austria) did the same in April 2020. France, Slovakia, Portugal, the UK, Ireland and Italy are expected to abandon coal by 2025 or earlier.

According to the latest data published by Eurostat, the European Union is systematically increasing the share of renewable energy sources in the energy mix. Since 2004, the share of RES in gross final energy consumption has increased significantly in 22 out of 28 Member States and reached an average level for all EU members 18.0% in 2018, which is more than twice as much as in 2004 (8.5%), when the data on this topic began to be collected. Now, 12 of the 28 EU Member States have already achieved an equal or higher share than their national binding targets for 2020. Those are: Bulgaria, the Czech Republic, Denmark, Estonia, Greece, Croatia, Italy, Latvia, Lithuania, Cyprus, Finland and Sweden. In four more ones, less than one percentage point is missing to achieve the RES 2020 target, and nine countries lack 1 to 4 percentage points. So far, the leader in the share of renewable energy in final energy consumption in EU is Sweden – 54.6% (2018). The Stockholm government has adopted a target of a 100% share of renewable energy in the electricity mix until 2040.

However, looking at the problem objectively, all efforts made in Europe to move away from burning fossil fuels and towards the use of alternative energy sources are a drop in the ocean against the background of the global situation.

The leading energy consumer in the world is China. China is now the highest producer and consumer of coal in the world, the second (after the USA) largest consumer of oil, and third in gas consumption after the USA and Russia. About 64% of the China's total energy consumption comes from coal but, in recent years, as much as 61% of China's oil needs were imported (Musa et al. 2018). Even though in 2019, Beijing declared its intention to limit the role of coal in the energy mix, the demand for coal in this huge society is steadily increasing so, at the same time, it has also agreed to increase the import of coal to the country. The China's dependence on the oil and gas imports is also growing. Therefore, the Middle Kingdom intensively supports the development of low-carbon energy sources. Currently, China has the world's largest wind energy production capacity and is the largest producer as well as exporter of wind turbines. They are also the largest producer and exporter of solar installations, and it is estimated that they will soon become one of the largest producers of energy from this source.

A summary of the global present situation of the development of renewable energy sources is presented in Table 2.1 and Figure 2.1.

Table 2.1. Global renewable energy generation capacity and share of renewable energy generation capacity by region (state for 2019). Self-reported data based on: Renewable capacity statistics 2020.

Region	Capacity [GW]	Global share [%]
Asia	1119	44,11
Europe	573	22,59
North America	391	15,41
South America	221	8,71
Eurasia	106	4,18
Africa	48	1,89
Oceania	40	1,58
Middle East	23	0,91
Central America and The Caribbean	16	0,63

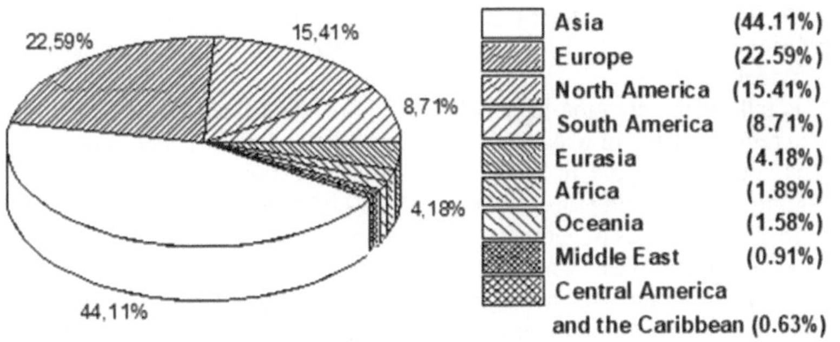

Figure 2.1. The share of renewable energy generation capacity by region in 2019. Self-reported data based on: Renewable capacity statistics 2020.

2.4 RENEWABLE ENERGY DEVELOPMENT IN POLAND

The instability and unpredictability of the renewable energy resources and their uneven distribution in space are natural barriers limiting the possibility of their permanent free use. Al-though the European Union strongly supports the development of renewable energy sources, in Poland, in recent years, despite the large potential, this development has slowed down. This is partly due to the historically conditioned relationship of the economy with coal and partly due to political conditions. Sometimes, in addition, this is difficult by environmental protection requirements, spatial development of the area and often simply unprofitable investment.

2.4.1 *Environmental conditions*

Considerations for the development of alternative energy sources in a given area should begin with the analysis of the environmental conditions on which the availability of renewable energy depends. One of the inherent features of renewable energy is the unevenness of its occurrence. In terms of the environment conditions, the areas more or less predestined to use specific forms of renewable energy can be selected also in Poland.

If the possibility of using kinetic energy of water is considered, the hydropower potential of Poland is characterized by uneven distribution across the country. It is favored by the mountainous terrain and the presence of watercourses with high flow rates. Poland is a largely lowland country, so the possibilities of constructing large dam reservoirs are therefore limited and rather exhausted in our country. In addition, compared to other European countries, the Poland's water resources are low, which results from both atmospheric factors and inadequate water management. The regions with the most favorable conditions for the development of hydroelectric power plants are in the south (mountain areas) and north (due to existing infrastructure). Currently, the future of hydropower in Poland is associated rather with the launch of small hydropower plants next to the existing dams, than with new, extremely costly investments difficult to implement without the State financial support.

Taking into account the wind energy potential, favorable conditions for this type of investment occur in most of the country, with particular emphasis on the coastal areas. Until recently, the largest plans have been made with this type of renewable energy (Vision of wind energy development in Poland 2010). In 2012, Poland could even boast with the two highest windmills in the world (tower with the height of 160 m). They were located in the village of Paproć near Nowy Tomyśl. In the following years, higher windmills appeared in the world.

At present, however, Poland is not among the European leaders in terms of the wind energy production and the investments in the wind energetics have almost disappeared. It can be presumed that the responsibility for this state of affairs is borne by the provisions of the Act of 2016 on investments in wind farms, which eliminated the possibility of locating most economically profitable power plants, not to mention higher ones that could successfully work in the areas, where the wind conditions below are insufficient.

There is talk of the creation of a Polish offshore wind farm in the Baltic Sea, but in the above-mentioned report (Vision of wind energy development in Poland 2010) it was anticipated that it would be connected to the network in 2018. It is currently still in the project phase, now the year 2025 is given (National energy and climate plan for 2021–2030 2019).

There are rich resources of geothermal energy in Poland. The geothermal energy technical potential is estimated at 1512 PJ/year, which is about 30% of the domestic heat demand (Zimny et al. 2015). It is also very important that the regions with optimal geothermal conditions largely overlap with the areas of high density of urban and rural agglomerations, highly industrialized areas as well as the regions of intensive agricultural and vegetable growing. However, so far only a few geothermal heating plants operate in Poland. Further development of this industry requires the devising and implementing new technologies with significant investment outlays, which, with relatively low prices of energy obtained from conventional sources, does not mobilize investors. The use of geothermal energy in Poland depends on the proper development of projects, guaranteeing its economic and ecological competitiveness compared to other energy carriers.

However, even in the cases where nature is conducive to the use of its energy resources, the fact that about 20% of Poland's land surface is covered by the European Natura 2000 program may be an additional difficulty. The regulations regarding Natura 2000 areas do not contain prohibitions on investment implementation, but oblige the investor to exclude a significant impact on the objects of protection of these areas. Therefore, the location of the investment in the Natura 2000 area or in its vicinity, apart from a few cases, does not prejudge the inability to implement it, provided that the impact assessment shows no significant detrimental effect or the possibility of eliminating it.

2.4.2 *Social conditions*

Many specialists emphasize the ability of socio-technical systems to imagine a common future in specific national contexts. They show how „hopes and desires" for the future are associated with difficult achievements of the past (Kuchler & Bridge 2018). Therefore, the role of the history of a given nation in its imagining the future cannot be overlooked.

For decades, Poles used to consider coal as a pillar of the country's energy security and trade balance. Not long ago – during the COP-24 in Katowice (one of the cities recognized as the capital of Polish smog) – the Polish president convinced the guests about the great attachment of Poles to the so-called "black gold". Hard coal has still played a significant role in Poland's economy and politics. It has become a basic national good, which is an essential energy resource in a highly energy-consuming economy. In the communist times, it was the only export commodity that provided the necessary foreign currency needed for the state purchases abroad. For a long time, this raw material have provided Poland with one of the highest indicators of energy independence in Europe (the energy dependence indicator in 2017 for Poland was 38%, with the EU average of 55% (National energy and climate plan for 2021–2030 2019). Such strong belief in the country's energy security persists in the consciousness of a large part of the society, even though the figures for many years contradict it.

In Poland, in the years 1990–2018, the geological hard coal resources decreased from 86.0 billion tonnes to 76.0 billion tonnes. Its annual production fell by around 57.8% (Wyszkowska et al. 2019). Currently, Poland is systematically becoming more and more dependent on the import of coal, and the situation looks even worse in relation to other energy carriers (oil, natural gas). The studies of the Polish Institute of Economics show that only around 20% of the domestic gas consumption is satisfied thanks to our own resources, while the domestic oil deposits account for only 3% of total demand. Russia remains the main supplier of all energy resources.

Figure 2.2 shows the changes in primary energy production obtained from domestic resources in Poland, compared to the amount of energy imported and to export of energy in 2010–2017. For comparison, total energy import dependency of Poland in 2000–2018 was also shown, expressed as the share of total energy needs of our country met by imports from other countries.

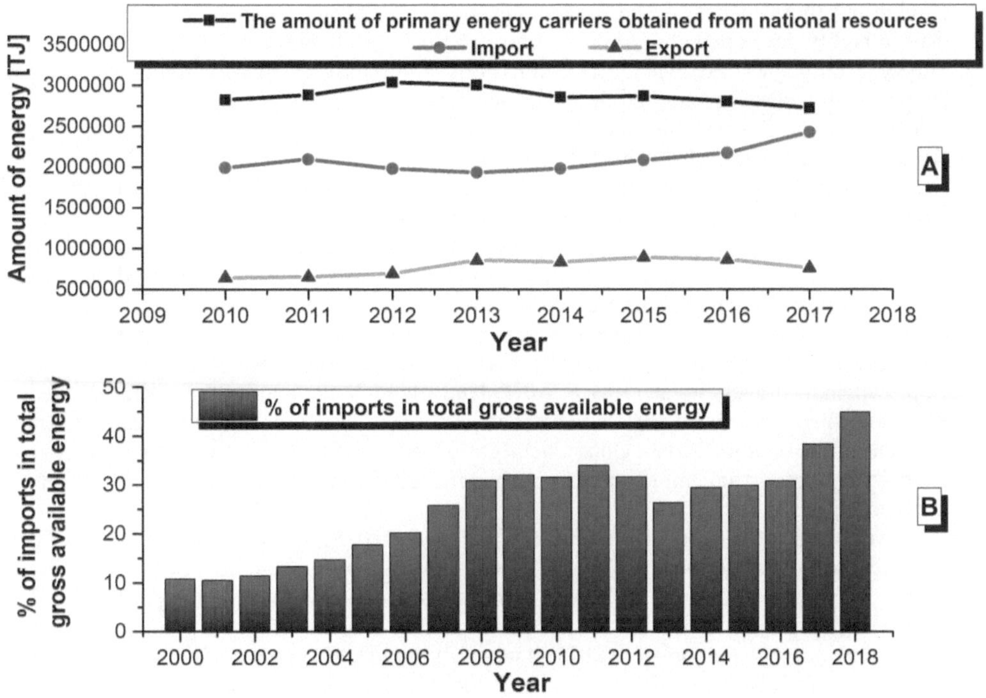

Figure 2.2. Summary of data on Poland's dependence on external energy supplies A) The quantity of primary energy obtained from domestic resources compared to the amount of energy imported and to export of energy. B) Total energy import dependency expressed in figures as the share of total energy needs of a country met by imports from other countries. It is calculated as net imports divided by the gross available energy. Energy dependence = (imports – exports) / gross available energy. Self-reported data based on: Statistics Poland (GUS) and Eurostat. The data applies to all carriers: solid fossil fuels, oil and petroleum products (excluding biofuel portion), natural gas.

The data shown in the figures indicate a worrying tendency to decrease in energy production in the country and increase in its import. The year 2018 was a record year for Poland in terms of hard coal imports, according to the Central Statistical Office data, as much as 19.3 million tons of this raw material came to our country at that time. This is an increase by as much as half, compared to the previous year. The data published by the Central Statistical Office in June 2019 shows that the ratio of energy imports to energy exports in Poland has been constantly growing over the past 20 years. Within a dozen or so years, from the commercial coal power, Poland has become an importer in need of coal from other countries.

However, no amount of coal will protect Poland against the situation that occurred in August 2015, when restrictions in power supply were introduced, due to the inability to properly cool the infrastructure of coal-fired power plants in hot weather.

Meanwhile, currently the greatest obstacle in the development of alternative energy sources in our country is the coal lobby, which by all means tries to extend the inevitable agony of mining in Poland and provide it with further financial support, despite the awareness of the impoverishment of the quantity and quality of domestic resources and catastrophically low air quality in Silesia.

2.4.3 *Economic conditions*

Poland has to move away from coal, aware that this process will cost it much more than other European countries. Due to high dependence on coal and the development of energy-consuming and emission-producing industries, Poland may painfully suffer from the need to meet the requirements of the EU climate and energy package. The energy intensity of the Polish economy is about 2.5 to 3 times higher than in highly developed countries; therefore, the increase in electricity prices after the introduction of new technologies to reduce the CO_2 emissions or purchase of emission rights will adversely affect the production costs of many products and services.

The analyses of experts and specialist analytical companies confirm these fears. In 2011 the World Bank Report (Jorgensen & Kasek 2011) pointed that the implementation of the European climate and energy package will cost the Polish economy a loss of 1.4% of GDP annually by 2020. The same indicator calculated by Żmijewski is even higher (Żmijewski 2011).

Experts also raise the problem of the possibility of the so-called "carbon leakage" phenomena, i.e. emigration of emission and energy-intensive industries outside the area covered by the restrictions of the European Climate and Energy Policy and the EU-ETS emissions trading system. This applies to various industries, in particular: the cement, mining, glass, chemical and steel industries. In the European Union, only two countries have a share of industries at risk of carbon leakage in total employment in industry greater than 8.5% (Poland and Finland), and another three (Sweden, Belgium, Romania) share greater than 5% with an average share in the European Union estimated at about 3% (ESPON & Innobasque 2010). Consequently, it is estimated that in these countries between five and ten percent of industrial employment could be affected if companies decide to relocate their activities to other regions.

However, during the already mentioned COP-24 (in 2018), the representatives of the World Bank claimed that, given the local and global context, moving away from coal in Poland is economically viable. A decrease of approx. 25% jobs in mining and quarrying would compensate for the creation of about 100,000 jobs in new sectors of the economy, related to the energy efficiency, e.g. thermo-modernization. In the report from 2019, it was also stated that the costs of pollution in our country are huge, and their reduction will pay off. The estimates show that the cost of air pollution in Poland is about 31–40 billion dollars a year – i.e. from 6.4 to 8.3% of GDP. What is more, these costs of climate change and extreme weather in the years 2021–2030 will reach PLN 120 billion.

2.4.4 *Legal conditions*

The development of renewable energy technologies gives Poland the opportunity to fulfill the obligations assumed under the so-called the climate and energy package (contained in Directive 2009/28 / EC of the European Parliament and of the Council of 23 April 2009 on the promotion of the use of energy from renewable sources). In the case of Poland, the target for the share of energy from renewable sources in gross final energy consumption, to be achieved in 2020, was set at 15%. At the same time, each Member State should achieve in 2020 is at least 10% of renewable energy in the final energy consumption in transport.

Figure 2.3 shows the changes in the share of energy from renewable resources in gross final energy consumption and in transport in Poland since the accession to the EU (2004–2018).

As can be seen, Poland has not achieved the EU target so far neither in gross final energy consumption nor in the transport sector. The data for 2019 is not available yet, but there is no indication that the situation will improve significantly in the near future. It is clearly visible the collapse of the growing trend around 2016 and since then the level has not changed significantly. The forecast of the Polish government contained in the "National Plan for Energy and Climate for 2021–2030" forwarded to the European Commission in December 2019, estimates that Poland will achieve the RES 2020 target, but later (National energy and climate plan for 2021–2030 2019).

Regarding the other legal aspects of renewable energy policy, accession to the European Union forced Poland to adapt its legislation to the European law. This of course also applies to the energy law, the including renewable energy. The Renewable Energy Act was created for a period of 5 years. In accordance with the EU requirements, the new RES Support Act, harmonizing Polish law with Directive 2009/28, was to come into force in December 2010 and in fact, it was only on May 4, 2015. Since then, it has been called the "renewable over and over Act", since at the early stage of work on this Act its amendment was preparing, 5 amendments and a dozen or so amended regulations have been created since 2015.

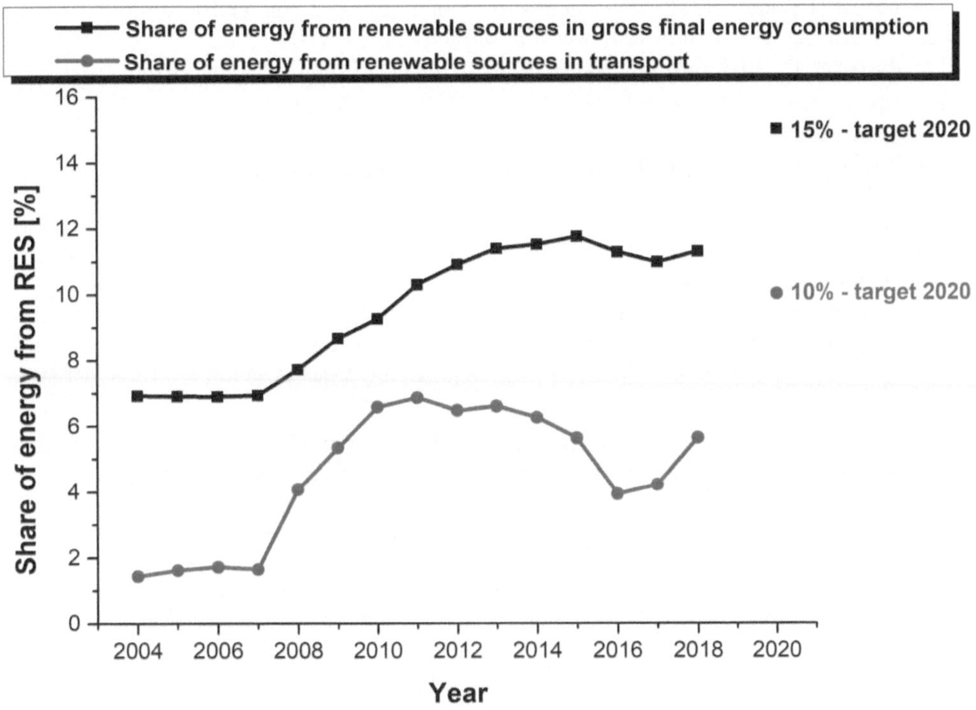

Figure 2.3. Changes in share of energy from renewable sources (RES) in gross final energy consumption and in transport in Poland in 2004–2018. Self-reported data based on: Eurostat.

In addition, at the same time as the RES Act (which was supposed to favor the development of renewable energy in Poland) was being developed, the Act on investments in wind farms (the so-called anti-wind mill act) was being worked out, which contains the regulations that significantly inhibit the development of energy wind power in Poland.

This act entered into force on July 16, 2016. This situation can give rise to understandable investor objections and the following question is reasonable:

Can the investors be expected to get involved in the renewable energy industry, when the investment process lasts a long time, and the legal bases can change several times during this time?

2.4.5 *Poland is smothered by smog and GHGs*

However, apart from any barriers to the development of renewable energy in Poland, one argument seems to be crucial in the matter of abandoning the combustion of fossil fuels in the country. The Polish air conditions are not very good, to put it mildly. The infamy of air pollution in Polish cities reaches not only Europe but the whole world. In recent years, there have been many recurring media reports such as:

- Poland has been listed second among the most polluted countries in Europe (Greenmatch, 2018),
- over 30 Polish cities have been on the list of the 50 most polluted cities in the EU (Euronews, 2017),
- on the WHO website, in the tab "Urban settings as a social determinant of health", a photo of smog in Krakow appeared as an example of the poorest quality of air (WHO, 2019),
- the highest level of air pollution in the world was recorded in Wrocław. The tenth place was taken by Krakow (IQAir, 2018).

However, high concentrations of dust during the heating period are not the only problem of Polish air. In Poland, the concentrations of carcinogenic benzo(a) pyrene are also highest among all EU countries. The dramatic situation results from the burning of coal and wood in old furnaces, which are the primary source of heat in over four million Polish houses.

Moreover, the EEA lists Poland according to the data from December 2019 as:

- one of the largest GHG issuers per GDP and per capita in the EU in 2017 (with values: 178 PPS, EU-28=100 and 11.0 tCO_2e per person) (The European environment – state and outlook 2020, EEA, 2019),
- one of the countries with the highest values of annual mean concentration PM2.5 and PM10 in the EU in 2017 (in order after Bulgaria and Estonia) (The European environment – state and outlook 2020, EEA, 2019),
- one of three countries (together with Spain and United Kingdom) that confirms the presence in the air of all recommended to detect VOCs, which are O_3 precursors (Air quality in Europe – 2019 report, EEA 2019; Annual European Union greenhouse gas inventory..., EEA 2019).

The available data is alarming. According to Eurostat, in 2019 Poland was the third largest CO_2 emitter in the European Union. The intensity of the CO_2 emissions in Polish electricity production is slightly higher than in China and more than twice as high as in EU-15.

Figure 2.4 presents a comparison between the GHG emissions in million tonnes CO_2 equivalent (excl. LULUCF) in Poland and EU-28 in 1990–2018.

The data shown in the graph presents a worrying stagnation in the total GHGs emission in Poland compared to Europe, where the emission value is systematically and clearly decreasing.

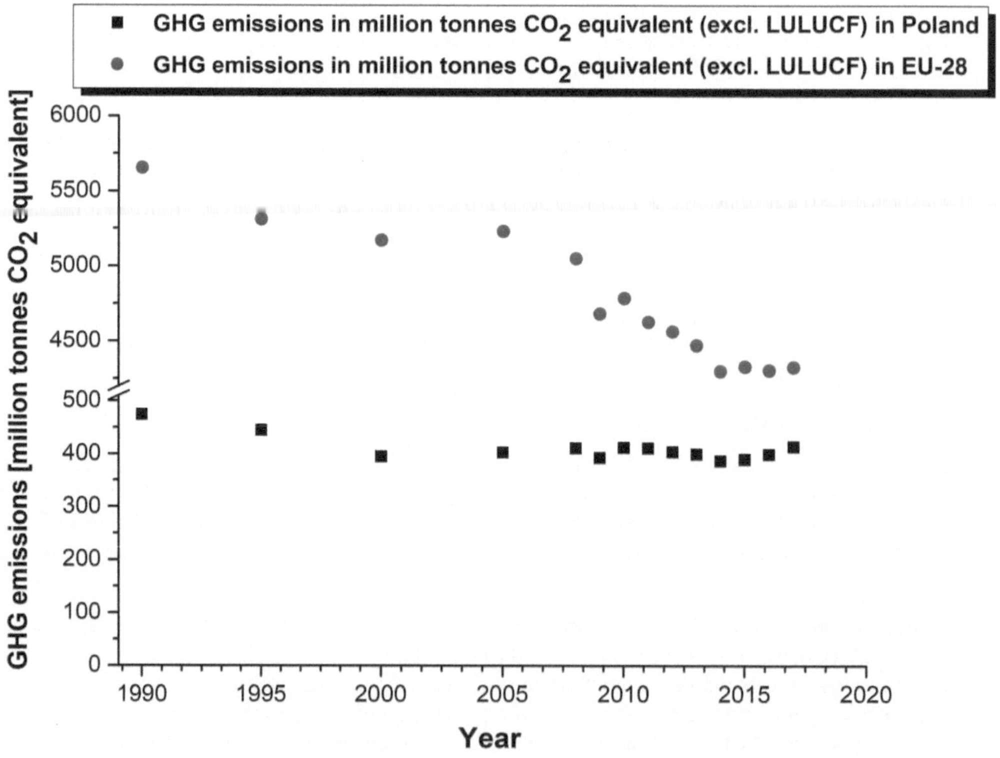

Figure 2.4. Changes in GHG emissions in million tonnes CO_2 equivalent (excl. LULUCF) in Poland and EU-28 in 1990–2018. Self-reported data from Annual European Union greenhouse gas inventory 1990–2017 and inventory report 2019.

2.5 CONCLUSIONS

In the face of limited, constantly diminishing resources of fossil fuels, the development of various forms of using alternative energy is looked upon with hope. The maintenance of a relative balance between the volume of non-renewable resources and their output is an important factor conditioning the sustainability of economic development. It is believed that the development of alternative power industry will allow us not only to maintain the security of energy supply but will also have a positive influence on the state of the natural environment.

Theoretically, there is an agreement among global decision-makers in this matter; however, having gone deeper into the detailed problems of renewable energy development in individual parts of the world, it turns out that the support for it is not so obvious everywhere. The infrastructure and economy conversion to alternative energy sources is difficult and time consuming. In order to create infrastructure like hydroelectric plants, solar panels, wind turbines, tidal and geothermal facilities, nuclear power plants, hydrogen cells and others advanced technologies and a lot of energy are needed – both directly in the energy processes and indirectly in the output of raw materials, transport and factory operations. If the fossil fuels run out completely, it may simply no longer be possible to produce them.

In the meantime, new problems regarding the use of renewable energy appear, such as:

- the economics of obtaining this energy,
- the need to provide reserve sources based on conventional energy, and
- real amounts of reducing greenhouse gas emissions and other environmental problems through the use of renewable energy throughout the life cycle.

As a rule, the latter are usually ignored (or omitted) in the publicly available analyses, while they should be thoroughly studied and optimized as much as possible. The fact that the elimination of fossil fuels and their replacement with infrastructure for the production of energy from renewable resources would require countless amounts of wind turbines, solar panels and other devices, for the production of which the non-renewable and fossil raw materials (for example rare-earth elements) are also needed, is not widely publicized.

At the same time, we are looking forward to new technologies that are currently at the research stage: green hydrogen, fuel cells, perovskites, graphene, quantum dot technology…Perhaps in the near future, they will constitute the basis for the energy production in the world, eliminating some of the disadvantages of the existing alternative technologies and significantly reducing the demand for energy from fossil fuels. We are at a crossroads, and how we proceed in the face of the impending energy crisis will most likely determine the future of next generations and our planet.

REFERENCES

Air quality in Europe – 2019 report, EEA Report No 10/2019; European Environment Agency; Luxembourg: Publications Office of the European Union, 2019, doi:10.2800/822355

Annual European Union greenhouse gas inventory 1990–2017 and inventory report. 2019, European Environment Agency, Copenhagen

Global Environment Outlook – GEO-6: Healthy Planet, Healthy People. 2019. UN Environment (Ed.) Cambridge University Press doi:10.1017/9781108627146

https://ec.europa.eu/eurostat/cache/infographs/energy/bloc-2c.html

https://ec.europa.eu/eurostat/statistics-explained/index.php?title=Energy_production_and _imports/pl

https://www.euronews.com/2017/11/30/poland-among-europe-s-worst-for-smog

https://www.greenmatch.co.uk/blog/2018/11/mapped-europes-most-and-least-polluted-countries#Methodology

https://www.iqair.com/

https://www.who.int/news- room/facts-in-pictures/detail/10-facts-on-urban-settings-as-a-social-determinant-of-health

Hubbert M.K. 1956. *Nuclear Energy and the Fossil Fuels.* Publication No 95 Shell Development Company Exploration and Production Research Division, Houston, Texas

Jorgensen E. & Kasek L. 2011. Transition to a low-emissions economy in Poland (English). Europe and Central Asia knowledge brief; issue no. 42 Washington, D.C.: World Bank Group.

http:// documents.worldbank.org/curated/en/951921468282911433/ Transition-to-a-low-emissions-economy-in-Poland

Kuchler M. & Bridge G. 2018. Down the black hole: Sustaining national socio-technical imaginaries of coal in Poland. *Energy Research & Social Science* (41): 136–147

Meadows D.H., Meadows D.L., Randers J. and Behrens III W.W. 1972. *The Limits to Growth. A report for the Club of Rome's project on the predicament of Mankind.* New York: Universe Books

Musa S.D., Zhonghua T., Ibrahim A.O., Habib M. 2018. China's energy status: A critical look at fossils and renewable options. *Renewable and Sustainable Energy Rev.* 81: 2281–90

National energy and climate plan for 2021–2030 vs. 4.1. z dn. 18.12.2019 published on: https://www.gov.pl/web/aktywa-panstwowe/krajowy-plan-na-rzecz-energii-i-klimatu-na-lata-2021–2030-przekazany-do-ke w dniu 30.12.2019 roku (in Polish)

Proceedings of a Conference of Governors held in the White House May 13–15, 1908; PUBLISHED Washington: G.P.O. 1909. (access on the website: http://lcweb4.loc.gov/cgi-bin/query/r?ammem/-consrv:@field(DOCID+@lit(amrvgvg16div19))

Renewable capacity highlights 2020, International Renewable Energy Agency (2020), available at: https://www.irena.org/-/media/Files/IRENA/Agency/Publication/2020/Mar/IRENA_RE_Capacity_ Highlights_2020.pdf?la=en&hash=B6BDF8C3306D271327729B9F9C9AF5F1274FE30B

Souza Mendonça A.K., Conradi Barni G.A., Moro M.F., Bornia A.C., Kupek E., Fernandes L. 2020. Hierarchical modeling of the 50 largest economies to verify the impact of GDP, population and renewable energy generation in CO_2 emissions. *Sustainable Production and Consumption*. 22: 58–67

The ESPON 2013 Programme ReRisk Regions at Risk of Energy Poverty Applied Research Project 2013/1/5 Draft Final Report, vs. 31/03/2010, ESPON & Innobasque, 2010

The European environment – state and outlook 2020. Knowledge for transition to a sustainable Europe. 2019. European Environment Agency; Luxembourg: Publications Office of the European Union. ISBN 978-92-9480-090-9

Vision of wind energy development in Poland until 2020. Summary. 2010. Instytut Energii Odnawialnej, https://www.cire.pl/pokaz-pdf-%252Fpliki%252F2%252Frap_pods_pl2.pdf (in Polish).

Wyszkowska D., Artemiuk H., Giziewska D., Godlewska A., Łapińska R., Rogalewska A., Słucka U., Szpaczko I. 2019. *Green economy indicators in Poland*. Warszawa, Białystok: Główny Urząd Statystyczny, Urząd Statystyczny w Białymstoku (in Polish)

Zimny, J., Struś, M. & Bielik, S. 2015. *Selected problems of renewable resources energetics*. Monografia Naukowa, tom 6, Polska Geotermalna Asocjacja, Stowarzyszenie Naukowo-Techniczne, Kraków (in Polish)

Żmijewski, K. 2011. *The threat of carbon leakage in Poland.* Warszawa: Instytut im. E. Kwiatkowskiego, https://www.cire.pl/pliki/2/IBS_carbon_leakage_PL.pdf (in Polish)

CHAPTER 3

The kinetics of biogas production in co-digestion of sewage sludge and agro-industrial wastes

A. Szaja
Faculty of Environmental Engineering, Lublin University of Technology, Poland

ABSTRACT: This study examined the influence of applying agro-industrial wastes on the kinetics of biogas production in the co-digestion with sewage sludge (SS). Brewery spent grain (BSG) and acid cheese whey (ACW) were used as co-substrates. The research was conducted in semi-flow anaerobic reactors. The results indicate that the addition of ACW resulted in an improvement of the constant of the biogas production rate (k), as compared to the SS mono-digestion. The average k values were 0.102 and 0.096 h^{-1} at HRT of 20 and 18 d, respectively. Therein, the untapped biogas potential (UBP) maintained 1.49 and 2.54 L at HRT of 20 and 18 d, respectively. In the BSG presence, the negative influence on its kinetics was noticed. The average k values were 0.060 and 0.051 h^{-1} at HRT of 20 and 18 d, respectively. While UBP values were 8.2 and 16.1 L at HRT of 20 and 18 d, respectively.

3.1 INTRODUCTION

In the recent years, an anaerobic digestion process (AD) has gained growing attention, as a cost-effective sustainable technology for the management of various organic biodegradable substrates and renewable energy source. Moreover, the energy crisis caused by the depletion of fossil fuel resources and progressive climate changes resulted in an increased interest in the improvement of biogas production in AD (Mata-Alvarez et al. 2010; Prajapati & Singh et al. 2020). Therefore, the simultaneous AD of at least two adequately selected substrates has been implemented on a large scale at many existing facilities (Mata-Alvarez et al. 2014). The co-digestion process (AcoD) offers several benefits over single substrate AD. In addition to increasing the biogas production, the improved process stability through diluting of inhibitory compounds, providing the necessary buffer capacity as well as enhancing nutrient balance in feedstock, can be achieved. Furthermore, the implementation of an additional substrate may provide the essential macro-and microelements in the AD process. However, the crucial element in effective AcoD is the selection of substrates with complementary composition. Various by-products, such as sewage sludge, manure and as well as lignocellulosic, industrial and municipal wastes have been evaluated (Zahan et al. 2018). Among them, agro-industrial by-products represent a promising group due to their significant availability and production. On the other hand, most of them are produced seasonally and require pre-treatment before implementation to digesters (Hillion et al. 2018). Additionally, these substrates constitute a diverse group. Their composition depends mainly on the source of origin (Álvarez et al. 2010).This group includes various by-products from the brewing, distillery, meat, fruit and vegetable and dairy industries. Most of them are characterized by significant organic load, high biogas potential and presence of different micro- and macro-elements. These features contribute to their widespread application in co-digestion with sewage sludge(Mata-Alvarez et al. 2014; Maragkaki et al. 2018).

Furthermore, the AD is a complex process involving different consortia of microorganisms and requiring different process conditions (Angelidaki & Batstone 2011). The implementation of the AcoD process requires consideration of many factors e.g. substrate proportion, C:N ratio, pH, presence of inhibitors/toxic compounds, biodegradable organic matter and dry matter content (Hartmann et al. 2003). For this reason, various kinetic models have been applied in recent years to simulate the biodegradation in AD process, becoming an indispensable tool in the design and operation of digesters

DOI 10.1201/9781003171669-3

(Zhan et al. 2018). In order to model the anaerobic digestion process, the kinetics of bacterial growth, substrate degradation and product formation have to be considered (Biswas et al. 2007).

This study examined the influence of applying agro-industrial wastes on the kinetics of biogas production in the co-digestion with sewage sludge (SS). Two by-products with different properties were investigated in the present work. Brewery spent grain (BSG) and acid cheese whey (ACW) were used as co-substrates. The first one is known as a hardly-biodegradable by-product of the brewing industry, while the second one as an easily biodegradable waste from the dairy industry. In order to evaluate the effect of components on kinetics, the constant of biogas production rate (k) and untapped biogas potential (UBP) were estimated. Moreover, the analysis of the biogas production curves was performed.

The results presented in this study were previously presented (Szaja & Montusiewicz 2019; Szaja et al. 2020). However, this work constitutes a collection of experiences in the terms of kinetics evaluation in semi-flow systems. Furthermore, a detailed analysis of the influence the co-substrate selection on kinetics of biogas production was presented.

3.2 MATERIALS AND METHODS

3.2.1 *Substrate characteristic*

The sewage sludge (SS) from municipal Puławy WWTP (Poland) was used as a main substrate. It was taken once a week from primary and secondary clarifiers. Under laboratory conditions, this sample was mixed at the recommended volume ratio of 60:40 (primary:waste sludge). Subsequently, it was homogenized and screened through a 3 mm sieve. The portioned sample was stored at 4°C in a laboratory refrigerator for no longer than one week. Two commonly obtained agro-industrial wastes from local region were applied as co-substrates. The first one, BSG, was taken once from a craft brewery Grodzka 15 in Lublin (Poland). In order to maintain an unchanged co-substrate composition, it was dried at 60°C for two hours in a laboratory dryer and then milled to a particulate size of 2.0 mm. The ACW was used as second co-substrate. It was the main liquid by-product of the District Dairy Cooperative in Piaski (Poland). This sample, as in the first case, to ensure a constant co-substrate characteristic, was frozen at a temperature of -25°C. Before feeding the reactors, the SS and ACW samples were kept for some time in the indoor air to reach 20°C. Then, the substrates in the adopted proportions were homogenized using a low-speed mixer. The composition of substrates is presented in Table 3.1.

Table 3.1. Composition of the substrates used in the experiments.

		SS		BSG	ACW
Parameter	Unit	Avg. value	Upp./low.95% mean	Avg. value ±SD	Avg. value ±SD
COD	mg L^{-1}	44227	40286/48168	72623 ± 3144	72259 ± 2072
SCOD	mg L^{-1}	2539	1685/3393	-	64720 ± 1794
VFA	mg L^{-1}	1143	694/1591	2095 ± 189	5536 ± 625
pH		6.19	5.93/6.45	6.19 ± 0.64	4.05 ± 0.71
Alkalinity	mg L^{-1}	843	753/933	2967 ± 139	-
TS	g kg^{-1}	37.8	35.2/40.4	223.9 ± 4.3	43.5 ± 1.1
VS	g kg^{-1}	28.3	26.4/30.2	215.1 ± 2.9	36 ± 0.9
TN	mg L^{-1}	3942	3431/4452	877 ± 359	4690 ± 454
TP	mg L^{-1}	1115	945/1285	171 ± 97	800 ± 96
NH_4^+-N	mg L^{-1}	54.9	36.6/73.3	22.1 ± 6.1	68.4 ± 10.5
PO_4^{3-}-P	mg L^{-1}	292.1	188.2/358.2	25.1 ± 6.9	614 ± 53

Table 3.2. Experimental settings in experiments.

Run	Feedstock composition	Component volume SS	Component volume ACW	Additive mass BSG	SS:ACW volumetric ratio / BSG mass: feedstock volume ratio*	HRT	OLR Avg.	OLR Upp./low. 95% mean
		L	L	g	g L^{-1*}	d	kg VS $m^{-3}d^{-1}$	
					Experiment 1			
R 1.1	SS (control)	2.0	-	-	100	20	1.35	1.23/1.46
R 1.2	SS + BSG	2.0	-	20	10:1*	20	1.73	1.68/1.78
R 1.3	SS + ACW	1.8	0.2	-	90:10	20	1.37	1.27/1.48
					Experiment 2			
R 2.1	SS (control)	2.0	-	-	100	20	1.49	1.41/1.58
R 2.2	SS + BSG	2.2	-	21.7	10:1*	18	1.98	1.84/2.13
R 2.3	SS + ACW	2.0	0.2	-	91:09	18	1.61	1.53/1.68

3.2.2 *Operational set-up*

The research was conducted in semi-flow anaerobic reactors with an active volume of 40 L each, operating under mesophilic conditions (35°C).Two experiments at different hydraulic retention times (HRT) of 18 and 20 d were performed (Table 3.2). In each one, the separate control run supplied daily with 2 L of SS and operated at HRT of 20 d was provided (R 1.1 and R 2.1). In the co-digestion runs, the constant co-substrate proportions were maintained. The BSG mass to the feed volume ratio was 1:10 (R 1.2 and R 2.2), while the ACW volume was retained at 0.2 L (R 1.3 and R 2.3).

The experiments lasted 90 days, including 30 d of acclimatization, and 60 d for measurements. The detailed experimental settings are presented in Table 3.2. The digested sludge for the mesophilic anaerobic digester (HRT = 25 d) from the Puławy WWTP was used as an inoculum. The adaptation of the digester biomass in the laboratory reactors was achieved after 30 d.

3.2.3 *Analytical methods*

The main substrate (SS) characteristics were monitored once a week after their delivery to the laboratory. In turn, the co-substrate compositions (ACW and BSG) were analyzed once immediately after collection. The following parameters were controlled: the total chemical oxygen demand (COD), total solids (TS), volatile solids (VS), total nitrogen (TN) and total phosphorus (TP), the soluble chemical oxygen demand (SCOD), VFA, alkalinity, pH level, ammonia nitrogen (NH_4^+-N) and orthophosphate phosphorus (PO_4^{3-}-P). The experimental analyses were performed using the Hach analytical methods by means of Hach Lange UV–VIS DR 5000. The pH values were controlled using a HQ 40D Hach-Lange multimeter. Total and volatile solids were determined according to the *Standard Methods for the Examination of Water and Wastewater* (APHA 2005).

3.2.4 *The kinetics of biogas production*

The kinetics of biogas production was estimated based on the constant of the biogas production rate and untapped biogas potential. The second kinetic parameter corresponds to the difference between the maximum biogas production (V_{max}) theoretically possible to achieve from a portion of feedstock supplied daily to the reactor (V_e) (Eq. 3.1).

$$UBP = V_{max} - V_e \tag{3.1}$$

The biogas production was determined every day involving an Aalborg (Orangeburg, NY,USA) digital mass flow meter. In turn, the biogas production curves were established on the basis of the averaged experimental data obtained from an XFM Control Terminal. A first-order kinetic equation was used to determine the biogas production rate (Eq. 3.2).

$$V_f = V_{max}[1 - \exp(-k \cdot t)] \quad (3.2)$$

where V_f = the biogas volume in time (L); V_{max} = a constant corresponding to the maximum daily biogas production it is theoretically possible to obtain from a portion of feedstock (L); k = a constant of the biogas production rate (h^{-1}) and t = the operational time (h).

This equation was commonly used to determine the kinetics in batch systems. Importantly, this method was efficiently adopted for semi-flow mode (Szaja & Montusiewicz 2019, Szaja et al. 2020).The high values of the determination coefficients (R^2) observed in the present study confirm the possibility of using this approach.

The ANOVA procedure involving Shapiro-Wilk's, Levene's and Tukey's tests was applied to perform the statistical analysis. The differences were assumed to be statistically significant at $p < 0.05$. Both kinetic constants (k and V_{max}) were determined using a nonlinear regression method. The strength of the relationships between the results was monitored involving Pearson's correlation coefficient (R) and determination coefficient (R^2). The Statsoft Statistica software (v 13) was used for the statistical analysis.

3.3 RESULTS AND DISCUSSION

As previously mentioned, the main substrate was SS, this sample was characterized by significant differences in terms of all analyzed parameters (Table 3.1). It resulted from the variable wastewater composition caused by a seasonal discharge of agro-industrial wastewater (fruit and vegetable processing companies) and temperature fluctuations throughout the year. Therefore, the separate control runs in each experiment have been provided (R 1.1 and R 2.1).

Two agro-industrial wastes (BSG and ACW) with different properties were applied as co-substrates. Comparing to the SS composition, both by-products were characterized by an increased organic matter content. However, for ACW, the beneficial sCOD/COD ratio with average value of 0.89 was observed. Additionally, it should be mentioned that this co-substrate contained significant VFA and nitrogen concentrations as well as the lowest pH. These factors could potentially inhibit the anaerobic digestion process (Fernández et al. 2014; Pilarska et al. 2016; Zielewicz et al. 2012). Additionally, the ACW used in this experiment was characterized by the highest concentration of orthophosphate phosphorus as compared to SS and BSG. The analyzed parameters in ACW reached a typical values for such substrate reported in literature (Azbar et al. 2009; Blonskaja & Vaalu 2006; Chatzipaschali & Stamatis 2012; Ferchichi et al. 2005). The exception was the nitrogen content, according to Carvalho et al. (2013); the concentration of this parameter reached the values between 200–1760 mg L^{-1}. In this study, the highest concentrations were observed, resulting from the milk composition used in the cheese production. The nitrogen content in this product mainly depends on the feed type used in breeding. This tendency is observed in the forage reached in protein. BSG showed the highest total and volatile solids, alkalinity as well as pH among other substrates. The supplementation of feedstock in BSG could provide the necessary buffer capacity and improve C/N ratio in feedstock resulting in an increased process stability (Chatzipaschali & Stamatis 2012). However, the presence of hardly degradable lignin as well as phenolic compounds content could have potentially affected the AD (Retfalvi et al. 2013; Sawatdeenarunat et al. 2015). Similarly to ACW, this substrate presented comparable chemical characteristic to the results found in previous studies (Bougrier et al. 2018; Dos Santos-Mathias et al. 2015; Poerschmann et al. 2014; Wang et al. 2015). Importantly, both co-substrates are known as wastes with significant biogas potential. Furthermore, the presence of crucial in AD process micro- and macro elements in ACW and BSG might enhance methanogens activity (Chatzipaschali & Stamatis 2012; Mussatto et al. 2006).

The results of kinetic evaluation are listed in Table 3.3. In turn, the biogas production curves are presented in Figure 3.1, the values presented on this graph were obtained as means from 30 measurement days forming the final phase of each run. As it was mentioned previously, reactors were

Table 3.3. The average values of kinetics constants as well as coefficients of determination (Szaja & Montusiewicz 2019; Szaja et al. 2020).

Parameter	Unit	R1	R 1.2	R 1.3	R 2.1	R 2.2	R 2.3
Constant of biogas production rate k	h^{-1}	0.076	0.060	0.102	0.078	0.051	0.096
Maximum biogas production V_{max}	L	24.3	36.3	25.0	33.1	54.8	34.3
Coefficient of determination R^2	-	0.9993	0.9997	0.9135	0.9997	0.9999	0.9228

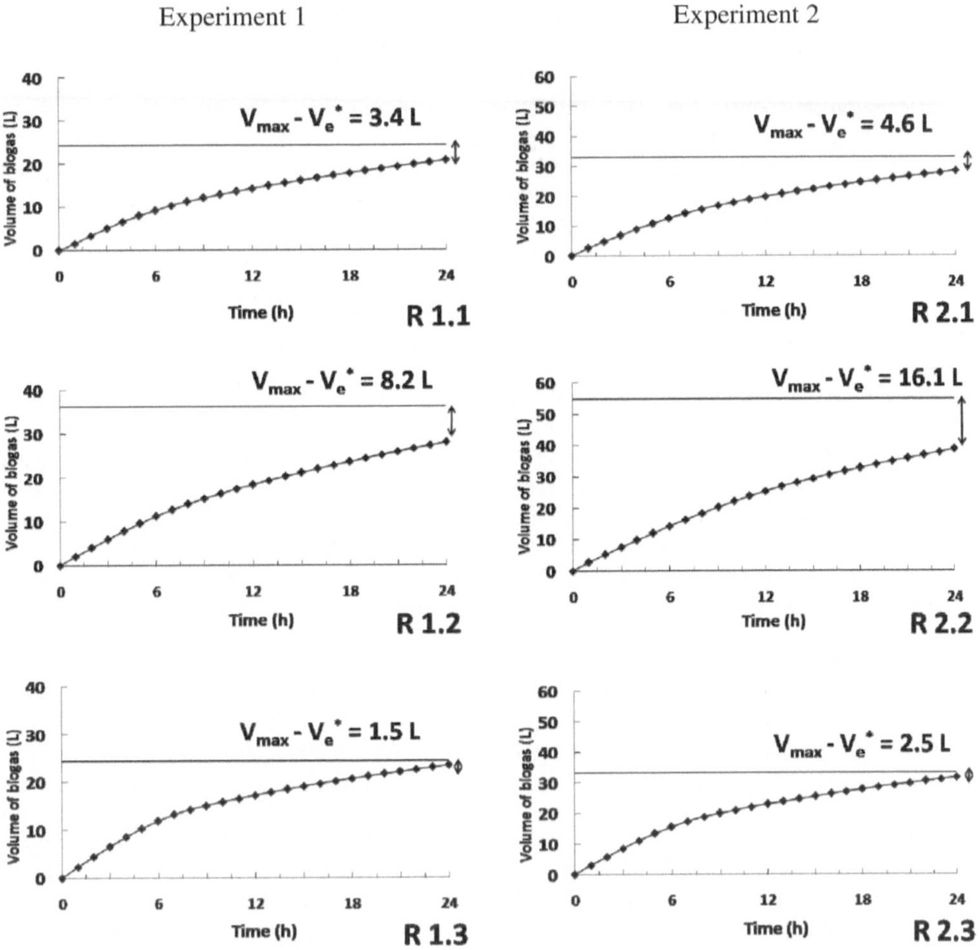

Figure 3.1. Biogas production in time, the average values from 30 measurement days are reported (Szaja & Montusiewicz 2019; Szaja et al. 2020).

operated in semi-flow mode. Therein, regularly once a day, the digester was supplied with the portion of feedstock and the same volume of digestate was removed from it.

Therefore, each tag on the graphs represented the average cumulative biogas volume (V_e) produced in time in semi-flow mode (i.e. subsequent hours from 0 to 24 h). In turn, the final V_e value (for 24 h) corresponded to the daily biogas production.

In the BSG presence, the negative effect on its kinetics was observed (R 1.2 and R 2.2). The supplementation of feedstock with this substrate contributed to a decline in the k value by 21 and

35% at HRT of 20 and 18 d, respectively, as compared to control (Table 3.3). Moreover, in the BSG occurrence, the highest UBP values were found, as compared to control and co-digestion with ACW. These were represented at approx. 23 and 29% at HRT of 20 and 18 d, respectively (Figure 3.1). The shortened HRT led to a decrease in this parameter, indicating the necessity to apply prolonged HRT for such co-substrate. The main reason of this effect might be the complex structure of BSG; importantly, this by-product might contain even up to 28% DM of lignin (Mussatto et al. 2006). These compounds are known as highly recalcitrant to biodegradation under anaerobic conditions (Ahring et al. 2015). An important factor is its cross-linking with the carbohydrates that reduces the surface area available for enzymes (Ahmed et al. 2019). Moreover, it was demonstrated that hydrolysis of lignin is a limiting step of AD (Ahring et al. 2015; Gonzalez-Estrella et al. 2017). Several studies on lignocelluloses biomass AD confirmed that such substrates were characterized by low first-order rate (Li et al. 2013). In the batch mode, Kafle et al. (2013) applied brewery grain and bred by-products in co-digestion with fish waste. Therein, in the presence of BSG (60% TS), the reduced methane production rate constant by 24% was found as compared to bred waste application. However, these difficulties this might be overcome by pretreatment of feedstock and its co-digestion with adequate substrate (Yang et al. 2015). It should be noticed that phenolic compounds that could potentially be generated through the pretreatment, might also inhibit AD process and result in kinetic deterioration (Panjičko et al. 2017; Paul & Dutta 2018; Yang et al. 2015). In the present study, BSG was dried and milled, which may have contributed to the formation of this compounds.

A different effect was observed for ACW (R 1.3 and R 2.3). The obtained results indicate that the addition of this co-substrate resulted in an improvement of the constant of the biogas production rate, as compared to the SS mono-digestion. However, more beneficial results were observed at prolonged HRT of 20 d (R 1.3). As compared to controls, the average k values were enhanced by approx. 34 and 23% at HRT of 20 and 18 d, respectively (Table 3.3). Moreover, in the ACW presence, the lowest UPB values were achieved. Therein, the UBP constituted only 6% and 7.4% at HRT of 20 d (R 1.3) and 18 d (R 2.3), respectively. For the SS mono-digestion, the increased values of this parameter were found. For both controls, it represented approx. 14% (Figure 3.1). These obtained values of UBP confirmed that the adopted operational conditions for such a co-digestion have been selected properly. As in the case of BSG, the prolonged HRT of 20 d is also recommended for this co-substrate. The shortened HRT of 18 d caused the faster washing out of the methanogens from the reactor, resulting in a lower growth of k and major value of UBP (Lee et al. 2011).This beneficial effect in ACW occurrence resulted from its composition. As was mention above, the ACW is characterized by significant content of biodegradable matter (lactose) that could be easily transformed to acids and subsequently to biogas (Chatzipaschali & Stamatis 2012; Carvalho et al. 2013; Stamatelatou et al. 2014). In this case, especially the hydrolysis rate is accelerated as compared to other substrates (Koch et al. 2015; Sanders et al. 2000). A similar tendency was observed for distillery waste (Syaichurrozi et al. 2013), food waste (Koch et al. 2015; Xie et al. 2017) and the organic fraction of municipal solid waste (Sosnowski et al. 2008). Moreover, the supplementation the feedstock in ACW rich in vitamins, mineral salts and proteins might also improve the kinetic parameters (Chatzipaschali & Stamatis 2012; Dereli et al. 2019). Recent studies have also shown that cheese whey indicated the stimulating effect on the population of methane-producing Archaea (Gonzalez-Martinez et al. 2016; Pagliano et al. 2018).

Interestingly, the highest maximum biogas production was found in the co-digestion with BSG. Therein, as compared to the SS mono-digestion, increases by 49 and 65% were achieved at HRT of 20 and 18 d, respectively. However, this tendency in the BSG presence did not correspond with its rates, mainly due to the composition and structure of this substrate. It was confirmed that AD of lignocellulose materials required the prolonged degradation time than other easily biodegradable substrates (Yadvika et al. 2004).

3.4 CONCLUSION

The presented results indicated the implementation of ACW resulted in an improvement of its kinetics. This effect resulted from the supplementation of feedstock with an easily biodegradable substrate rich in vitamins, mineral salts and proteins. However, a different tendency occurred in the BSG presence.

Despite the highest maximum biogas production, the negative influence on its kinetics was noticed. It was attributed to the application of hardly-biodegradable substrate, which is characterized by a complex structure. Furthermore, the prolonged HRT of 20 d is recommended for both wastes.

REFERENCES

Ahmed, B., Aboudi, K., Tyagi, V.K., Álvarez-Gallego, C.J., Fernández-Güelfo, L.A., Romero-García, L.I. & Kazmi, A.A. 2019. Improvement of Anaerobic Digestion of Lignocellulosic Biomass by Hydrothermal Pretreatment. *Appl. Sci.* 9: 3853.

Ahring, B. K., Biswas, R., Ahamed, A., Teller, P. J. & Uellendahl, H. 2015. Making lignin accessible for anaerobic digestion by wet-explosion pretreatment. *Bioresour. Technol.* 175: 182–188.

Angelidaki, I., & Batstone, D. J. 2011. Anaerobic Digestion: Process. *In Solid Waste Technology and Management* (Vol. 2.), Chichester, West Sussex, UK: Wiley.

Azbar, N., Çetinkaya-Dokgöz, F.T., Keskin, T., Korkmaz, K.S. & Syed, H.M. 2009. Continuous fermentative hydrogen production from cheese whey wastewater under thermophilic anaerobic conditions. *Int. J. Hydrog. Energy* 34: 7441–7447.

Biswas, J., Chowdhury, R. & Bhattacharya, P. 2007. Mathematical modeling for the prediction of biogas generation characteristics of an anaerobic digester based on food/vegetable residues. *Biomass and Bioenergy*. 31: 80–86.

Blonskaja, V. & Vaalu, T. 2006. Investigation of different schemes for anaerobic. treatment of food industry wastes in Estonia. *Proc. Est. Acad. Sci*. 55 (1): 14–28.

Bougrier, C., Dognin, D., Laroche, C. & Cacho Rivero, J.A. 2018. Use of trace elements addition for anaerobic digestion of brewer's spent grains. *J. Environ. Manage*. 223(1): 101–107.

Carvalho, F., Prazeres, A.R. & Rivas, J. 2013. Cheese whey wastewater: characterization and treatment. *Sci. Total. Environ*. 445–446: 385–396.

Chatzipaschali, A.A. & Stamatis, A.G. 2012. Biotechnological Utilization with a Focus on Anaerobic Treatment of Cheese Whey: Current Status and Prospects. *Energies* 5: 3492–3525.

Dereli, R.K., van der Zee, F.P., Ozturk, I. & van Lier, J.B. 2019. Treatment of cheese whey by a cross-flow anaerobic membrane bioreactor: Biological and filtration performance. *Environ. Res*. 168: 109–117.

Dos Santos-Mathias, T. R., Moretzsohn, de Mello, P.P. & Camporerese-Sérvulo E. F. 2014. Solid wastes in brewing process: A review. *J. Brew. Distilling*. 5 (1): 1–9.

Ferchichi, M., Crabbe, E., Hintz, W., Gil, G-H. & Almadidy, A. 2005. Influence of culture parameters on biological hydrogen production by Clostridium saccharoperbutylacetonicum ATCC 27021. *World J. Microbiol. Biotechnol*. 21: 855–862.

Fernández, C., Cuetos, M., Martínez, E., & Gómez, X. 2015. Thermophilic anaerobic digestion of cheese whey: Coupling H2 and CH4 production. *Biomass and bioenergy*, 81: 55–62.

Gonzalez-Estrella, J, Asato, C.M., Jerke, A.C., Stone, J.J. & Gilcrease, PC. 2017. Effect of structural carbohydrates and lignin content on the anaerobic digestion of paper and paper board materials by anaerobic granular sludge. *Biotechnol Bioeng.* 114(5): 951–960.

Hartmann, H., Angelidaki, I. & Arhing, B.K. 2003. Co-digestion of the organic fraction of municipal waste with other waste types. In: Mata-Alvarez, J. (Ed.), Biomethanization of the Organic Fraction of Municipal Solid Wastes. IWA Publishing, UK.

Hillion, M.L., Moscoviz, R., Trably, E., Leblanc, Y., Bernet, N., Torrijos, M. & Escudié, R. 2018. Co-ensiling as a new technique for long-term storage of agro-industrial waste with low sugar content prior to anaerobic digestion. *Waste Management* 71: 147–155.

Kafle, G.K, Kim, S.H. & Sung, K.I. 2013. Ensiling of fish industry waste for biogas production: a lab scale evaluation of biochemical methane potential (BMP) and kinetics. *Bioresour. Technol.* 127:326–336.

Koch, K., Helmreich, B. & Drewes, J.E. 2015. Co-digestion of food waste in municipal wastewater treatment plants: Effect of different mixtures on methane yield and hydrolysis rate constant. *Applied Energy* 137: 250–255.

Lee, I.S., Parameswaran, P. & Rittmann, B.E. 2011. Effects of solids retention time on methanogenesis in anaerobic digestion of thickened mixed sludge. *Bioresour Technol*. 102: 10266–10272.

Li, Y., Zhang, R., Liu, G., Chen, C., He, Y. & Liu, X. 2013. Comparison of methane production potential, biodegradability, and kinetics of different organic substrates. *Bioresour. Technol.*, 149: 565–569.

Maragkaki, A.E., Fountoulakis, M., Kyriakou, A., Lasaridi, K. & Manios, T. 2018. Boosting biogas production from sewage sludge by adding small amount of agro-industrial by-products and food waste residues. *Waste Manage*. 71: 605–611.

Mata-Alvarez, J., Dosta, J., Romero-Güiza, M.S., Fonoll, X., Peces, M. & Astals, S. 2014. A critical review on anaerobic co-digestion achievements between 2010 and 2013. *Renewable and Sustainable Energy Reviews* 36: 412–427.

Mata-Alvarez, J., Mace, S. & Llabres, P. 2000. Anaerobic digestion of organic solid wastes: an overview of research achievements and perspectives. *Bioresour. Technol.* 74 (1): 3–16.

Mussatto, S.I., Dragone G. & Roberto I.C. 2006. Brewers' spent grain: generation, characteristics and potential applications. *Journal of Cereal Science* 43(1): 1–14.

Panjičko, M., Zupančič, G.D., Fanedl, L., Logar, R. M. Tišma, M. & Zelić, B. 2017. Biogas production from brewery spent grain as a mono-substrate in a two-stage process composed of solid-state anaerobic digestion. *Journal of Cleaner Production* 166: 519–529.

Paul, S. & Dutta, A. 2018. Challenges and opportunities of lignocellulosic biomass for anaerobic digestion. *Resources, Conservation and Recycling* 130: 164–174.

Pilarska, A.A., Pilarski, K., Witaszek, K., Waliszewska, H., Zborowska, M., Waliszewska, B., Kolasiński, M., & Szwarc-Rzepka, K. 2016. Treatment of dairy waste by anaerobic co-digestion with sewage sludge, *Ecological Chemistry and Engineering S* 23(1): 99–115.

Poerschmann, J., Weiner, B., Wedwitschka, H., Baskyr, I., Koehler, R. & Kopinke, F.D. 2014. Characterization of biocoals and dissolved organic matter phases obtained upon hydrothermal carbonization of brewer's spent grain. *Bioresour. Technol.* 164, 162–169.

Prajapati, K.B. & Singh, R. 2020. Enhancement of biogas production in bio-electrochemical digester from agricultural waste mixed with wastewater. *Renewable Energy* 146: 460–468.

Retfalvi, T., Tukacs-Hajos, A. & Szabo, P. 2013. Effects of artificial overdosing of p-cresol and phenylacetic acid on the anaerobic fermentation of sugar beet pulp. *International Biodeterioration & Biodegradation* 83: 112–118.

Sanders, W.T.M., Geerink, M., Zeeman, G. & Lettinga, G. 2000. Anaerobic hydrolysis kinetics of particulate substrates. *Water Science and Technology* 41: 17–24.

Sawatdeenarunat, C., Surendra, K.C., Takara, D., Oechsner, H. & Khanal, S.K. 2015. Anaerobic digestion of lignocellulosic biomass: challenges and opportunities. *Bioresour. Technol.* 178: 178–186.

Sežun, M., Grilc, V., Zupančič, G.D. & Marinšek-Logar, R. 2011. Anaerobic digestion of brewery spent grain in a semi-continuous bioreactor: inhibition by phenolic degradation products. *ActaChimicaSlovenica* 58(1): 158–66.

Sosnowski, P., Wieczorek, A. & Ledakowicz, S. 2003.Anaerobic co-digestion of sewage sludge and organic fraction of municipal solid wastes. *Advances in Environmental Research* 7(3): 609–616.

Stamatelatou, K., Giantsiou, N., Diamantis, V., Alexandridis, C., Alexandridis, A. & Aivasidis, A. 2014. Biogas production from cheese whey wastewater: laboratory- and full-scale studies. *Water Science and Technology* 69: 1320–5.

Syaichurrozi, I., Budiyono & Sumardiono, S. 2013. Predicting kinetic model of biogas production and biodegradability organic materials: biogas production from vinasse at variation of COD/N ratio. *Bioresour Technol.* 149: 390–397.

Szaja, A. & Montusiewicz, A. 2019. Enhancing the co-digestion efficiency of sewage sludge and cheese whey using brewery spent grain as an additional substrate. *Bioresour. Technol.* 291: 1–9.

Szaja, A., Montusiewicz, A., Lebiocka, M. & Bis, M. 2020. The effect of brewery spent grain application on biogas yields and kinetics in co-digestion with sewage sludge. *PeerJ* (in press).

Wang, K., Yin, J., Shen, D. & Li, N. 2014. Anaerobic digestion of food waste for volatile fatty acids (VFAs) production with different types of inoculum: effect of pH. *Bioresour. Technol.* 161: 395–401.

Xie, S., Wickham, R. & Nghiem L.D. 2017. Synergistic effect from anaerobic co-digestion of sewage sludge and organic wastes. *Int. Biodeter. Biodegr.* 116: 191–197.

Yadvika, S., Sreekrishnan, T. R., Kohli ,S. & Rana V. 2004. Enhancement of biogas production from solid substrates using different techniques - a review. *Bioresour. Technol.* 95, 1–10.

Yang, L., Xu, F., Ge, X., & Li, Y. 2015. Challenges and strategies for solid-state anaerobic digestion of lignocellulosic biomass. *Renewable and Sustainable Energy Reviews* 44: 824–834.

Zahan, Z, Othman, M.Z. & Muster, T.H. 2018. Anaerobic digestion/co-digestion kinetic potentials of different agro-industrial wastes: A comparative batch study for C/N optimisation. *Waste Management* 71:663–674.

Zielewicz, E., Tytła, M., Liszczyk, G. 2012. Possibility of sewage sludge and acid whey co-digestion process. *Arch. Civil Eng. Environ.* 5 (1): 87–92.

CHAPTER 4

Application of the smoothing methods to the PV power output data for the evaluation of the performance photovoltaic system

S. Gułkowski
Department of Renewable Energy Engineering, Faculty of Environmental Engineering, Lublin University of Technology, Lublin, Poland

ABSTRACT: The analysis of the output data generated by large-scale photovoltaic power plant is of great importance not only from the scientific but also from the investment perspective, due to the possibility of system efficiency evaluation depending on many parameters related to the weather conditions, devices (PV modules, inverters) degradation or module soiling. Moreover, long-term system performance monitoring helps to detect system failures, and thus to evaluate the quality of the installation by system and array losses computations. The PV system performance evaluation requires the analysis of large sets of output data which is a complex and challenging task due to their dynamic behavior caused by a number of factors (irradiance, operating temperature, wind, partial shading by moving clouds). The purpose of this paper is the effectiveness analysis of the smoothing algorithms applied to the daily PV power output data collected during partly cloudy days. Such data is characterized by the highest power fluctuations. Three methods have been studied: locally weighted non-parametric regression (LOESS), moving average and Savitzky-Golay filtering. On the basis of the modified data of power production, the daily energy yields were computed and compared with the experimental results to find the smoothing technique of the highest accuracy. Relative mean error (%RE) was calculated. The results show an excellent agreement of the smoothed data with experimental measurements in terms of LOESS and Savitzky-Golay filtering.

4.1 INTRODUCTION

World today is faced with a few critical issues related to the environment degradation due to the increase of energy demands which leads to fossil fuels consumption and thus to the growing emission of carbon dioxide (CO_2) and other greenhouse gases (GHG) (Braungardt et al. 2019; Chepeliev et al. 2020; Shahnazi et al. 2020). The energy production without a negative impact on the environment is one of the most challenging tasks nowadays (Ahmad et al. 2020). According to European Directive 2009/28/EC on the Promotion of the Use of Energy from Renewable Sources and to further Directive 2015/2013 all European Countries are required to increase the share of renewable sources in the final consumption of energy (Directive 2009/28/ec, Directive (EU) 2015/1513). Moreover, the proposed scenario assumes not only a 20% increase of renewable energy consumption in 2020 compared to 1990 but also a 20% reduction of CO_2 emission. It is worth mentioning that many EU countries achieved the targets for 2020 much earlier, in 2015. However, in that year, the renewable energy shares in the EU member states varied from 5% to 54%. The assumption of the European REmap is a significant reduction of these differences (Janeiro et al. 2018). Among the environmentally friendly technologies which enable access to clean energy production, photovoltaics is one of the most promising mostly due to the inexhaustible energy source, little maintenance costs, and flexibility related to energy demands. In the last years, photovoltaic market experienced fast growth reaching 488 GWp of total installed power capacity all over the World (end of 2018). In the European Union this value was found to be 119 GWp including 46 GWp in Germany (Photovoltaics Report 2017). The total installed power of PV systems in Poland was found to be about 0.7 GWp in April 2019 (Photovoltaics Report 2020). This value cannot be considered impressive, comparing to the largest European members; however, it

DOI 10.1201/9781003171669-4

should be emphasized that the most of the PV systems were installed in the last few years. The forecast for the future growth of PV market in Poland is even more optimistic.

With the worldwide development of photovoltaic solar systems, performance evaluation under different climate conditions are of high importance. It enables to control the installation quality by analyzing the PV system losses (Kymakis et al. 2009; Zdyb & Gulkowski 2020). Site specific evaluation studies allow investigating the influence of climate conditions on the performance of PV modules and thus determining the most suitable technology for a given location (Sharma et al. 2013; Zdyb et al. 2017). The impact of the temperature on the efficiency of different technology modules, spectral sensitivity or annual degradation rate analyses help to predict the long-term energy production (Huld et al. 2010; Jordan et al. 2018; Solís-Alemán et al. 2019). Another important field of research involves the modeling studies of the photovoltaic modules related to their I-V characterization under given weather conditions. It should be emphasized that the characteristic parameters of the PV device provided by manufacturer are measured under standard test conditions which are, in most cases, different than the real working conditions. The simulations of the I-V curves based on single or two-diode equivalent circuit enable to extract the real parameters of the module and thus to provide its behavior under any irradiance or temperature conditions (Gulkowski et al. 2019). Comprehensive studies of photovoltaic systems provide a source of knowledge not only for researchers but also for software developers, who can improve their algorithms for higher accuracy forecast of the PV power production. Many commercial software products (e.g. PVSyst, PVSol, DDS CAD, Homer) are successfully used by the PV installers to design the highest efficiency systems for small households as well as large PV plants (Ahsan et al. 2016; Krawczak 2018; Krawczak 2019; Pabasara et al. 2019). In all cases, the economic and environmental aspects play a fundamental role (Gradziuk et al. 2019; Zdyb et al. 2019).

This article presents the comparison of the averaging approach to the PV power output measured during partly cloudy days, which are characterized by the highest dynamic changes of power. Three methods (LOESS, moving average and Savitzky-Golay) were studied with a time span ranging from 5% to 40%, to analyze their sensitivity to power changes. Relative mean errors were computed to evaluate the accuracy of the methods. The previously published findings related to the application of the Savitzky-Golay (S-G) filtering (Gulkowski 2015) showed a good agreement between the computed and measured daily energy production of the PV system. The results presented in this work confirmed the low relative mean error of S-G filtering for a wide range of time span.

However, the LOESS method revealed better accuracy at higher values of time span. The precision of the moving average algorithm was found to be much lower.

4.2 EXPERIMENTAL FACILITY AND DATA ACQUISITION

The photovoltaic system used in this research is shown in Figure 4.1. It consists of 85 polycrystalline silicon (pc-Si) modules of 250 Wp each and is a part of a large-scale 1.4 MWp power plant located in East Poland (latitude 51° 51′N, longitude 23° 10′E). The modules are south-oriented. The optimal inclination angle was chosen (34°). The total power capacity of the investigated installation is 21.25 kWp. In order to avoid the shading effects, the modules were installed 0.5 m above the ground. The distance between rows is 6.3 m.

The selected string is grid-connected via inverter with nominal power equal to 20 kW and relatively high efficiency equals 98%. The inverter is equipped with two inputs of maximum power point tracking systems. The PV data is collected by the central data-logging computer system connected to the inverter. Beside the power output and energy data, the irradiance at the location of the experimental system and temperature of the modules were also registered synchronously every 5 minutes during the measurement day.

Figure 4.2 illustrates the daily PV power output profile normalized to the 1 kWp measured in June during the sunny and partly cloudy day for comparison.

As can be seen, the PV output power (in June) is available from 5.30 a.m. until 8 p.m. Within this range, 150 experimental points were collected. In the location of the solar plant (temperate climate)

Figure 4.1. Experimental setup consisted of one string of pc-Si modules.

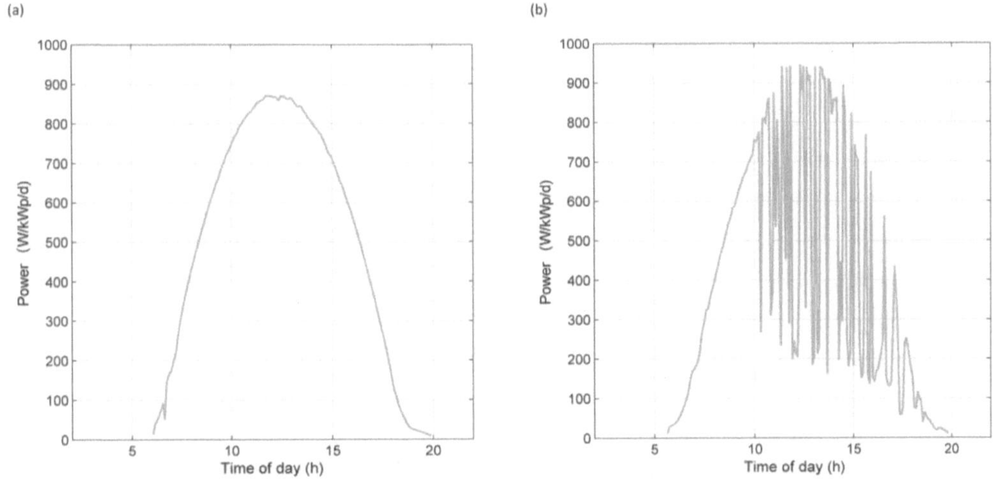

Figure 4.2. PV power output profile measured during sunny (a) and partly cloudy (b) day.

many days of the year are partly cloudy which are characterized by a high level of fluctuations of the output power (Figure 4.2b). Due to highly fluctuating data, it is difficult to forecast the average level of power production and thus the efficiency of the system. The application of the smoothing methods, especially for highly oscillating data, enhances the average level of power and energy estimation.

The basic idea of such methods is to replace the data points by some kinds of a local average of surrounding data points (Brandt 2014). For example, Locally-Weighted Regression (LOESS) uses a linear regression model for a given data point (power in the case of this study). The significant parameter of each analyzed method is the time span parameter which determines the number of the nearest neighbors taken into account for the estimation of a given data point. The smoothing methods used in this research are well known and widely used in many applications. A detailed description can be found in (Cleveland et al. 1988; Press et al. 2007; Schafer et al. 2011).

For each analyzed method, the energy calculated as an area under the averaged power output data was compared with the experimental value registered by the inverter. Relative mean error (RE) of daily energy defined by the eq. (4.1) was calculated.

$$RE\ (\%) = \frac{|E_c - E_M|}{E_M} \tag{4.1}$$

Subscripts 'C' and 'M' denote the computed and measured energy values. All computations were carried out using the Matlab/Simulink software with the implemented averaging methods.

4.3 RESULTS AND DISCUSSION

Figure 4.3 shows the results of the smoothing algorithms application to the PV power output data collected in one day of June. As can be seen from the raw data presented in Figure 4.3a, the day was characterized by a relatively high PV power production with instantaneously power drop between 10 a.m. and 1 p.m., which led to decrease of energy production. Another aspect of lower energy

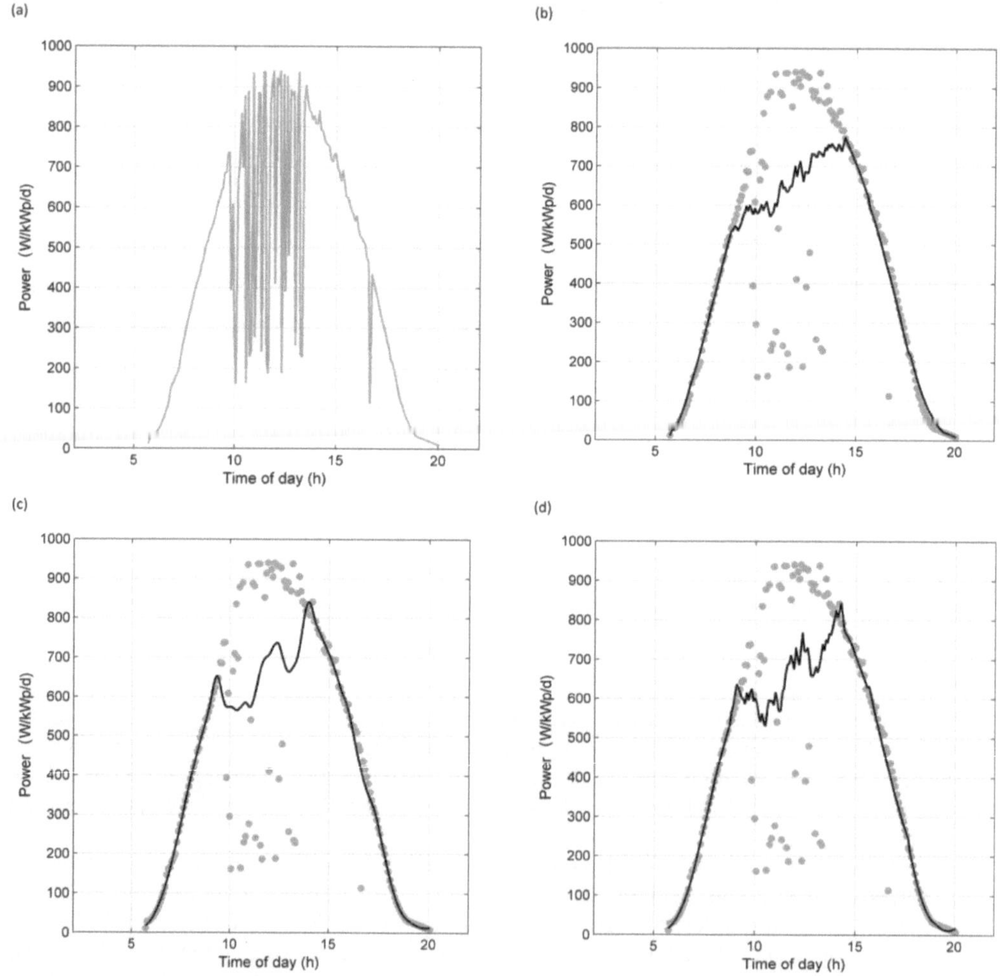

Figure 4.3. a) PV power output data registered during partly cloudy day and the results of smoothing with the use of b) moving average. c) locally weighted regression (LOESS) and d) Savitzky-Golay methods with 20% of span.

production is the temperature influence on the efficiency of the PV modules as it was discussed in previous research (Gulkowski et al. 2019). The energy registered in the analyzed day was 6.36 kWh/kWp/d. The results of power data after filtering with the use of three analyzed methods were illustrated in Figures 4.3b-d. The span parameter was equal to 20 % of the data points. It should be noticed that 10% of the span is corresponding to 15 experimental points and thus to 75 minutes of the measurement day. As can be seen, each method is characterized by a different sensitivity on the PV power changes. The best smoothing effect was found for the LOESS algorithm. However, the differences in energies computed on the basis of average power values (after filtering) were relatively low. Each value was found in the range between 6.29 kWh/kWp/d and 6.36 kWh/kWp/d. Relative errors of locally-weighted regression and Savitzky–Golay filtering were 0.017% and 0.021%, respectively. In terms of moving average, the RE(%) was higher and equaled to 1.14%. Figure 4.4 illustrates the results of smoothing the PV power data using the LOESS algorithm with higher values of span (up to 40%).

As can be seen in Figure 4.4b with 40% (5 hours) of the nearest neighbors taken into account for smoothing the fluctuations of power output data are almost invisible. Moreover, further analysis showed that the amount of energy calculated from the smoothed data with higher span values was not changed significantly. The values of 6.38 kWh/kWp/d with the span of 30% and 6.40 kWh/kWp/d with the span of 40 % were noticed. Percentage relative error was found to be 0.22 % and 0.69% respectively. It leads to the conclusion that the locally-weighted regression can be used for the PV power output data modeling without considerable changes of the energy yields. These findings can be significant for the PV studies on real system efficiency as well as system losses calculations.

Figures 4.5 shows the span parameter influence on relative error (%RE) for all three methods under study. Another partly cloudy day with the total energy production of 4.68 kWh/kWp/d was chosen in this case. As can be seen from the Figure 4.5a, normalized power output was dynamically changing during the whole day in the range from 150 W to about 950 W.

The calculated %RE value increased along with the increase of span value. For span up to 10% all three studied methods revealed high precision of the daily energy computation based on the smoothed data. However, in terms of the moving average method, at higher levels data, filtering led to a significant increase of the relative error even up to 6%. The calculated energy production was found to be 4.39 kWh/kWp/d. Such modification of the energy data is not acceptable for the system performance analysis. Much better results were obtained for two other methods for which relative error was below 1%. With the use of Savitzky-Golay filtering, the daily energy was found to be 4.71 kWh/kWp/d with a 40% span (0.74% of RE). Locally-weighted regression revealed the best agreement of the calculated energy with the energy collected by the inverter (4.69 kWh/kWp/d with 0.43 % of RE).

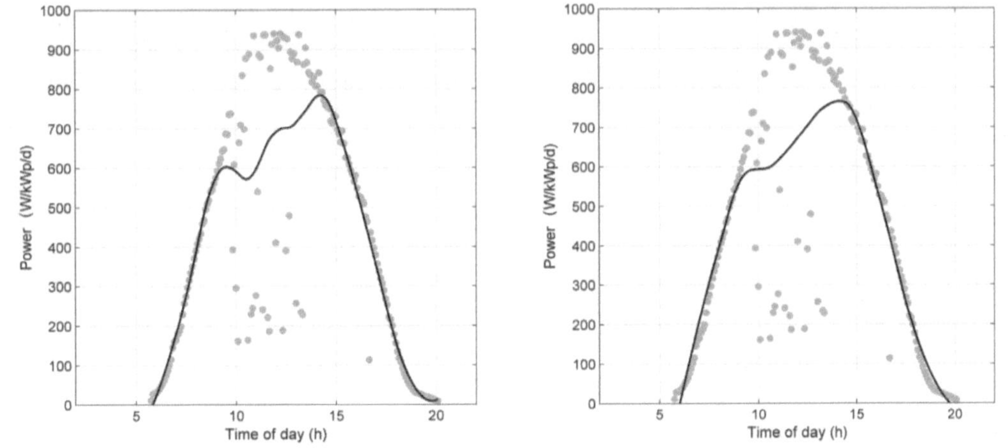

Figure 4.4. The effect of PV power output data averaging using LOESS method with span equals a) 30 % and b) 40 %

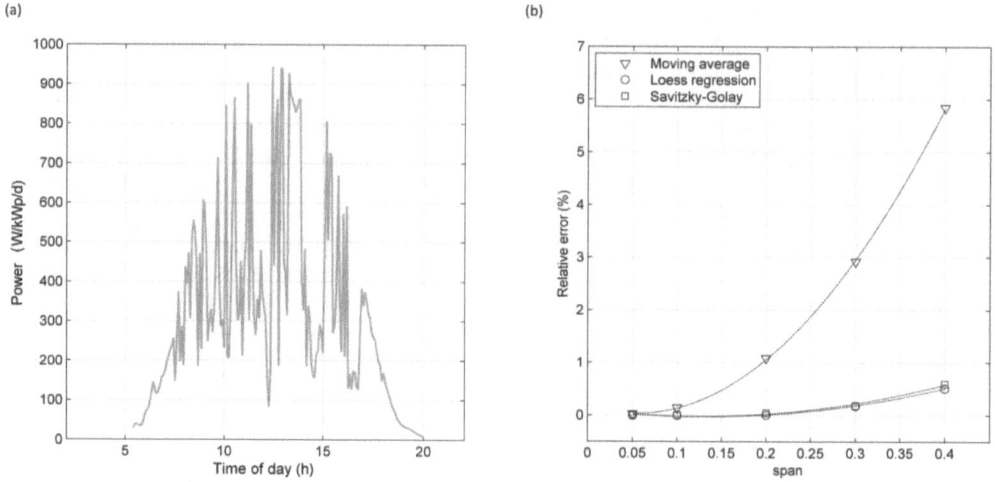

Figure 4.5. a) PV power output data profile and b) the influence of span on relative mean error of smoothing methods.

4.4 CONCLUSION

Large-scale photovoltaic power plant, aside from energy production, generates large amount of useful data which can be used for system failures detection as well as performance evaluation. Some of the data, not only PV power output but also irradiance or wind speed, has to be averaged, especially under temperate climate conditions. This research showed the results of the PV power output data filtering using moving average algorithm, locally-weighted regression (LOESS) and Savitzky-Golay methods for the efficiency evaluation purposes. On the basis of averaged power data, daily energy was computed and compared with the experimental values measured by the inverter. Relative mean error was calculated to evaluate the precision of each smoothing method in terms of the span parameter. The results showed that for the most fluctuating power data collected during a partly cloudy day, the LOESS method was characterized by the lowest relative error (below 0.7 % with the span up to 40%). The best smoothing effect (sensitivity on power changes) was also confirmed for this method. The relative error for Savitzky-Golay filtering was below 0.75% for the same conditions. On the contrary, total daily energy calculated on the basis of modified data using the moving average was changed significantly. RE (%) was found to be even 6% with the span equal to 40%. This observation brings the conclusion that both the LOESS and Savitzky-Golay methods were precise enough for the fluctuating PV output data filtering without significant changes of the final energy yield. The moving average algorithm can be useful with low values of span (below 10%).

REFERENCES

Ahmad, T., Zhang, H., Yan, B., 2020. A review on renewable energy and electricity requirement forecasting models for smart grid and buildings. *Sustainable Cities and Society* 55: 102052.

Ahsan, S., Javed, K., Rana, S., Zeeshan, M. 2016. Design and cost analysis of 1kW photovoltaic system based on actual performance in Indian scenario, *Perspectives in Science* 8: 642–644.

Brandt, S. 1999. Data Analysis. Statistical and Computational Methods for Scientists and Engineers. Springer Verlag, New York, USA.

Braungardt, S., Van den Bergh, J., Dunlop, T. 2019. Fossil fuel divestment and climate change: Reviewing contested arguments, *Energy Research & Social Science* 50: 191–200.

Chepeliev, M., Mensbrugghe, D. 2020. Global fossil-fuel subsidy reform and Paris Agreement, *Energy Economics* 85: 104598.

Cleveland, W.S., Devlin, J.S. 1988. Locally Weighted Regression: An Approach to Regression Analysis by Local Fitting, Journal of the American Statistical Association 83(403): 596–610.

Directive 2009/28/ec of the European Parliament and of the Council of 23 April 2009 on the Promotion of the Use of Energy from Renewable Sources. Available online:https://eur-lex.europa.eu/LexUriServ/LexUriServ.do?uri=OJ:L:2009:140:0016:0062:EN:PDF (accessed on April 2020).

Gradziuk, P., Gradziuk, B. 2019. Economic profitability of investment in a photovoltaic plant in south-east Poland. *Annals of the Polish Association of Agricultural and Agribusiness Economists* 21: 124.

Gulkowski, S. 2015. Application of the Savitzky-Golay method for PV power output data set. Task QUARTERLY 19(1): 25–34.

Gulkowski, S., Zdyb, A., Dragan, P. 2019. Experimental Efficiency Analysis of a Photovoltaic System with Different Module Technologies under Temperate Climate Conditions. *Applied Sciences* 9(1): 141.

Gulkowski, S., Muñoz Díez, J. V., Tejero Aguilera, J., Nofuentes, G. 2019. Computational modeling and experimental analysis of heterojunction with intrinsic thin-layer photovoltaic module under different environmental conditions, *Energy* 172: 380–390.

Huld, T., Gottschalg, R., Beyer, H., Topič, M. 2010. Mapping the performance of PV modules, effects of module type and data averaging, *Solar Energy* 84(2): 324–338.

Institute for Renewable Energy (IEO). 2019. Report on PV Market in Poland – 2020. Warsaw, Poland.

Janeiro, L., Gutierrez, L., Prakash, G., Saygin, D. 2018. IRENA REmap EU 2018 - Renewable energy prospects for the European Union, REmap analysis to 2030.

Jordan, D., Deline, C., Kurtz, R., Kimball, G. Anderson, M. 2018. Robust PV Degradation Methodology and Application, *IEEE Journal of Photovoltaics* 8(2): 525–531.

Krawczak, E. 2018. Energy, economical and ecological analysis of a single-family house using photovoltaic installation. E3S Web Conf. 49.

Krawczak, E. 2019. Studies on PV power plant designing to fulfil the energy demand of small community in Poland. E3S Web Conf. 116.

Kymakis, E., Kalykakis, S., Papazoglou, T. 2009. Performance analysis of a grid connected photovoltaic park on the island of Crete, *Energy Conversion and Management* 50(3): 433–438.

Press, W.H., Vetterling, W.T., Teukolsky, S.A., Flannery B.P. 2007. Numerical Recipes in C:The Art of Scientific Computing, Third Edition, Cambridge University Press, Cambridge, United Kingdom.

Pabasara, W.M., Wijeratne, U., Yang, R., Too, E., Wakefield, R. 2019. Design and development of distributed solar PV systems: Do the current tools work?, *Sustainable Cities and Society* 45: 553–578.

Photovoltaics Report. 2017. Fraunhofer Institute for Solar Energy Systems, ISE with support of PSE AG; Freiburg, Germany.

Schafer, R.W. 2011. What Is a Savitzky-Golay Filter?, *IEEE Signal Processing Magazine* 28(4): 111–117.

Sharma, V., Kumar, A., Sastry, O, Chandel, S. 2013. Performance assessment of different solar photovoltaic technologies under similar outdoor conditions. *Energy* 58: 511–518.

Shahnazi, R., Shabani, Z. 2020. Do renewable energy production spillovers matter in the EU?, Renewable Energy 150: 786–796.

Solís-Alemán, E., De La Casa, J., Romero-Fiances, I., Silva, J. P., Nofuentes, G. 2019. A study on the degradation rates and the linearity of the performance decline of various thin film PV technologies, *Solar Energy* 188: 813–824.

Zdyb, A., Żelazna, A., Krawczak, E. 2019. Photovoltaic System Integrated Into the Noise Barrier – Energy Performance and Life Cycle Assessment, *Journal of Ecological Engineering* 20(10):183–188.

Zdyb, A., Gulkowski, S. 2020. Performance Assessment of Four Different Photovoltaic Technologies in Poland. *Energies* 13(1): 196.

Zdyb A., Krawczak, E. 2017. The influence of external conditions on the photovoltaic modules performance. In Pawłowska M., Pawłowski (ed.), Environmental Engineering V, CRC Press Taylor&Francis Group, 261–266.

CHAPTER 5

Catechol dye as a sensitizer in dye sensitized solar cells

A. Zdyb & E. Krawczak
Lublin University of Technology, Lublin, Poland

ABSTRACT: The main trends in the research on dye-sensitized solar cells include investigation of new compounds that can act as sensitizers which exhibit a broad range of light absorption. This work presents the evaluation of the photovoltaic performance of catechol in dye cells. The energies of HOMO (highest occupied molecular orbital) and LUMO (lowest occupied molecular orbital) in a catechol molecule were determined and compared with the conductive band edge of the used semiconductor and redox potential of the electrolyte. The photoelectrical properties of catechol were studied based on the calculated chemical reactivity parameters. The cells sensitized with catechol were prepared and tested under standard conditions.

5.1 INTRODUCTION

Solar energy can be harnessed in all parts of the globe, regardless of the geopolitical relations. The irradiation is of the highest value (1300–2500 kWh/m^2) in the regions within the tropical belt, especially in the mountains; however, in high latitude countries, where the irradiation is in the range of 900–1200 kWh/m^2, conversion of solar radiation into other forms of energy also gains popularity. Photovoltaics (PV) is the technology that enables to use the solar radiation and produce the most desired energy form, which is electricity. Over the recent years a growth of investments in the research on photovoltaic cells is observed. Numerous studies on the photovoltaic performance have also been performed both at the locations characterized by good insolation (Belluardo 2015; Gaglia 2017; Romero-Fiances 2019) as well as in the places located at higher latitude, where significant differences between the summer and winter months are observed (Gulkowski 2019; Louwen 2017).

In general, photovoltaic cells can be categorized into three generations:

- first generation – monocrystalline, polycrystalline and microcrystalline silicon cells,
- second generation – thin film cells based on amorphous silicon, cadmium telluride or copper indium gallium diselenide,
- third generation– emerging technologies including dye-sensitized solar cells (DSSC), devices based on organic semiconductors (OPV) or perovskits (Hagfeldt 2018).

The dye-sensitized solar cells that are the object of the presented study display low materials use, easy assembly and recycling process, the efficiency of 11.9% for the single cell, 10.7% for minimodule and 8.8% for submodule (Green 2020). DSSC can be applied in BIPV due to semitransparency and possible different colors. DSSC market is expected to grow rapidly, since this technology has the potential to reduce the price below \$0.5/Wp (Mariani 2015; Sharifi 2014). Additionally, the energy-payback time for DSSC is less than 1 year while it is 1 to over 2 years for other photovoltaic technologies (Report PV 2019).

DSSC, as a photoelectrochemical PV device, consists of two electrodes and electrolyte between them (Archer & Nozik 2010; Grätzel 2001; Sharma 2018). The most important processes that influence the DSSC performance occur within the illuminated electrode which is conductive glass covered by a mesoporous layer of titanium dioxide nanoparticles with dye molecules adsorbed on the surface. The photons of light can be absorbed by the dye molecules that work as a sensitizer which broadens the absorption spectrum, since TiO_2as a wide bandgap semiconductor absorbs only the UV light. Upon

DOI 10.1201/9781003171669-5

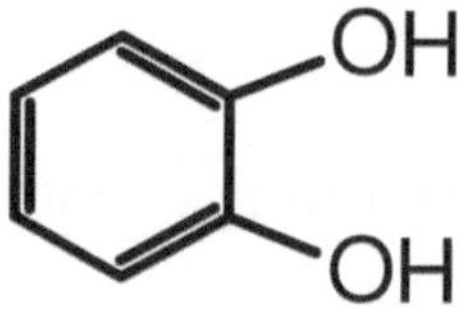

Figure 5.1. Scheme of a catechol molecule.

photoexcitation of the dye molecule, the electrons are transferred from the excited state of the dye into the conduction band of TiO_2 (Krawczyk et al. 2018; Listorti et al. 2011), collected by conductive layer of fluorine-doped indium tin oxide (FTO) and then pass through the external circuit delivering electrical power. The illuminated electrode (photoanode) is in contact with electrolyte that provides electrons to regenerate the dye ground state.

The heart of the dye cell is the sensitizer that has to fulfill some requirements to perform well: it has to pose anchoring groups (e.g. carboxylate, phosphonate) to firmly graft to the surface of semiconductor nanoparticles, absorb solar light in a wide range, its HOMO (highest occupied molecular orbital) energy level should be higher than the bottom of the TiO_2 conduction band, LUMO (lowest occupied molecular orbital) has to be located below the redox level of electrolyte and finally it should be resistant to contact with the electrolyte (Hagfeldt & Grätzel 2000). The process of fundamental importance to DSSC performance is the charge transfer from the excited dye molecules to the semiconductor (Zdyb 2014; Zdyb 2016). This process is influenced by the mode of binding and the rate of the electron injection (Nieto-Pescador et al. 2014; Zigler et al. 2016).

The search for efficient sensitizers, both metal-containing (Ambre et al. 2015; Kaewin et al. 2017) as well as metal-free compounds (Naik et al. 2018; Naik et al. 2017) and natural dyes among them (Kabir 2019; Krawczak & Zdyb 2019; Richhariya 2017; Shalini 2016), is a topic widely addressed in literature. Very good photovoltaic performance was achieved with the complexes of ruthenium (Nazeeruddin 2005); however, low abundance and high price of ruthenium lead to the investigations aiming at the development of different, new sensitizing substances. One of the metal-free dyes is catechol, which can be a good candidate for a sensitizer in DSSC, according to the theoretical and experimental investigations (Sánchez-de-Armas 2011). Catechol is an enediol compound with the molecular formula $C_6H_4(OH)_2$ that offers two hydroxyl groups attached to one of the carbon double bonds in the benzene ring (Figure 5.1) and can bind to the TiO_2 nanoparticle in three different modes (Redfern 2003). Good electronic coupling is observed between TiO_2 and catechol, which was confirmed by UV-Vis absorption measurements. Adsorption of catechol on titanium dioxide nanoparticles results in the appearance of new absorption band ascribed to the charge transfer process, which can be beneficial for the dye cell operation (Nawrocka 2009). This new band is broad and enables to absorb solar radiation of the wavelength over 500 nm while free catechol, before the adsorption process absorbs only in UV with a narrow maximum around 275 nm.

The promising combination of the described features of catechol was a source of motivation for applying this dye in DSSC device. At first, in this study the HOMO and LUMO energies in catechol/TiO_2 assembly were determined and their position in relation to the TiO_2 conduction band was evaluated. Then, the dye cells sensitized with catechol were prepared and their characteristic parameters were derived from the experimental measurements.

5.2 METHODOLOGY

The TiO_2 covered photoanodes and counter electrodes were bought from the Dyesol company. In order to carry out the dye adsorption process on the TiO_2 electrode, the 1 mM solution of catechol was prepared and then the electrode was immersed in the ethanolic solution for 4 h. Both electrodes were assembled using the Surlyn sealing foil and heated. The I^-/I_3^- electrolyte was introduced into the cell through the small hole in the counter electrode. The active area of the cell was equal to 0.7 cm^2. The chemicals used in dye cells were purchased from Sigma and used as received. In order to perform the measurements of the current-voltage characteristics of the prepared cells solar simulator class AAA, Abet Technologies was used. The simulator is equipped with a xenon lamp that provides the irradiance of 1000 W/m^2.

5.3 RESULTS AND DISCUSSION

5.3.1 *Energy levels in TiO2/catechol/electrolyte interface*

Charge separation and electrons movement in a dye cell is driven by the proper position of energy levels in TiO_2/dye/electrolyte interface. The HOMO level (E_{HOMO}) of catechol is equal to 1.18 V vs. NHE scale (Hod 2010), thus it is -5.62 eV on the absolute scale. The HOMO position is thus higher of 0.902 eV than the redox potential of the electrolyte redox couple (E_{el}) I^-/I_3^-, which is equal to -4.718 eV. It means that the HOMO level of catechol is compatible with the redox level of the electrolyte which enables regeneration of the catechol ground state during the solar cell operation.

In order to determine the LUMO level (E_{LUMO}), the transition energy of the absorption maximum has to be considered. According to the previous investigations (Nawrocka et al. 2009), the maximum of catechol/TiO_2 absorption occurs at 400 nm, so the optical transition energy can be calculated according to the following formula:

$$E = \frac{hc}{\lambda}, \tag{5.1}$$

in which h – Planck constant, c– speed of light, λ– wavelength.

The energy of transition equals 3.1 eV thus the LUMO level can be calculated as: -5.62 eV $+3.1$ eV $= -2.52$ eV according to the rule of determination of LUMO level. The LUMO level for catechol is located 1.78 eV higher than the TiO_2 conduction band (E_{CB}) which is at -4.3 eV. This kind of relation between LUMO of the sensitizer and the semiconductor conduction band is required to ensure the efficient electron transfer into the semiconductor conductionband. A schematic presentation of ground and excited energy levels of the dye, the electrolyte redox potential, as well as the electron transfer mechanism in DSSC are shown in Figure 5.2.

The electron transfer reactions are described quantitatively by free energy driving force Δ_{Ginj} and electron regeneration driving force ΔG_{reg} defined by the following equations:

$$\Delta G_{inj} = E_{LUMO} - E_{CB} \tag{5.2}$$

$$\Delta G_{reg} = E_{HOMO} - E_{el} \tag{5.3}$$

For greater ΔG_{inj} more effective electron transfer is observed, contrary to ΔG_{reg} for which lower values are beneficial (Galappaththi et al. 2018, Islam et al. 2003).

The photoelectrical properties of the dye can also be evaluated based on the chemical reactivity parameters such as: chemical potential μ, global hardness η andelectrophilicity index ω. The better photovoltaic performance is expected for higher electroaccepting power and electrophilicity as well as lower chemical hardness (Soto-Rojo 2015; Soto-Rojo 2016). The descriptors of reactivity can be determined according to the following expressions (Parr 1999; Parthasarati 2004; Ruiz-Anchondo 2010):

$$\mu \approx \frac{1}{2}\left(E_{LUMO} + E_{HOMO}\right) \tag{5.4}$$

$$\eta \approx \frac{1}{2}\left(E_{LUMO} - E_{HOMO}\right) \tag{5.5}$$

$$\omega = \frac{\mu^2}{2\eta} \approx \frac{\left(E_{LUMO} + E_{HOMO}\right)^2}{4\left(E_{LUMO} - E_{HOMO}\right)}. \tag{5.6}$$

Table 5.1 summarized the values of the reactivity parameters as well as theoretical maximum open circuit voltage in eV calculated for the catechol sensitized dye cell as: $V_{OC} = E_{HOMO} - E_{CB}$.

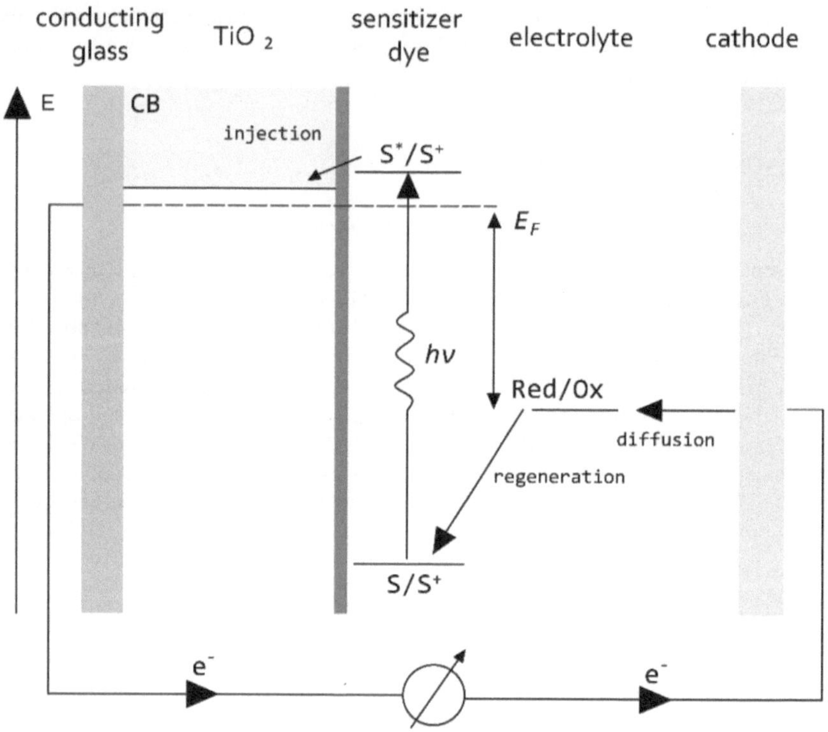

Figure 5.2. Scheme of working mechanism of catechol-sensitized solar cell. The energy levels and electron transfer reactions are indicated.

Table 5.1. The chemical potential μ, global hardness η and electrophilicity index ω of catechol dye.

Dye	ΔG_{inj} (eV)	ΔG_{reg} (eV)	V_{OC} (eV)	μ (eV)	η (eV)	ω (eV)
catechol	1.78	0.902	1.93	4.07	1.55	5.34

The determined values of chemical potential, global hardness and electrophilicity index are similar to those found for the carotenoid dyes qualified as good sensitizers upon a theoretical study (Ruiz-Anchondo 2010).

5.3.2 *Photovoltaic characteristic*

The overall photovoltaic conversion of a dye cell is determined by current-voltage ($I - V$) characteristic and electric parameters such as: short circuit current I_{SC} and open circuit voltage V_{OC}. The measure of the squareness of the $I - V$ curve that increases for larger maximum power from the solar cell, is the fill factor (Sharma 2018):

$$FF = \frac{P_M}{V_{OC} I_{SC}} = \frac{V_M I_M}{V_{OC} I_{SC}}, \tag{5.7}$$

where P_M– the maximum power, V_M – voltage at the maximum power point, I_M – current at the maximum power point.

The efficiency of the cell can be determined according to the equation:

$$\eta = \frac{P_M}{P_{in}} = \frac{FFV_{OC}I_{SC}}{P_{in}}, \tag{5.8}$$

where P_{in} – the incident light power.

Figure 5.3 shows a current–voltage curve for the cell based on the TiO2 prefabricated Dysol electrodes using catechol as a sensitizer.

The photovoltaic performance of the cell sensitized with the catechol dye is depicted in Figure 5.3 and the corresponding parameters derived from the presented current-voltage characteristic are collected in Table 5.2.

The performance of catechol can be juxtaposed with the pyranoflavylium dyes possessing a catechol linkage unit in their structure. The reported the V_{OC} and V_M values obtained for this kind of dyes are in the range of 252–338 mV and 172–233 mV respectively, which is the same order as the values achieved in the presented work. However, the current parameters (I_{SC} and I_M) obtained for pure catechol are not satisfactory and indicate a problem with series resistance of the prepared cells. The findings of this study include also fill factor $FF = 30.65\%$ and efficiency $\eta = 0.003\%$, determined based on current and voltage data, according to Eq. 5.7, 5.8. The obtained fill factor is comparable with the values of 26–69% presented in recent experimental work dedicated for use of natural dyes (Ferreira et al. 2020).

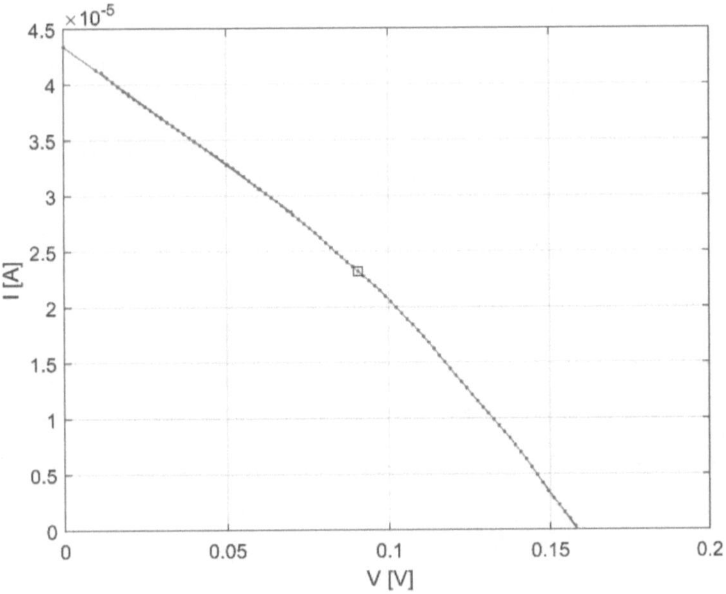

Figure 5.3. Current-voltage characteristic of the catechol sensitized cell.

Table 5.2. Photovoltaic parameters of catechol sensitized cell.

Characteristic parameters of catechol sensitized cell	I_{SC} [mA]	V_{OC}[mV]	I_M [mA]	V_M [mV]	P_M [mW]
	0.0434	158	0.0232	90.66	0.0021

5.4 CONCLUSIONS

Catechol is the compound that adsorbs on the titanium dioxide nanoparticles creating a chelate bond through two hydroxy groups and in consequence broadens the range of the absorbed solar radiation. In this paper, the assessment of catechol as a sensitizing compound in the dye-sensitized photovoltaic cells was presented. The location of the excited state of catechol molecule after adsorption on theTiO_2 nanoparticles and the differences between particular energy levels in the catechol-sensitized DSSC indicate that catechol HOMO and LUMO positions are compatible for efficient charge transfer in the cell. The obtained values of chemical reactivity parameters confirm the usefulness of catechol as a sensitizing dye. Photovoltaic characterization providing $I - V$ curve and the parameters of the cell exhibit satisfactory values of voltage. However, current and – by extention – efficiency, need further improvement through the optimization of catechol solution concentration and the method of the electric contacts deposition.

REFERENCES

Ambre, R.B., Mane, S.B., Chang, G.F., Hung, C.H., 2015. Effects of number and position of meta and para carboxyphenyl groups of zinc porphyrins in dye-sensitized solar cells: structure-performance relationship. ACS Appl. Mater. Interfaces 7: 1879–1891. https://doi.org/10.1021/am507503d.

Archer, M.D., Nozik, A.J. 2010. Nanostructured and photoelectrochemical systems for solar photon conversion, Imperial College Press, London.

Belluardo, G., Ingenhoven, P., Sparber, W., Wagner, J., Weihs, P., Moser, D. 2015. Novel method for the improvement in the evaluation of outdoor performance loss rate in different PV technologies and comparison with two other methods. Sol. Energy 117: 139–152. doi:10.1016/j.solener.2015.04.030.

Ferreira, F.C., Babua, R.S., de Barros, A.L.F., Raja, S., da Conceição, L.R.B., Mattoso, L.H.C. 2020. Photoelectric performance evaluation of DSSCs using the dye extractedfrom different color petals of Leucanthemum vulgare flowers as novel sensitizers. Spectrochimica Acta Part A: Molecular and Biomolecular Spectroscopy 233: 118198–118206. doi.org/10.1016/j.saa.2020.118198.

Gaglia, A.G., Lykoudis, S., Argiriou, A.A., Balaras, C.A., Dialynas, E. 2017. Energy efficiency of PV panels under real outdoor conditions—An experimental assessment in Athens, Greece. Renew. Energy 101: 236–243. doi:10.1016/j.renene.2016.08.051.

Gunappathi, R., Lim A., Ekanayake, P., Petra, M.I. 2018. A rational design of high efficient and low-cost dye sensitizer with exceptional absorptions: computational study of cyanidin based organic sensitizer. Sol. Energy 161: 83–89. https://doi.org/10.1016/j.solener.2017.12.027.

Grätzel, M. 2001. Photoelectrochemical cells, Nature 414: 338–344.

Green, M.A., Dunlop, E.D., Hohl-Ebinger J., Yoshita M., Kopidakis N., Ho-Baillie A.W.Y. 2020. Solar cell efficiency tables, Progress in Photovoltaics 28: 3–15. https://doi.org/10.1002/pip.3228.

Gułkowski, S., Zdyb, A., Dragan, P. 2019. Experimental Efficiency Analysis of a Photovoltaic System with Different Module Technologies under Temperate Climate Conditions. Appl. Sci. 9: 141–154. https://doi.org/10.3390/app9010141.

Hagfeldt, A., Cappel, U.B., Boschloo, G., Sun, L., Kloo, L., Pettersson, H., Gibson, E.A. 2012. Dye-Sensitized Photoelectrochemical Cells (pp. 501–560). In Kalogirou, S.A (ed.) *McEvoy's Handbook of Photovoltaics: Fundamentals and Applications*. Elsevier 2018. Academic Press. London.

Hagfeldt, A., Grätzel, M. 2000. Molecular Photovoltaics. Acc. Chem. Res. 33: 269–277. https://doi.org/10.1021/ar980112j.

Islam, A., Sugihara, H., Arakawa, H. 2003. Molecular design of ruthenium(II) polypyridyl photosensitizers for efficient nanocrystalline TiO2 solar cells. J. Photochem. Photobiol. A 158: 131–138. doi:10.1016/S1010-6030(03)00027-3.

Keawin, T., Tarsang, R., Sirithip, K., Prachumrak, N., Sudyoadsuk, T., Namuangruk, S.,

Roncali, J., Kungwan, N., Promarak, V., Jungsuttiwong, S., 2017. Anchoring number performance relationship of zinc-porphyrin sensitizers for dye-sensitized solar cells: A combined experimental and theoretical study. Dyes Pigm. 136: 697–706. https://doi.org/10.1016/j.dyepig.2016.09.035.

Kabir, F., Sakib, S.N., Matin, N. 2019. Stability study of natural green dye based DSSC. Optik — International Journal for Light and Electron Optics 181: 458–464. https://doi.org/10.1016/j.ijleo.2018.12.077.

Krawczak, E., Zdyb, A. 2019. The influence of the dye adsorption time on the DSSC performance. E3S Web of Conferences 100: 1–8. doi.org/1051/e3sconf/201910000040.

Listorti, A., O'Regan, B., Durrant, J.R. 2011. Electron Transfer Dynamics in Dye-Sensitized Solar Cells. Chemistry of Materials 23: 3381–3399. DOI: 10.1021/cm200651e.

Louwen, A., de Waal, A.C., Schropp, R.E.I., Faaij, A.P.C., van Sark, W.G.J.H.M. 2017. Comprehensive characterization and analysis of PV module performance under real operating conditions. Prog. Photovolt. Res. Appl. 25: 218–232. doi:10.1002/pip.2848.

Krawczyk, S., Nawrocka, A., Zdyb, A. 2018. Charge-transfer excited state in pyrene-1-carboxylic acids adsorbed on titanium dioxide nanoparticles. Spectrochimica Acta Part A: Molecular and Biomolecular Spectroscopy 198: 19–26. doi.org/10.1016/j.saa.2018.02.0611386-1425.

Mariani, P., Vesce, L., Di Carlo, A. 2015. The role of printing techniques for large area dye-sensitized solar cells. Semicond. Sci. Technol. 30: 104003-104019. doi:10.1088/0268-1242/30/10/104003.

Naik, P., Elmorsy, M.R., Su, R., Babu, D.D., El-Shafei, A., Adhikari, A.V. 2017. New carbazole based metal-free organic dyes with D-π-A-π-A architecture for DSSCs: Synthesis, theoretical and cell performance studies. Solar Energy 153: 600–610. https://doi.org/10.1016/j.solener.2017.05.088.

Naik, P., Su, R., Elmorsy, M.R., El-Shafei, A., Adhikari, A.V. 2018. Investigation of new carbazole based metal-free dyes as active photosensitizers/co-sensitizers for DSSCs. Dyes and Pigments 149: 177–187. doi.org/10.1016/j.dyepig.2017.09.068.

Nawrocka, A., Zdyb, A., Krawczyk, S. 2009. Stark Spectroscopy of Charge-Transfer Transitions in Catechol-Sensitized TiO2 Nanoparticles. Chemical Physics Letters 475: 272–276. doi:10.1016/j.cplett.2009.05.060.

Nieto-Pescador, J., Abraham, B., Gundlach, L. 2014.Photoinduced ultrafast heterogeneous electron transfer at molecule-semiconductor interfaces. J. Phys. Chem. Lett. 5: 3498–3507. DOI:10.1021/jz501541a.

Nazeeruddin, M.K., Angelis, F.D., Fantacci, S., Selloni, A., Viscardi, G., Liska, P., Ito, S., Takeru, B., Grätzel, M. 2005. Combined Experimental and DFT-TDDFT Computational Study of Photoelectrochemical Cell Ruthenium Sensitizers. J. Am. Chem. Soc. 127: 16835–16847. https://doi.org/10.1021/ja052467l.

Parr, R.G., Szentpály, L.V., Liu, S. 1999. Electrophilicity index, J. Am. Chem. Soc. 121: 1922–1924. https://doi.org/10.1021/ja983494x.

Parthasarathi, R., Padmanabhan, J., Elango, M., Subramanian, V., Chattaraj, P.K. 2004. Intermolecular reactivity through the generalized philicity concept. Chemical Physics Letters 394: 225–230. doi:10.1016/j.cplett.2004.07.002.

Redfern, P.C., Zapol, P., Curtiss, L.A., Rajh, T., Thurnauer, M.C. 2003. Computational Studies of Catechol and Water Interactions with Titanium Oxide Nanoparticles. J. Phys. Chem. B 107: 11419–11427. https://doi.org/10.1021/jp0303669.

Richhariya, G., Kumara, A., Tekasakul, P., Gupta, B. 2017. Natural dyes for dye sensitized solar cell: A review. Renewable and Sustainable Energy Reviews 69: 705–718. http://dx.doi.org/10.1016/j.rser.2016.11.198.

Report PV 2019. https://www.ise.fraunhofer.de/content/dam/ise/de/documents/publications/studies/Photovoltaics-Report.pdf.

Romero-Fiances, I., Muñoz-Cerón, E., Espinoza-Paredes, R., Nofuentes, G., de la Casa, J. 2019. Analysis of the Performance of Various PV Module Technologies in Peru. Energies 12: 186–205. doi:10.3390/en12010186.

Ruiz-Anchondo, T., Flores-Holguín, N.,Glossman-Mitnik, D. 2010. Natural Carotenoids as Nanomaterial Precursors for Molecular Photovoltaics: A Computational DFT Study. Molecules 15: 4490–4510. doi:10.3390/molecules15074490.

Sánchez-de-Armas, R., San-Miguel, M.A., Oviedo, J., Fdez. Sanz, J. 2011. Direct vs. indirect mechanisms for electron injection in DSSC: Catechol and alizarin Computational and Theoretical Chemistry. J. Phys. Chem. C 115: 11293–11301. https://doi.org/10.1021/jp201233y.

Shalini, S., Balasundaraprabhu, R., Kumar, T.S., Prabavathy, N., Senthilarasu, S., Prasanna, S. 2016. Status and outlook of sensitizers/dyes used in dyesensitized solar cells (DSSC): a review. Sensitizers for DSSC. Int. J. Energy Res. 40: 1303–1320. DOI:10.1002/er.3538.

Sharifi, N., Tajabadi, F., Taghavinia, N. 2014. Recent developments in dye-sensitized solar cells. Chem. Phys. Chem. 15: 3902–3927. https://doi.org/10.1002/cphc.201402299.

Sharma, K., Sharma, V., Sharma S.S. 2018. Dye-Sensitized Solar Cells: Fundamentals and Current Status, Nanoscale Research Letters 13: 381–427. https://doi.org/10.1186/s11671-018-2760-6.

Soto-Rojo, R., Baldenebro-Lopez, J., Glossman-Mitnik, D. 2015. Study of chemical reactivity in relation to experimental parameters of efficiency in coumarin derivatives for dye sensitized solar cells using DFT. Phys. Chem. Chem. Phys. 17: 14122–14129. DOI:10.1039/c5cp01387a.

Soto-Rojo, R., Baldenebro-Lopez, J., Glossman-Mitnik, D., 2016. Computational study ofthe influence of the pi-bridge conjugation order of novel molecular derivatives of coumarins for dye-sensitized solar cells using DFT. Theor. Chem. Acc. 135: 68. doi:10.1007/s00214-016-1826-8.

Zdyb, A., Krawczyk, S. 2014. Adsorption and electronic states of morin on TiO2 nanoparticles. Chemical Physics 443: 61–66. doi.org/10.1016/j.chemphys.2014.08.009.

Zdyb, A., Krawczyk, S. 2016. Characterization of adsorption and electronic excited states of quercetin on titanium dioxide nanoparticles. Spectrochimica Acta Part A: Molecular and Biomolecular Spectroscopy 157: 197–203. doi.org/10.1016/j.saa.2016.01.006.

Zigler, D.F., Morseth, Z.A., Wang, L., Ashford, D.L., Brennaman, M.K., Grumstrup, E.M., Brigham, E.C., Gish, M.K., Dillon, R.J., Alibabaei, L., Meyer, G.J., Meyer, T.J., Papanikolas, J.N. 2016. Disentangling the physical processes responsible for the kinetic complexity in interfacial electron transfer of excited Ru(II) polypyridyl dyes on TiO2. J.Am. Chem. Soc. 138: 4426–4438. DOI: 10.1021/jacs.5b12996.

Sewage sludge management

CHAPTER 6

Influence of the processing method on chemical forms of heavy metals in sewage sludge

J. Wiater & D. Łapiński
Department of Technology in Environmental Engineering, Bialystok University of Technology, Poland

ABSTRACT: The aim of the study was to analyze the general content and mobility of heavy metals at particular stages of production and processing of sewage sludge and to assess the possibility of their natural management on the basis of the determined parameters. The studies were carried out on the samples of sewage sludge collected from the treatment plant in Białystok. The samples were collected in three study cycles: autumn (November), winter (February) and spring (April) in 2017 and 2018. In studies was analyzed four types of sludge: primary sludge, excess sludge after dewatering by hydraulic press, fermented sludge and dewatered fermented sludge. The sewage sludge was subjected to sequential extraction and the content of heavy metals in them and in general forms was determined using Thermo Scientific AAS ICE 3000 atomic absorption spectrometer.

6.1 INTRODUCTION

Nowadays, a number of municipal wastewater treatment plants are struggling with the constantly increasing amount of sludge produced (Bień 2012 Bień et al. 2014). According to the forecasts of the National Waste Management Plan 2022 (M.P. 2016, item 784), in 2014, this volume amounted to 556 thousand tons of dry matter, and each year the amount of municipal sewage sludge converted into dry matter will increase by 2–3%. Each treatment plant selects the method of treatment and final disposal of sewage sludge based on its own capabilities. Stabilized sewage sludge, treated as waste of code 19 may be used for the natural or energetic purposes (Journal of Laws 2014, item 1923). However, it should be emphasized that any waste can be recovered, making it a useful product again (Rosik–Dulewska 2015). This also applies to sewage sludge, which – after processing – can be used for natural purposes. The possibilities of natural use of sediments are limited by the presence of pollutants in the form of heavy metals, which, when released, may have a negative impact on the environment and ultimately on humans (Ociepa–Kubicka, Ociepa 2012). The natural use of sludge may cause these pollutants to accumulate in the soil and migrate into waters. Regulation of the Minister of the Environment of 6 February 2015 on municipal sewage sludge (Journal of Laws 2015, item 257) specifies what parameters must be met by the sewage sludge used for natural purposes, including agriculture. The sewage sludge produced in wastewater treatment plants may contain small amounts of heavy metals and meet the requirements in accordance with the RM, but in the case of natural use, their solubility and assimilability to living organisms is important.

The aim of the study was to analyze the general content and mobility of heavy metals at particular stages of production and processing of sewage sludge and to assess the possibility of their natural management on the basis of the determined parameters.

6.2 MATERIALS AND METHODS

The studies were carried out on the samples of sewage sludge collected from the treatment plant in Białystok. The samples were collected in three study cycles: autumn (November), winter (February) and spring (April) in 2017 and 2018. Four types of sludge samples were analyzed in each cycle:

- primary sludge;
- excess sludge after dewatering by hydraulic press;

DOI 10.1201/9781003171669-6

Table 6.1. Conditions for extraction of metals from soils using an ultrasonic probe.

Fraction	Extractor	Probe operating time	Probe amplitude/power
F I	40 ml, 0.11 mol/l CH_3COOH	7 min	70% / 15W
F II	40 ml, 0.5 mol/l $NH_2OH{\cdot}HCl$	10 min	
F III	10 ml, 30% H_2O_2 evaporation	4 min	
	50 ml, 1 mol/l CH_3COONH_4	6 min	

- fermented sludge;
- dewatered fermented sludge.

All samples were studied in the Department of Environmental Engineering Technology and Systems. In order to determine the total heavy metal content, the sludge samples were dried at 105°C to determine their dry matter and then digested in BUTCHI Speed Digester K-425 using the following reagents: nitric acid – 65% HNO_3 – 8 cm^3 and perhydrol – 30% H_2O_2 – 10 cm^3. This process resulted in the thermal decomposition of the samples and transition of complex organic compounds to simple inorganic compounds. In this way the samples were prepared for the quantitative analysis of total metal content.

Sequential extraction of heavy metals was performed with a modified three-stage BCR method (proposed by the European Community Bureau of Reference), (Szumska & Gworek 2009). The following reagents were used to separate individual forms of metal binding:

F I – exchangeable fraction (ion exchangeable and carbonate) – 0.11 mol/l CH_3COOH – 40 ml;
F II – reducible fraction (oxide and hydroxide) – 0,5 mol/l NH_2OH-HCl – 40 ml;
F III – oxidisable fraction (organic and sulfide) – 30% H_2O_2 – 10 ml; after evaporation 1 mol/l CH_3COONH_4 – 50 ml.

In order to speed up the extraction, the process has been assisted by an ultrasonic probe. The extraction procedure including operating time, amplitude and power of the probe is shown in Table 6.1.

The total content of metals and the content of individual metal fractions were determined using Thermo Scientific AAS ICE 3000 atomic absorption spectrometer. Depending on the expected amount of a given metal in the samples, two atomization methods were applied:

- FAAS flame atomization method – for Zn, Cu, Cr, Ni, Pb;
- furnace atomization method with a graphite cuvette – for Cd.

The fraction F IV – residual, was determined by subtracting the sum of the contents of the first three fractions from the total contents of a given metal. The results of total metal content were given in milligrams per kilogram of dry matter. The percentages of individual fractions in the total content of each metal were calculated. The changes occurring in the arrangement of variables were analyzed by means of the Fisher's least significant differences test. As a qualitative factor in the analysis of variance, a series of studies (sampling) and the type of the studied sludge were assumed. All variables accepted for analysis were characterized by normal distribution according to the Saphiro-Wilk test and uniformity of variance according to Bartlett's test.

6.3 RESULTS AND DISCUSSION

In Table 6.2 presents the content of heavy metals at particular fractions in the tested sewage.

Table 6.2. Total concentration of metals in the examined sludge and the limit values of metals given in Polish law [mg/kg d.m.].

Lp	Type of sludge	series	Zn	Cu	Cr	Ni	Pb	Cd
1	Primary sludge	I	1400.2	166.2	35.9	14.5	8.0	0.17
		II	985.80	113.0	26.6	17.7	8.1	0.60
		III	1169.0	84.0	10.2	7.0	6.3	0.78
		average	1185.0	121.1	24.2	13.1	7.5	0.52
2	Excess sludge dewatered by hydraulic press	I	1705.0	264.4	71.1	18.5	24.5	0.20
		II	988.1	119.2	29.1	16.6	9.3	0.61
		III	1252.8	118.2	23.8	15.4	6.6	0.59
		average	1315.3	167.3	41.3	16.8	13.4	0.47
3	Fermented sludge	I	1503.9	202.6	64.8	11.9	14.9	0.12
		II	1098.0	155.9	40.3	21.8	15.8	0.60
		III	1515.4	158.9	34.7	20.3	13.3	0.61
		average	1372.4	172.4	46.6	18.0	14.7	0.44
4	Dewatered fermented sludge	I	1548.1	262.8	74.0	29.0	19.0	0.14
		II	1109.3	169.3	41.0	22.2	16.6	0.71
		III	1677.7	162.1	34.6	18.1	12.9	0.67
		average	1445.0	198.1	49.87	23.1	16.2	0.50
5	Limit values for municipal sewage sludge approved for agricultural use (Dz.U. 2015 poz. 257)		2500.0	1000.0	500.0	300.0	750.0	20.00

The total zinc content in the studied sludge ranged from 985.8 to 1705.0 mg/kg of sludge d.m. Statistical differences were found in the case of total zinc, depending on the date of sludge collection, but no differences were found depending on the type of sludge. A wide range of zinc content in the sludge is confirmed by the results from the years 2011–2013 (Wiater & Butarewicz 2013). The obtained content is significantly lower than the maximum content of this element in the sludge from the Podlaskie Voivodeship in the years 1998–2000, which was even over 2000 mg/kg d.m. (Boruszko et al. 2005).

The total copper content in the studied sludges varied between 84.0 and 264.4 mg/kg d.m. and depended on the date of the sludge sampling and less on its type. In the case of this element, its maximum content was higher than that obtained for sludge from wastewater treatment plants in the Podlaskie Province, which were examined in the years 1998–2000 and averaged 136 mg/kg d.m. (Boruszko et al. 2005). Additionally, the sludge analyses carried out in the studied wastewater treatment plant in the years 2011–2013 indicate a lower copper content in this period and amounted to 195 – 200 mg/kg d.m. (Wiater & Butarewicz 2014). Thus, on average, a higher copper content was recorded than the average for the sludge of the Podlaskie Voivodeship.

The chromium content in the studied sludges was at the level of 10.2– 74.0 mg/kg d.m. and differed significantly depending on the date of their collection and the type of sewage sludge. It is lower than the results obtained for the sludges analyzed in Białystok a few years ago, where the minimum chromium content was 69.3 mg/kg d.m. and the maximum 70.7 mg/kg d.m. (Wiater & Butarewicz 2014). The lowest chromium content was observed for preliminary sludge and the highest for dehydrated fermented sludge. In the years 1998–2000, the content of chromium in the sludge from the treatment plants in Podlaskie Voivodeship was many times higher and reached even 1000 mg/kg d.m. of sludge (Boruszko et al. 2005), which resulted from the inflow of industrial wastewater in those years.

The total nickel content in the primary sludge s varied between 7.02 and 29.00 mg/kg d.m., but no statistical differences between the sampling date and sludge type were proven. The least amount of nickel was contained in the preliminary and the most dehydrated sludge. In the years 2011–2013, the content of this metal in the sludge of the studied treatment plant was 23.5 - 27.3 mg/kg d.m. (Wiater & Butarewicz 2014). These values were similar for all treatment plants in the Podlaskie Province – in the years 1998–2000, the maximum nickel content in Podlaskie treatment plants was 25 mg/kg d.m. (Boruszko et al. 2005).

The total cadmium content in the studied sludge varied between 0.12 and 0.78 mg/kg d.m. of sludge. It was more differentiated by the sampling date than by the type of sludge analyzed. In the years 2011–2013, the content of cadmium in the sludge from the studied treatment plant was higher and amounted to 1.08–1.36 mg/kg d.m. (Wiater & Butarewicz 2014). Over the next few years, the content of this toxic element at the Wastewater Treatment Plant in Białystok decreased. The obtained values were much lower than the maximum cadmium content in the sludge of the treatment plant in the years 1998–2000, which averaged 4.9 mg/kg d.m. (Boruszko et al. 2005)

The lead content in the studied sludge ranged from 8.0–24.5 mg/kg d.m., but the content did not differ significantly from the sampling date. This is much lower than that observed in the years 1998–2000. During this period, the maximum lead content in the sludge from the treatment plants in Podlasie reached 194 mg/kg d.m. (Boruszko et al. 2005). In 2011, the content of this metal in the sludge of the Białystok Treatment Plant was 24.4 mg/kg d.m., and already in 2013 it decreased to 21.2 mg/kg d.m. (Wiater & Butarewicz 2014). Thus, in the years 2011–2013, the lead content in the sludge was similar to the results obtained. The least amount of lead was in the preliminary sludge, while in the excessive sludge it doubled and more than doubled. No significant differences in the lead content were found between the excessive sludge, after fermentation and after dehydration.

The sewage sludge produced at the Białystok Wastewater Treatment Plant can be used in nature, including agriculture, and contains significantly less metals than the sludge studied by Wilk and Gworek (2009). The content of heavy metals is significantly below the standard given in the Regulation of the Minister of Environment of 6 February 2015 on municipal sewage sludge (Journal of Laws 2015, item 257).

The share of mobile and non-mobile zinc fractions in the samples of the analyzed sludge varied depending on the stage of processing and date of sampling (Figure 6.1). Approximately half of the total amount of zinc compounds in the preliminary and excessive sludge are bioavailable and potentially bioavailable. On the other hand, the share of both fractions in the sludge after fermentation decreased and the share of the fraction bound to the organic substance and the residual fraction increased.

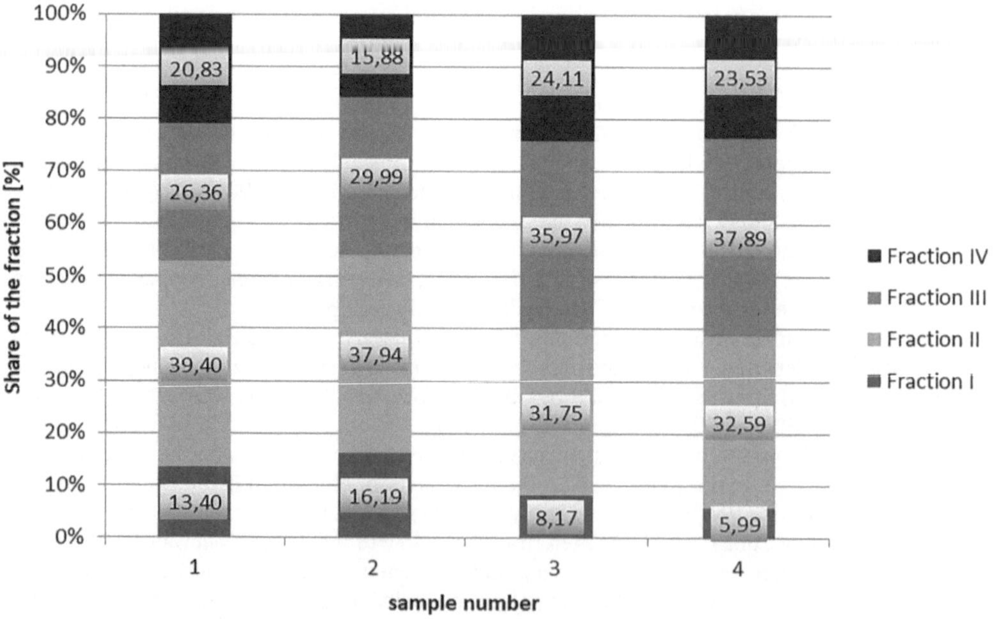

Figure 6.1. Percentage fraction of zinc in the total content.

These results differ from the results obtained by other authors, who showed a predominance of fractions I and II in the sludge (Boruszko 2013), fractions III and IV (Gawdzik 2012) or fractions II and IV (Ignatowicz et al. 2011). In the studied sewage sludge, zinc is most strongly associated with aluminum and iron oxides and hydroxides and organic matter. The fermentation and dehydration process has contributed to the reduction of mobility of this metal.

In the studied sludge, a clear predominance of non-mobile copper fractions was observed from 97.27% for excess sludge to 99.51% for dehydrated digested sludge (Figure 6.2). The largest share in all the sludges, over 50% is the fraction bound to organic matter. Similar results were obtained by Gawdzik (2012) and Ignatowicz et al. (2011), in their studies the content of non-mobile fractions of copper was 97.14% and 89.19%, respectively. The prevalence of the third fraction is characteristic for copper, as this metal forms permanent compounds with organic matter.

The share of this fraction increased in the sludge after fermentation and dehydration. This was partly at the expense of the residual fraction. In the case of copper, there is practically no exchangeable and reducible fraction, so it is a metal contained in the sewage sludge that is very mobile.

The chromium contained in the sludge turned out to be not very mobile metal, because in the majority of cases it occurred in the residual fraction, i.e. in stable and hardly accessible chemical compounds (Figure 6.3). The IV fraction alone represents as much as 76.05–88.89% of the total chromium content. Gawdzik (2013) received a similar result for the sludge treatment plant in Ostrowiec Świętokrzyski, where its content in the residual fraction was 73.4% of total content. In the research on the sludge from the Sokółka WWTP conducted by Ignatowicz et al. (2011), chromium was present to the greatest extent in the organic fraction for the sludge stabilized after press (75.99%). Such discrepancies in results may be affected by the share of industrial wastewater flowing into the treatment plant. The changes in the share of fractions in particular types of sludge depended mainly on the date of its collection. In precipitation, the share of this metal in fraction III was twice as high in relation to the remaining sludge. The reverse dependence occurred in the case of fraction IV. The share of exchangeable and reducible fractions decreased after the fermentation process. Dehydration contributed to a renewed increase in the exchangeable fraction.

Out of all the heavy metals determined, the share of individual fractions of nickel in the processing of sludge changed most dynamically (Figure 6.4). For fractions I and IV, the greatest differences occurred between the precipitate and excess sludge – the share of fraction I increased by 23.71% and

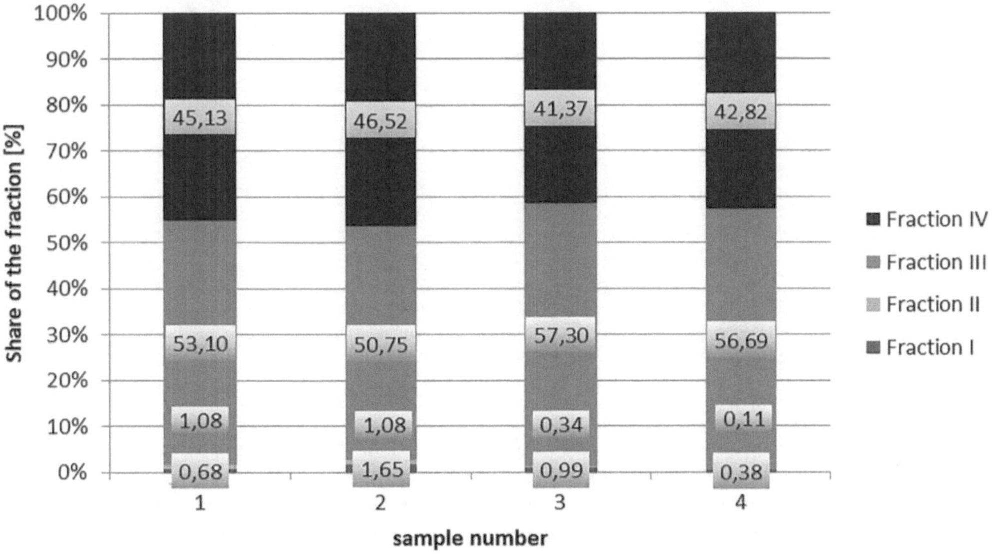

Figure 6.2. Percentage fraction of copper in total content.

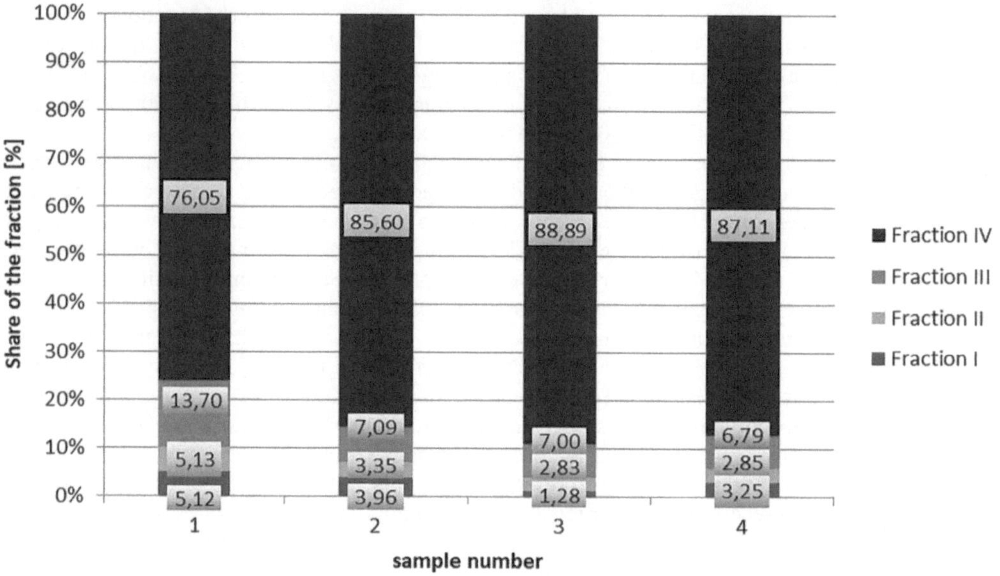

Figure 6.3. The percentage of chromium fraction in the total content.

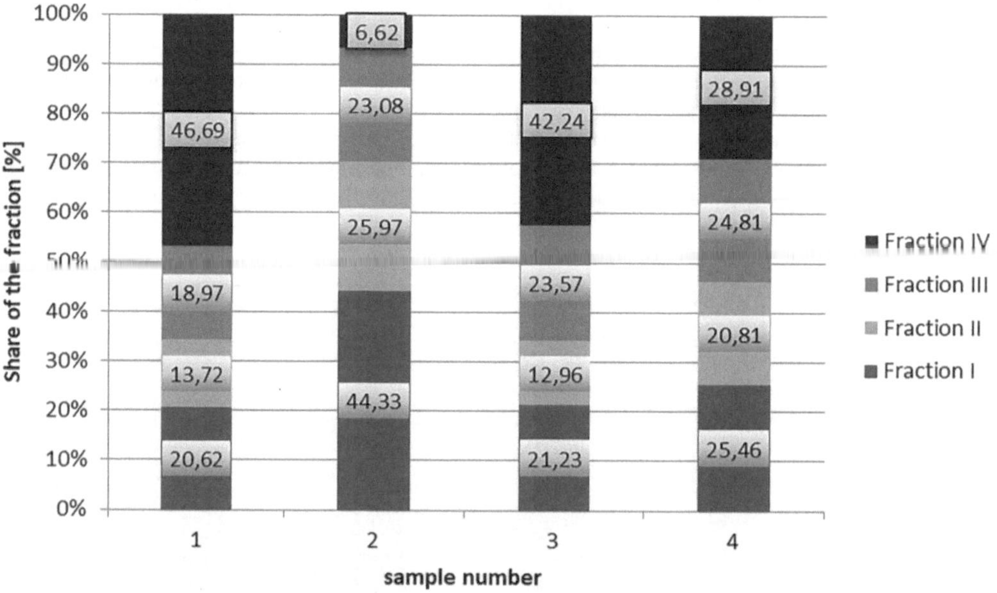

Figure 6.4. Percentage fraction of nickel in the total content.

fraction IV decreased by as much as 40.07%. After the fermentation process, the share of fraction II decreased by 13.01%. The smallest fluctuations in particular types of sludge were found for fraction III. Nickel was the most mobile in excess sludge after press, and the least in the digestate. The sum of the share of exchangeable and reductive fractions in these sludge samples was 70.30% and 34.19%, respectively. Thus, the fermentation process reduced the mobility of this element. Boruszko (2013), in the studies on the dairy and municipal sludge, also obtained different results of individual nickel

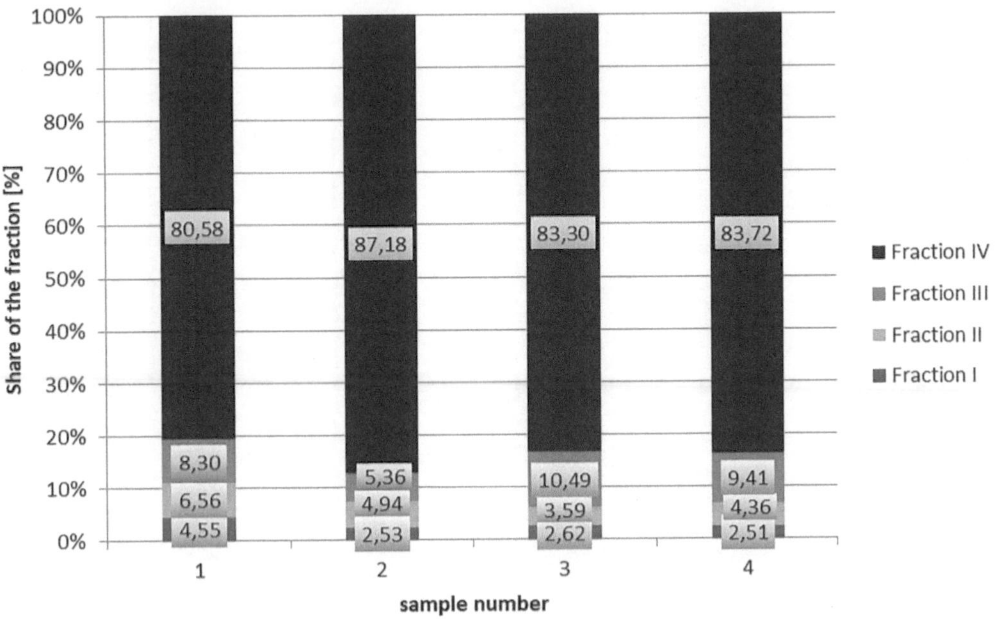

Figure 6.5. Percentage fraction of lead in the total content.

fractions. They varied between 1.1–11.0% (fraction I), 6.0–45.9% (fraction II), 7.8–60.4% (fraction III) and 29.1–83.8% (fraction IV) depending on the method of sludge processing.

The share of nickel in individual fractions did not differ significantly depending on the date of individual sludge samples collection.

As in the case of chromium, most of the lead present in the samples is contained in the residual fraction (Figure 6.5). The share of fraction IV in the total lead content was similar for all studied sludge samples (80.58– 87.18%). The fermentation process contributed to a slight increase in fraction III. In the sludge from the Ostrowiec Świętokrzyski WWTP this value was 75.6% (Gawdzik 2013). Similar results were obtained by Dąbrowska and Nowak (2014) in the studies of the sludge from Częstochowa and Łężyca near Zielona Góra. The share of bound lead in the residual fraction of sludge originating from these objects ranged from 85 to 96%. The share of lead in the remaining fractions did not differ much between sludge types. Like copper and chromium, lead is a poorly mobile metal in sewage sludge.

The share of the analyzed cadmium fractions in the sludge depended on its type (Figure 6.6). The largest share in relation to total cadmium content was the residual fraction, which increased after the sludge fermentation process. This metal was also to a large extent related to organic matter – the share of fraction III, which decreased by over 10% in comparison with the excessive sludge. Similar results were obtained by Gawdzik (2010), in whose studies the share of fractions IV and III was 72.8% and 19.4% respectively. In the Dąbrowska's and Nowak's studies (2014), fraction III (71%) had the advantage. Both of these fractions are considered immobile, so that cadmium was mostly concentrated in fractions III and IV. Fractions I and II accounted for over 22% of the preliminary sludge; their share in the excessive and digested sludge decreased. After the dewatering process, the exchangeable fraction I significantly increased, at the expense of the reducible fraction II, which made cadmium in the dewatered sludge the most mobile with respect to the excessive sludge. Changes were also found in the share of mainly mobile fractions depending on the date of sampling.

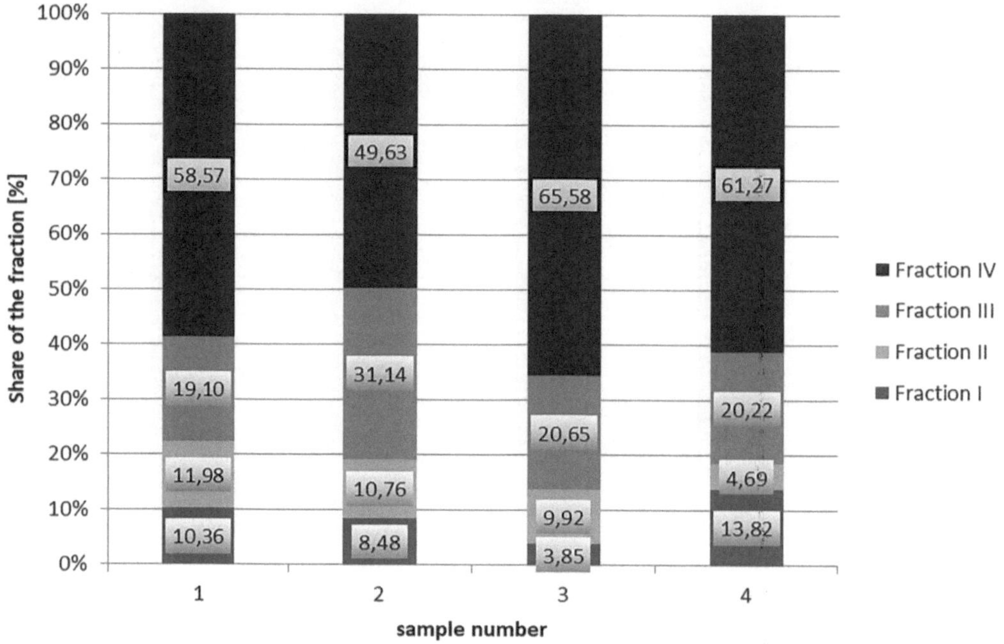

Figure 6.6. Share of the cadmium percentage fraction in the total content.

6.4 CONCLUSIONS

1. The total content of heavy metals in sewage sludge depended on the time of its collection to a greater extent than on the type of sludge. The least amounts of metals were contained in the primary sludge, while in the others the content was higher and similar to each other.
2. All types of the sludge met the requirements for the heavy metal content set out in the Ordinance of the Minister of Environment of 6 February 2015 on municipal sewage sludge when used in agriculture and for land reclamation for agricultural purposes.
3. The most mobile metals were zinc and nickel, and the least mobile were chrome and lead. In the studied sewage sludge, zinc was most closely related to oxides and hydroxides and organic compounds. The highest content of copper and cadmium was observed in the organic and residual fractions. Chromium and lead were most strongly bound in the residual fraction. The distribution of nickel in the fractions was very varied.
4. Sludge processing contributed to a reduction in the mobility of the analyzed metals, which is indicated by an increase in the share of non-mobile fractions at the expense of mobile fractions.
5. The analysis of the sludge from the treatment plant in Białystok, in comparison with the literature data from previous years, indicates a decrease in the pollution of the environment with heavy metals.

REFERENCES

Bień J.B., Pająk T., Wystalska K. (2014). Disposal of municipal sewage sludge. Częstochowa University of Technology Publishing House, Częstochowa.

Bień J. D. (2012). Management of municipal sewage sludge by thermal methods. *Engineering and environmental protection*, 15/2012, 439–449.

Boruszko D. (2013). Fractionation of selected heavy metals in sewage sludge processed by low input methods. *Environmental Protection Yearbook*, 15/2013, 1787–1803.

Boruszko D., Butarewicz A., Dąbrowski W., Magrel L. (2005). Research on the final use of dewatered sewage sludge for non-industrial use. Białystok University of Technology. Białystok

Dábrowska L., Nowak R. (2014). Chemical fractions of heavy metals in sewage sludge and in the solid residue after its incineration. *Environmental Engineering and Protection*, 3/2014, 403–414.

Gawdzik J. (2010). Heavy metal speciation in sewage sludge on the example of a selected municipal treatment plant. *Environmental Protection*, 4/2010, 15–19.

Gawdzik J. (2012). Mobility of heavy metals in sewage sludge on the example of a selected wastewater treatment plant. *Engineering and Environmental Protection*, 1/2012, 5–15.

Gawdzik J. (2013). Mobility of selected heavy metals in sewage sludge. Swietokrzyska University of Technology Publishing House, Kielce.

Ignatowicz K., Garlicka K., Breńko T. (2011). Effect of sewage sludge composting on the content of selected metals and their fractions. *Ecological Engineering*, 25/2011, 231–241.

Ociepa-Kubicka A., Ociepa E. (2012). Toxic effects of heavy metals on plants, animals and humans. *Engineering and Environmental Protection*, 15/2012, 169–180.

Regulation of the Minister of Environment of 6 February 2015 on municipal sewage sludge (Journal of Laws 2015, item 257).

Regulation of the Minister of Environment of 9 December 2014 on the waste catalog (Journal of Laws 2014, item 1923).

Resolution No. 88 of the Council of Ministers of 1 July 2016 on the National Waste Management Plan 2022 (M.P. 2016 item 784).

Rosik-Dulewska C. (2015). Fundamentals of waste management. Scientific Publishing PWN, Warsaw.

Szumska M., Gworek B. (2009). Methods for determination of heavy metal fractions in sewage sludge. *Protection of the Environment and Natural Resources*, 41/2009, 42–63.

Wiater J., Butarewicz A. (2014). Ways of using sludge from the Wastewater Treatment Plant in Białystok. *Engineering and Environmental Protection*, 17/2014, 281–291.

Wilk M., Gworek B. (2009). Heavy metals in sewage sludge. Environmental and Natural Resources Protection, 39/2009, 40–59.

CHAPTER 7

Composting of sewage sludge in technical scale: The influence of straw added mass of the humification process

R. Sidełko, B. Walendzik, B. Janowska, K. Szymański, A. Leśniańska & R. Królak
Faculty of Civil Engineering, Environmental and Geodetic Sciences, Koszalin University of Technology, Koszalin, Poland

ABSTRACT: The main objective of presented research work was the assessment of impact of reduced straw content, as one of the components of the compost mixture and organic carbon source, on the course of sewage sludge composting process. During the research work performed in technical conditions, the composting process going in periodically overturned windrows differing in mass proportion of dewatered sludge, straw and bulking agent (wood chips and matured compost) being 4:1:1 and 8:1:2 respectively, was observed. The consequence of increase of sludge concentration with relation to straw was decrease of C:N ratio in the input material from 11.5 to 8.5. The following indices were analysed as indicators for the assessment of the composting process: contents of fulvic acids (FA), humic acids (HA) as well as absorbance in UV/VIS ($\lambda = 280$, 465 and 665 nm) range. The results obtained have indicated that the increase of sludge content extends the elevated temperature ($T > 50°C$) period from 42 days to approximately 65 days. Our tests did not confirm that limitation of straw content added to sewage sludge had any adverse effect on the course of composting. Value of PI (HA/FA) > 3.6 index qualifying compost as mature in the first case – No 1, was determined on the 83rd day, whereas, in the second case No 2, on the 48th day.

7.1 INTRODUCTION

Due to high content of macroelements, mainly organic carbon, nitrogen and phosphorus, the mechanically dehydrated sludge makes a valuable material for production of compost complying with the requirements set for soil improvers and materials substituting soil for crop production (Publications Office of the European Union 2001). Composting of sewage sludge from municipal wastewater treatment plants is practically a single method of biological treatment, which can provide a product that can be used in agriculture (Szymański et al. 2007). Whereas application of methane fermentation considerably improves effectiveness of the entire sludge processing process, application of a post fermentation material composting module ultimately determines quality of the final product, i.e. compost (Curtis & Claassen 2009; Carrizo et al. 2015). High concentration of total nitrogen in mechanically dehydrated sludge, generally falling within the 2÷7% dry mass range (Sidełko et al. 2010; Świerczek et al. 2018) and its high humidity amounting to 85÷75% (Kacprzak et al. 2017) cause that composting of sewage sludge requires addition, at the compost mass formation stage, of supplementary material with high concentration of organic carbon and low concentration of nitrogen. The optimum C/N ratio value at the moment of composting commencement should fall, according to various sources, into the 20÷35 interval (Bernal et al. 2009; Sweeten & Auvermann 2008) whereas value of this parameter in dehydrated sewage sludge generally does not exceed C/N = 7 (Kacprzak et al. 2017). Therefore, a statement can be made, that composting of sewage sludge without addition of any material reducing concentration of nitrogen due to the risk of NH_3 formation (Hellebrand & Kalk 2001) is practically impossible.

Furthermore, addition of various supplements is a factor modifying the carbon to nitrogen proportion in a way assuring proper C/N ratio (Alvi et al. 2017; Doublet et al. 2010; Sanchez et al. 2017).

DOI 10.1201/9781003171669-7

An important element of the analysis of the course of the entire composting process is evaluation of the compost maturity level. Immature compost may contain phytotoxic substances mostly in form of short-chain organic acids, ammonia, hydrogen sulphide, phenol and other compounds, occurrence of which, has adverse effect on plant germination and growth processes. To prevent the adverse effects caused by application of immature compost, physico-chemical, microbiological and fertilising criteria of its maturity are used. Such physical features as colour, odour and temperature provide a general view of the composting process advancement level (Bernal et al. 2009; Sidełko et al. 2017). Chemical methods allow for determination of variation of biogenic elements content and, consequently, values of various indices describing speed of organic and mineral components transformation during composting. From agricultural point of view an important element is not only the total carbon content, but also the form of its occurrence. Transformation of organic matter during compost maturing phase leads to generation of large molecular compounds in form of organic polymers. This process is called humification and the generated substances having exceptionally complex molecular form are called humus. The forerunner of humic compounds are mainly lignin, cellulose and hemicellulose from which originate various products such as, among other things, amino acids and phenol, making a basic component of humic acids generated due to occurrence of various enzymatic reactions (Yuan et al. 2017). During the preliminary phase of organic matter transformation, in the humification process dominate fulvic acids (FA), which are transformed, over time, into humic acids (HA) (Kononowa 1968). Change of forms of organic carbon, occurring in form of the so-called specific and non-specific humic compounds, during composting process, makes a basis for determination of indices allowing for assessment of the humification process progress defined as HA/FA and HA/TOC (Bustamante et al. 2008; Hsu & Lo 1999; Sanchez-Monedero et al. 1999). The first index described by the content of humic acids carbon (C_{HA}) to fulvic acids carbon (C_{FA}) ratio, is indicated by PI (Polymerisation Index) abbreviation. The second index expresses the percentage of humic acids carbon content in total organic carbon (TOC); it is expressed as HI (Humification Index).

There are many examples in scientific literature describing research on composting of mixtures of sewage sludge within broad scope of C/N = 15÷28.9 ratio (Głąb et al. 2018; Gonzalez et al. 2019; Kulikowska & Sindrewicz 2018; Li et al. 2017; Sidełko et al 2017; Zheng et al. 2018).

Generally, straw being a supplementary source of organic carbon, causes increase of C/N ratio value up to the level considered as optimal. However, straw is currently used more and more frequently in other sectors of economy, which means that it is in short supply on the market. Therefore, taking up any research work on the course of composting with reduced straw content, in relation to the amount originating from the balance of mass of carbon and nitrogen in the context of recommended C/N ratio, is justified.

The main objective of this research work was determination of impact of increased sewage sludge share in the mixture with straw on the course of humification of organic matter during composting. Our research comprised mostly an analysis of variation of humic substance content expressed by occurrence of fulvic and humic acids. Additionally contents of such organic compounds as lignin, cellulose and hemicelullose, occurrence of which may have impact on the values of humification indices, were determined.

7.2 MATERIAL AND METHODS

Our field research was performed at the site of Goleniów (Poland) wastewater treatment plant. Processing of the sludge generated in the municipal wastewater treatment process consist in its mechanical dewatered and then composting with addition of barley straw, wood chips and mature compost inoculum. Composting proceeds in roofed windrows approximately 70 m long, with dimensions of trapezoid transversal cross-section being: 3 m bottom base width and 1.5 m height.

The windrows are mechanically overturned from time to time, twice a week during the first three weeks of composting, and, in subsequent weeks – once per week on average. Composting takes 4–5 months depending on the external conditions.

The field research was performed in two series (S1 and S2). In the first run - S1, two windrows varying in the mass proportion of their components being respectively for windrow No 1 – 4:1:1 and windrow No 2 – 8:1:2 where examined. These proportions means the mass share of the components in the compost mixture, i.e. sewage sludge: barley straw: half wood chips and inoculum (matured compost). In the second run – S2, which was performed after completion of S1, the method of accomplishment of the field research, including the composition of the composted mixture, i.e. windrows No 3–4:1:1 and windrows No 4–8:1:2, was the same. During the tests, changes of temperature in all windrows were monitored. From each windrow, samples of compost were taken to determine the following indicators: dry mass (d.m.) after sample drying at 105°C, total organic carbon (TOC) in elemental analysis by application of VarioMAX CN according to PN-Z-15011-3 standard and total nitrogen (N_{tot}) through elemental analysis having prepared samples in accordance with PN-R-04006 standard. During TOC determination using the VarioMAX CN analyzer, the analytical procedure for determination of only organic carbon forms was employed. The procedure is based on a modification done in the apparatus combustion system compared to the standard CN operating mode and is compatible with DIN/ISO 10694 standard. Also contents of organic substances, total phosphorus and selected heavy metals, with definition of their chemical forms, where being determined during the tests.

Contents of lignin, cellulose and hemicellulose were determined in composts using the filtration bags technique in Ankom A200. Content of neutral detergent fibre (NDF) was determined based on Van Soest method (Van Soest et al. 1991), content of acid detergent fiber (ADF) and content of lignin (ADL) were determined too. The cellulose content was determined based on the difference between the cumulative lignin and cellulose (ADF) amount and content of lignin itself (ADL), whereas content of hemicellulose was established based on the difference of NDF and ADF fraction shares.

Extraction of the sum of humus acids (HS) was performed using 0.5 M NaOH applying the modified IHSS method (Swift 1996). Fulvic acids (FA) were determined in solution after acidification using HCl to pH = 2. Carbon in alkaline extracts (C_{HS}) and acidic extract (C_{FA}) was determined in VARIOMAX CN. Carbon in humic acids (C_{HA}) was determined as a difference between C_{HS} and C_{FA} carbon contents.

Absorbance in humus acids and humic acids contents was measured three times at the following wavelengths: $\lambda = 280$ nm (A2), $\lambda = 465$ nm (A4) and 665 nm (A6) (Sapek & Sapek 1986). Then, absorbance indicators defining the degree of humification A2/A4 and A4/A6 were calculated.

Results were worked out using Statistica 12 software from StatSoft. The analysed physico-chemical parameters data were presented as mean arithmetic values of three samples. The data reduction procedure was performed through the primary components analysis (PCA) using XLSTAT software from Addinsoft.

If your text starts with a heading, place the cursor on the I of INTRODUCTION and type the correct text for the heading. Now delete the word INTRODUCTION and start with the text after a return. This text should have the tag First paragraph.

If your text starts without a heading you should place the cursor on the I of INTRODUCTION, change the tag to First paragraph and type your text after deleting the word INTRODUCTION, but not the return at the end.

7.3 RESULTS AND DISCUSSION

Test results are shown in Table 7.1. Values of N_{tot}, TOC, HA, FA, lignin, cellulose, hemicellulose, A2/A4 and A4/A6 constitute a mean value of the assays taken in two runs i.e. S1 and S2. The mean error of a single carbon and nitrogen assay calculated based on the elementary analysis using the VarioMAX analyzer was: TOC $\pm$ 0.11 and $N_{tot} \pm 0.025$. The detailed data, including that used for validation of the methodology, have been presented in the report available from the site of the STEP project (Interreg South Baltic 2018), which did not comprise IR spectral tests and lignin, cellulose and hemicellulose contents tests.

The test results analysis was performed based on changes in the determined indicators values constituting mean values of assays for two windrows featuring the same proportions of windrow components i.e. windrow No 1 and 3– case No 1 and windrow No 2 and 4– case No 2. Figure 7.1 illustrates changes of the mean temperature values for the two above-defined cases.

Table 7.1. Evolution of some parameters and indexes during composting

Composting time day (no.)	N_{tot} [g/kg d.m.]	TOC [g/kg d.m.]	HA [g/kg d.m.]	FA [g/kg d.m.]	Lignin [g/kg d.m.]	Celluloze [g/kg d.m.]	Hemicell. [g/kg d.m.]	PI	C/N	A2/A4	A4/A6
case no. 1 (windrows no. 1 and 3) – 4:1:1 (sewage sludge/straw/wood chips and innoculum)											
3 (1/1)	30.48	413.43	153.74	58.17	18.28	9.37	15.21	2.64	14	5.66	16.74
10 (2/1)	27.36	397.08	148.51	47.22	30.25	10.83	11.14	3.15	15	4.37	14.03
16 (3/1)	28.68	389.93	148.00	42.51	29.48	14.40	7.55	3.48	14	3.69	14.13
24 (4/1)	29.07	378.05	149.46	48.52	34.26	11.56	12.87	3.08	13	4.53	14.06
29 (5/1)	33.25	378.48	161.60	78.47	22.94	10.56	8.57	1.90	11	3.11	16.35
48 (6/1)	35.05	365.30	163.01	46.12	29.41	11.54	10.31	3.50	10	2.58	14.20
62 (7/1)	35.79	367.28	156.00	71.46	24.90	8.12	9.63	2.28	10	2.76	14.71
83 (8/1)	34.51	379.68	160.78	44.05	27.61	10.11	3.56	3.54	11	1.77	13.78
111 (9/1)	34.41	358.55	163.92	42.02	30.83	11.94	2.49	3.83	10	1.79	13.74
133 (10/1)	35.53	362.18	165.51	55.03	29.87	10.41	0.68	2.98	10	1.82	13.04
case no. 2 (windrows no. 2 and 4) – 8:1:2 (sewage sludge/straw/wood chips and innoculum)											
3 (1/2)	40.74	390.78	146.84	74.77	25.18	11.76	5.75	1.96	10	4.99	15.76
10 (2/2)	27.35	386.85	158.45	46.96	27.29	11.23	9.49	3.37	14	5.08	13.93
16 (3/2)	27.84	378.00	142.11	43.75	30.61	13.86	10.56	3.25	14	4.08	14.76
24 (4/2)	28.63	371.23	142.57	53.69	27.68	11.53	9.43	2.66	13	4.20	14.65
29 (5/2)	35.59	358.35	143.01	46.20	26.45	13.53	16.43	3.10	10	3.42	14.23
48 (6/2)	31.22	362.80	152.76	28.50	31.83	10.81	4.38	5.36	12	2.19	13.83
62 (7/2)	33.93	358.03	157.28	36.63	32.30	8.50	5.54	4.29	11	2.36	13.00
83 (8/2)	34.75	363.43	154.75	44.49	26.96	6.10	4.14	3.48	10	2.45	12.63
111 (9/2)	36.73	358.08	152.54	28.10	26.96	5.40	2.27	5.43	10	2.59	12.12
133 (10/2)	35.53	337.05	163.68	45.50	26.31	5.76	2.58	3.60	9	2.65	11.71
sewage sludge	74.80	339.15	169.73	105.38	14.27	3.53	13.64	1.61	5	–	–

Already on the third day of composting, temperature in excess of 50°C (Figure 7.1) was noted in both cases. Such high temperature indicates intensive decomposition of organic matter during composting of both windrows. However, the thermophilic phase period in windrows of different compositions was not the same. In case No 2 (Figure 7.1) it was approx. 65 days, when the highest temperature was noted on approximately the 8th day of composting. In case No 1 (Figure 7.1) the thermophilic phase lasted in excess of 42 days, and the highest temperature was noted on approximately the 11th day of composting. Such situation may have originated from higher content of the sewage sludge fraction in windrow No 2 and 4. Insignificantly higher values for total nitrogen noted in these cases indicate higher accessibility of nutrients for development of thermophilic bacteria during composting. Temperature in excess of 50°C remained for a period ca. 50 and 20 days as shows curves for cases No 1 and 2 respectively (Figure 7.1).

The process of degradation of organic matter during composting was analysed based on TOC variations, which gradually decreased for two cases. The process conditions in the thermophilic phase promoted intensification of mineralisation, which resulted in 12% (case No 1) and almost 14% (case No 2) TOC reduction, by the end of composting of both windrows, with relation to its initial value (Figure 7.2).

Higher intensity of organic matter biodegradation process in windrows No 2 and 4 were probably related to longer period of the thermophilic phase as well as higher composting temperature compared to windrows No 1 and 3. Such situation could contribute, in the opinion of Meng et al. (2017), to intense decomposition of cellulose and hemicellulose in the composed matter of case No 2.

Variations of HS concentration during composting were observed. In case No 1 approx. 27% reduction of FA content with relation to this parameter value at the beginning of the composting process was noted. In case No 2 this decrease was more pronounced and was in excess 58%. The herein described

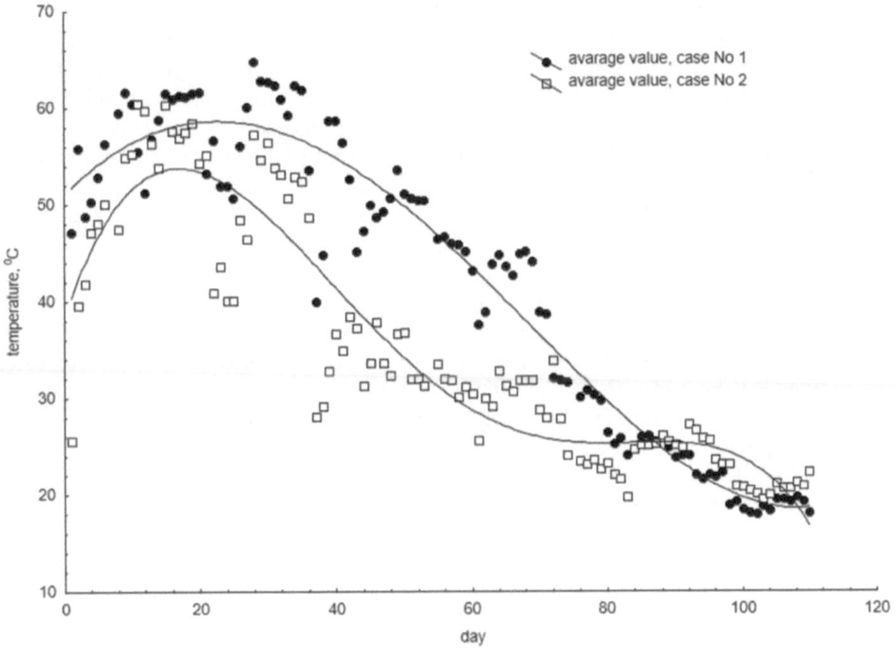

Figure 7.1. Variations of average temperature values in both cases during composting.

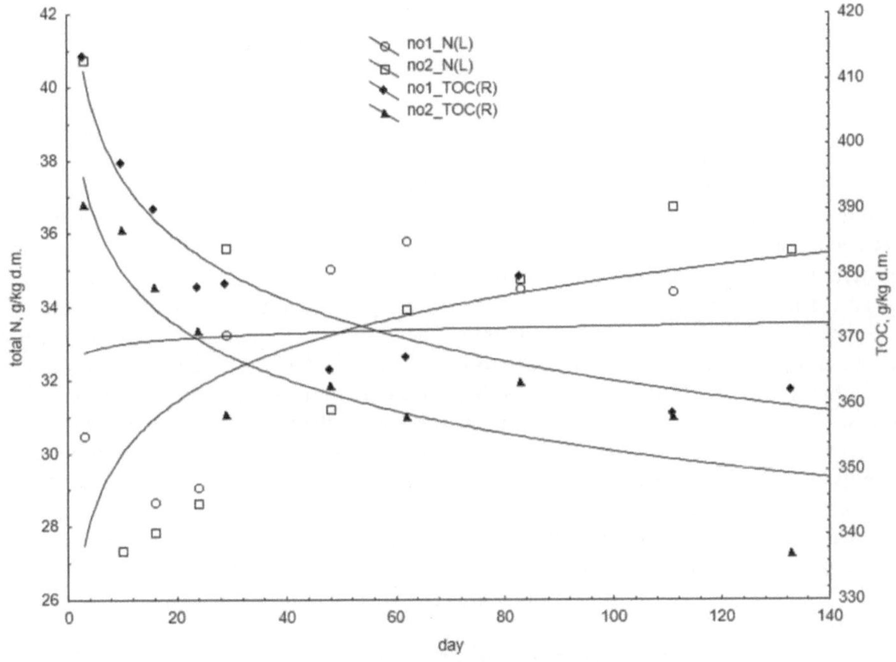

Figure 7.2. Change in TOC and total nitrogen concentration during the field research.

structures feature low molecular mass and high content of hydroxylic and carboxylic functional groups (Sellami et al. 2008). Faster decrease of FA content in case No 2 indicates better achievement of the stability level by organic matter; easily accessible carbon occurring in the composted matter was subjected to considerable decomposition. Also increase of HA content in both cases indicated achievement of sufficient stability by the tested composts. However, in case No 2, increase of HA was higher and appeared to be close to 12%. Such situation indicates more intense transformation of low-molecular substances into high-molecular ones and it may suggest, consequently, faster achievement of satisfactory level of maturity by compost with higher content of sewage sludge (Zhou & Selvam 2014). Humic acids could additionally be generated from other humic substances available in the compost, including fulvic acids (Kulikowska 2016).

The windrow composting process was analysed in terms of lignin, cellulose and hemicellulose contents. The research in the above scope was carried out only during the runs No 1 (windrows No 1 and 2). Contents of those substances in windrow No 1 decreased by 1.37%, 18.83%, 95.53%, and in windrow No 2 by – 7.11%, 51.02%, 79.76% respectively. A significant reduction of contents was noted only in the case of the two last parameters. Such situations confirms the fact that high process temperature favours decomposition of complex hydrocarbon structures i.e. cellulose and hemicellulose. However, occurrence of lignin hampered the organic matter decomposition processes. Its degradation proceeded mainly in acid environment and with participation of fungal enzymes. Higher temperature in windrow No 2 indicated higher activity of microorganisms in the thermophilic phase and contributed to effective degradation of the cellulose and hemicellulose fractions during composting.

PI index is extensively used for description of relative transformation of humic substance during composting and index reflects intensity of generation of complex humic acids molecules from less complex (Czekała 2008; Domeizel et al. 2004). The values of humification index for the tested windrows increased, but not in the same way. PI index for case No 1 increased almost by 14%, and for case No 2 – by 83%. However, mean initial values of this index were similar for both windrows and were: for case No 1- PI = 2.64 and for case No 2 – PI = 1.96 respectively. These values tell us that at the preliminary composting phase, low-molecular fractions of fulvic acids dominated in both windrows over the high-molecular fraction of humic acids. According to Zhou & Selvam (2014), HA/FA ratio within the interval of 3.6÷6.2 indicates generation of mature compost. The very moment in which windrows with lower sludge content (windrows No 1 and 3) reached the stable value of 3.6 occurred at the 83rd day of the process counting from the date of composting commencement, whereas in windrows No 2 and 4 it was the 48th day. At the same time, the humification process in case No 2 proceeded in a more stable way, which was indicated by the dynamics of PI index value change. PI value for both cases was maintained above 3.6 until the end of the composting process. The significant impact on the humification process has the timing of the thermophilic phase, particularly its final stage. The thermophilic phase period was definitely longer for case No 2 and allowed for generation of fully mature compost at the 40th day.

At the next stage of tests A2/A4 and A4/A6 indices were determined; they were calculated from absorbance of humic acids in UV-VIS range. The first index indicates the content of organic substance at the preliminary decomposition stage (Sapek & Sapek 1986). At the preliminary phase of composting, this index clearly decreased in both compost windrows, which was associated with intensely proceeding processes of depolymerisation associated with microbiological decomposition of complex organic structures, mainly hemicellulose and cellulose. Said non-humic compounds usually absorb energy of radiation near UV ($\lambda = 280$ nm) (Zbytniewski & Buszewski 2005a,b). After the 48th day of composting, value of A2/A4 index for case No 2 started to slightly increase untill the end of the process. This can be explained by decrease of intensity of organic matter decomposition process. It can be supposed that at the same time an increase in the share of the phenol and benzenecarboxylic groups in the humic substances structure has occurred (Veeken et al. 2000). Value of this index for case No 1 decreased until the 83rd day of the process. During composting, the organic material transformed gradually into structures of highly aromatic degree and high-molecular mass (Lv et al. 2013; Zhang et al. 2015). Such situation was confirmed by decrease of A4/A6 index for both windrows. During 133 days of composting, A4/A6 index decreased for case No 1 by approximately 22%, and for case No 2 – by approximately 26%. Such difference was probably caused by more intense decomposition

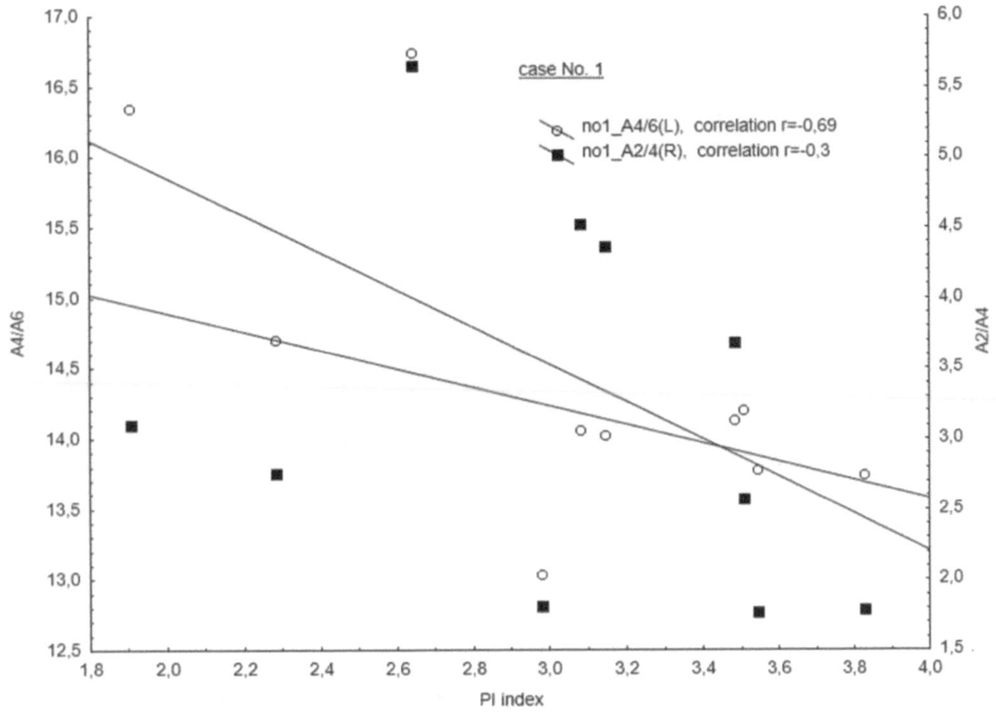

Figure 7.3. Correlation between absorbance coefficients A4/A6 and A2/A4 vs. PI index- case no.1

of the organic matter in case No 2, which had a favourable effect on the process of humification of composted materials (Yuan et al. 2016). The above results indicate that compost with lower straw content may not only get to the intended stability, but may also mature faster.

The changes in the value of the PI index and the absorbance indices of electromagnetic radiation in the range of visible light (VIS) and ultraviolet (UV) are identified with the structural reconstruction of organic matter. The strong relationship between these indices is evidenced by the high negative Parson correlation coefficient- r, calculated separately for individual pairs of indexes presented in Figures 7.3 and 7.4. It was found that in the case of the correlation between A2/A4 and PI the value of the coefficient r in compost with a lower ratio C/N- the case No. 2 compared to the case No. 1 is more than double and equals: -0.3 and -0.74, respectively. The value of the A2/A4 parameter related to the presence of organic compounds being a precursor of humic substances during composting gradually decreases (Table 7.1). At the same time, the value of the PI index is rising as a result of the HA increase and FA decrease. Lower value of the correlation coefficient A2/A4 vs. PI, $r = -0.74$ in the case of No 2, so when composting sewage sludge with a lower straw content, it may prove that the increased content of sewage sludge in the composted mixture has a positive effect on the humification process.

7.4 CONCLUSIONS

Quality of compost produced in industrial conditions depends on, among other things, share of the straw as an additional source of organic carbon. Unfortunately, this material is currently in short supply, therefore, costly. The tests we performed revealed that reduction of barley straw share in sewage sludge composting process may be, to some extent, favourable for the process. Such situation creates better conditions for the microorganism being a component of sewage sludge. Their occurrence may have favourable impact on decomposition and transformation of organic matter. Reducing barley straw content in the sewage sludge composting process, one can intensify the biodegradation process thus

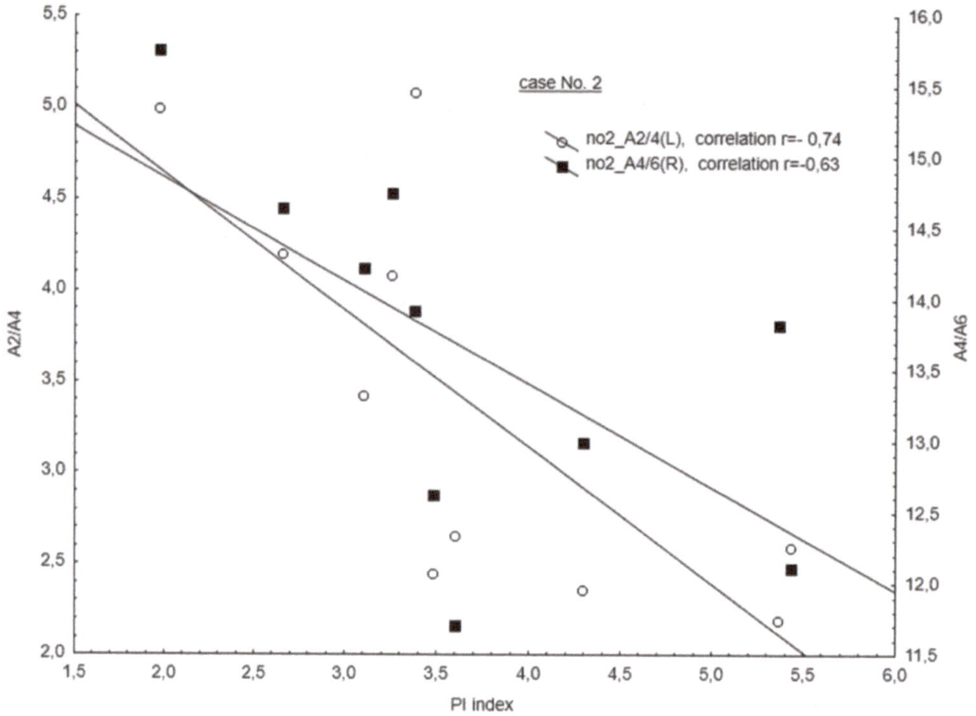

Figure 7.4. Correlation between absorbance coefficients A2/A4 and A4/A6 vs. PI index- case no.2.

extending the termophilic phase and, at the same time, increasing the degree of humification of the final product. However, during the ensuing tests, the minimum straw dose must be defined because excessive reduction of straw share in composted matter may inhibit the maturing processes, therefore, extend the entire composting operation.

Also the issue of transformation of heavy metals, the quantity of which increases with increased share of sewage sludge in the composted mass, is important. The issue of transformation of heavy metals compounds during composting, that was documented in numerous research works (Janowska et al. 2017; Liu et al. 2019; Robledo-Mahón et al. 2019), is extremely important in the context of evaluation of the real hazard, particularly in the case of use of the sewage sludge based compost in agriculture. My own research of this matter is continued.

ACKNOWLEDGEMENTS

This research was partly funded by the European Union Interreg South Baltic Programme, „Sludge Technological Ecological Progress – increasing the quality and reuse of sewage sludge" number STHB.02.02.00-32-0110/17.

REFERENCES

Alavi, N., Daneshpajou, M., Shirmardi, M., Goudarzi, G., Neisi, A. & Babaei, A.A. 2017. Investigating the efficiency of co-composting and vermicomposting of vinasse with the mixture of cow manure wastes, bagasse, and natural zeolite. *Waste Management* 69: 117–126.

Bernal, M.P., Alburquerque, J.A. & Moral, R. 2009. Composting of animal manures and chemical criteria for compost maturity assessment. A review. *Bioresource Technology* 100(22): 5444–5453.

Bustamante, M.A., Paredes, C., Marhuenda-Egea, F.C., Pérez-Espinoza, A., Bernal, M.P. & Moral, R. 2008. Co-composting of distillery with animal manures: carbon and nitrogen transformations in the evaluation of compost stability. *Chemosphere* 72(4): 551–557.

Carrizo, M.E., Alesso, C.A., Cosentino, D. & Imhoff, S. 2015. Aggregation agents and structural stability in soils with different texture and organic carbon content. *Scientia Agricola* 72(1): 75–82.

Curtis, M.J. & Claassen, V.P. 2009. Regenerating topsoil functionality in four drastically disturbed soil types by compost incorporation. *Restoration Ecology* 17(1): 24–32.

Czekała, J. 2008. Chemical properties of a compost produced on the basis of sewage sludge and different biowastes (in Polish). *Journal of Research and Applications in Agricultural Engineering* 53(3): 35–41.

Domeizel, M., Khalil, A. & Prudent, P. 2004. UV spectroscopy: a tool for monitoring humification and for proposing an index of the maturity of compost. *Bioresource Technology* 94(2): 177–184.

Doublet, J., Francou, F., Poitrenaud, M. & Houot, S. 2010. Sewage sludge composting: Influence of initial mixtures on organic matter evolution and N availability in the final composts. *Waste Management* 30(10): 1922–1930.

Głąb, T., Żabiński A., Sadowska, U., Gondek, K., Kopeć, M., Mierzwa–Hersztek, M. & Taborc, S. 2018. Effects of co-composted maize, sewage sludge, and biochar mixtures on hydrological and physical qualities of sandy soil. *Geoderma* 315: 27-35.

Gonzalez, D., Colon, J. Gabriel, D. & Sanchez, A. 2019. The effect of the composting time on the gaseous emissions and the compost stability in a full-scale sewage sludge composting plant. *Science of the Total Environment* 654: 311–323.

Hellebrand, H.J. & Kalk, W.D. 2001. Emission of methane, nitrous oxide and ammonia from dung windrows. *Nutrient Cycling in Agroecosystems* 60: 83–87.

Hsu, J.H. & Lo, S.L. 1999. Chemical and spectroscopic analysis of organic matter transformations during composting of pig manure. *Environmental Pollution* 104(2): 189–196.

Interreg South Baltic 2018. STEP. Sludge Technological Ecological Progress – increasing the quality and reuse of sewage sludge. Project no. STHB.02.02.00-32-0110/17, (https://www.step-interreg.eu/pl/)

Janowska, B., Szymański, K., Sidełko, R., Walendzik, B. & Siebielska, I. 2017. Assessment of mobility and bioavailability of mercury compounds in sewage sludge and composts. *Environmental Research* 156: 394–403.

Kacprzak, K., Neczaj, E., Fijałkowski, K., Grobelaka, A., Grosser, A., Worwag, M., Rorat, A., Brattebo, H., Almas, A. & Singh, B.R. 2017. Sewage sludge disposal strategies for sustainable development. *Environmental Research* 156: 39–46.

Kulikowska, D. & Sindrewicz, S. 2018. Effect of barley straw and coniferous bark on humification process during sewage sludge composting. *Waste Management* 79: 207–213.

Kulikowska, D. 2016. Kinetics of organic matter removal and humification progress during sewage sludge composting. *Waste Management* 49: 196–203.

Kononowa, M.M. 1968. *Substancje chemiczne gleby, ich budowa, właściwości i metody badań (Soil chemicals, their structure, properties and research methods)* (in Polish). Warszawa: Państwowe Wydawnictwo Rolnicze i Leśne.

Li, S., Li, D., Li, J., Li, G. & Zhan, B. 2017. Evaluation of humic substances during co-composting of sewage sludge and corn stalk under different aeration rates. *Bioresource Technology* 245: 1299–1302.

Liu, L., Wang, S., Guo, X.P. & Wang, H.G. 2019. Comparison of the effects of different maturity composts on soil nutrient, plant growth and heavy metal mobility in the contaminated soil. *Journal of Environmental Management* 250: 109525.

Lv, B., Xing, M., Yang, J., Qi, W. & Lu, Y. 2013. Chemical and spectroscopic characterization of water extractable organic matter during vermicomposting of cattle dung. *Bioresource Technology* 132: 320–326.

Meng, L., Li, W., Zhang, S., Wu, C. & Lv, L. 2017. Feasibility of co-composting of sewage sludge, spent mushroom substrate and wheat straw. *Bioresource Technology* 226: 39–45.

Publications Office of the European Union 2001. Commission Decision. Establishing ecological criteria for the award of the community eco-label to soil improvers and growing media. 2001/688/EC, (https://op.europa.eu/en/publication-detail/-/publication/9781b13a-e8aa- 4be4-9d38-f55e448fc03b).

Robledo-Mahón, T., Martín, M.A., Gutiérrez, M.C., Toledo, M., González, I., Aranda, E., Chica, A.F. & Calvo, C. 2019. Sewage sludge composting under semi-permeable film at full-scale: Evaluation of odour emissions and relationships between microbiological activities and physico-chemical variables. *Environmental Research* 177: 108624.

Sanchez, O.J., Ospina, D.A. & Montoya, S. 2017. Compost supplementation with nutrients and microorganisms in composting process. *Waste Management* 69: 136–153.

Sanchez-Monedero, M.A., Roig, A., Cegarra, J. & Bernal, M.P. 1999. Relationship between water-soluble carbohydrate and phenol fraction and the humification indices of different organic waste during composting. *Bioresources Technology* 70(2): 193–201.

Sapek, B. & Sapek, A. 1986. The use of 0.5 M sodium hydroxide extract for characterizing humic substances from organic formations. *Soil Science Annual* 37(2–3): 139–147.

Sellami, F., Hachicha, S., Chtourou, M., Medhioub, K. & Ammar, E. 2008. Maturity assessment of composted olive mill wastes using UV spectra and humification parameters. *Bioresource Technology* 99(15): 6900–6907.

Sidełko, R., Siebielska, I., Janowska, B., Skubała, A. 2017._Assessment of biological stability of organic waste processed under aerobic conditions. *Journal of Cleaner Production* 164: 1563–1570.

Sidełko, R., Janowska, B., Walendzik, B. & Siebielska, I. 2010.Two composting phases running In different process conditions timing relationship. *Bioresources Technology* 101(17): 6692–6698.

Sweeten, J.M. & Auvermann, B.W. 2008. Composting Manure and Sludge. *AgrilifeExtension* E-479: 06–08.

Swift, R.S. 1996. Organic Matter Characterization. In: D.L. Sparks, A.L. Page, P.A. Helmke & R.H. Loeppert (eds), *Methods of Soil Analysis Part 3—Chemical Methods*: 1011–1069. Madison Wis.: Soil Science Society of America.

Szymański, K., Sidełko, R., Janowska, B. & Siebielska, I. 2007. Monitoring of waste landfills (in Polish). *Scientific Papers of the Faculty of Civil and Environmental Engineering* 23:75–133.

Świerczek, L., Cieślik, B.M. & Konieczka, P. 2018. The potential of raw sewage sludge in construction industry-A review. *Journal of Cleaner Production* 200: 342–356.

Van Soest, P.J., Robertson, J.B. & Lewis, B.A. 1991. Methods for dietary fiber, neutral detergent fiber, and nonstarch polysaccharides in relation to animal nutrition. *Journal of Dairy Science* 74(10): 583–3597.

Veeken, A., Nierop, K., Wilde, V.D. & Hamelers, B. 2000. Characterisation of NaOH-extracted humic acids during composting of a biowaste. *Bioresource Technology* 72(1): 33–41.

Yuan, J., Chadwick, D., Zhang, D., Li, G., Chen, S., Luo, W., Du, L., He, S., & Peng, S. (2016). Effects of aeration rate on maturity and gaseous emissions during sewage sludge composting. *Waste Management*, 56, pp. 403–410, DOI: 10.1016/j.wasman.2016.07.017.

Yuan, Y., Xi, B., He, X., Tan, W., Gao, R., Zhang, H., Yang, Ch., Zhao, X., Huang, C. & Li, D. 2017. Compost-derived humic acids as regulators for reductive degradation of nitrobenzene *Journal of Hazardous Materials* 339: 378–384.

Zbytniewski, R. & Buszewski, B. 2005a. Characterization of natural organic matter (NOM) derived from sewage sludge compost. Part 1: chemical and spectroscopic properties. *Bioresource Technology* 96(4): 471–478.

Zbytniewski, R. & Buszewski, B. 2005b. Characterization of natural organic matter (NOM) derived from sewage sludge compost. Part 2: multivariate techniques in the study of compost maturation. *Bioresource Technology* 96(4): 479–484.

Zhang, J., Lv, B., Xing, M. & Yang, J. 2015. Tracking the composition and transformation of humic and fulvic acids during vermicomposting of sewage sludge by elemental analysis and fluorescence excitation–emission matrix. *Waste Management* 39: 111–118.

Zheng, G., Wang, T., Niu, M., Chen, X., Liu, Ch., Wang, Y. & Chen, T. 2018. Biodegradation of nonylphenol during aerobic composting of sewage sludge under two intermittent aeration treatments in a full-scale plant. *Environmental Pollution* 238: 783–791.

Zhou, Y., Selvam, A. & Wong, J.W.C. 2014. Evaluation of humic substances during co-composting of food waste, sawdust and Chinese medicinal herbal residues. *Bioresource Technology* 168: 229–234.

CHAPTER 8

Phosphorus recovery from sewage sludge

A.M. Czechowska-Kosacka
Faculty of Environmental Engineering, Lublin University of Technology, Nadbystrzycka, Lublin, Poland

G. Niedbała
Department of Biosystems Engineering, Faculty of Environmental Engineering and Mechanical Engineering, Poznań University of Life Sciences, Poznań, Poland

P. Kolarzyk
THORNFIELD SP Z O. O., Kraków, Poland

J. Ristvej
Faculty of Security Engineering, University of Žilina, Žilina, Slovakia

ABSTRACT: Phosphorus is an essential element for all living organisms. In the recent years a growing awareness of the limited resources of phosphorus has been observed. As phosphate rock is a non-renewable resource and is in a serious danger of depletion, sustainable phosphorus management has gained a great importance. It is estimated that at the current level of phosphorus consumption, 80% of all phosphate rock reserves will be consumed by 2070. A part of the mined phosphate rock is used to manufacture fertilizers and chemical products. This perspective increases the interest in the recycling of phosphorus. The ashes generated in the processes of thermal sewage sludge treatment are characterized by the highest content of phosphorus. The study results showed that thermal processing of sewage sludge, namely incineration, pyrolysis and hydrothermal treatment, can transform phosphorus in various species to increase its bioavailability. The chemical and physical properties of thermal treatment of phosphorus influence the effectiveness of its further recovery and reuse. The aim of this study is to summarize the latest progress made in the field of phosphorus recovery from sewage sludge.

8.1 INTRODUCTION

Phosphorus is an indispensable element of the life process and constitutes the building block of all living organisms. It is found in all living organisms and accounts for approximately 2–4% of the dry weight of most cells (Desmidt et al. 2015; Kahiluoto et al. 2014).

Phosphorus is also present in the genetic structure of DNA and RNA and plays the key role in inheriting traits and reproduction of living organisms. It is also crucial for energy production, transfer and storage in any of the biological processes being the basic component of ATP and ADP. Additionally, it is integral to the metabolic processes in plants, including photosynthesis.

The study results showed that phosphorus accumulates in the fastest-growing parts of plants. Plants uptake phosphorus and its majority accumulates in seeds and fruits, which makes the element so important. It is estimated that for the population of approximately 9 billion people, food production will have to be increased by roughly 30%. At present, more than 82% of the mined phosphorus is used in agriculture (Koppelaar & Weikard 2013; Suh & Yee 2011), approximately 7% in animal feed production, while the remaining 11% in the chemical, textile an d pharmaceutical industry (Cordell et al. 2009; SøRensen et al. 2015;.

Considering the continuous growth in population, the rapidly growing demand for food plants and consequently for phosphate fertilizers will be observed. The supply of phosphates is closely linked to the sustainable development of agriculture. As the primary industry, agriculture provides

DOI 10.1201/9781003171669-8

the most fundamental guarantee for other sectors of economy. Therefore, the supply of sufficient phosphorus resources guarantees the security and adequate level of global food resources (Koppelaar & Weikard 2013; Suh & Yee 2011). Unlike nitrogen, phosphorus exists in nature in the form of phosphate rocks of organic and non-organic origin and in the animal fossils and excretory products, e.g. phosphorite, struvite and other phosphate ore. However, the majority of phosphorus in the biosphere is not recoverable.

8.2 PHOSPHORUS RESOURCES

Phosphorus occurs in nature in the form of phosphate ore. There are 120 other minerals containing phosphorus that occur naturally, but because of their quality and quantity only a few can be exploited. When it comes to the usefulness of those materials for the industry mainly apatite, containing approximately 95% of phosphorus, is used. Other minerals exploited for industrial purposes are: phosphorite, svanbergite, struvite, and bluestone.

According to some sources, the economic resources of phosphate rock amounted to 17 billion tons while the primary resources – approximately to 50 billion tons (Dawson & Hilton 2012).

At present, phosphorus is produced in a few countries only. In Europe, except for Finland and Russia where phosphorus is produced in small volumes, the demand for this element is covered in full with imported produce. According to International Fertilizer Centre World Phosphate rock reserves and resources (IFDC 2010) 2/3 of the phosphate rock reserves come from Morocco and Western Sahara, China and US. In longer perspective, the future volume of supply and the demand for phosphorus are rather difficult to predict.

The annual global production of phosphorites exceeds 100 million tons and it is constantly increasing. If the annual consumption of phosphate rock is to grow at the rate of 3% per year, the annual consumption of phosphate rock will reach the level of 170 million tons in 2050. It is predicted that the phosphate rock reserves will become depleted by 2065. In the recent years, as a result of agricultural and industrial production development, the demand for phosphate ores and its consumption have been growing successively. Considering the statistical data for the period from 1990s to date, the demand for phosphate rock has been growing at the rate of 2.5% annually. Thus, taking into account the total volume of global phosphorus resources and the tempo of their exploitation, the global phosphate rock resources may deplete in approximately 100 years (Gilber 2009).

The phosphate rock reserves will continuously decrease and may soon be insufficient to meet the needs of the global development. Therefore, searching for alternative and renewable resources of phosphorus has become the most urgent necessity. Phosphorus can be recovered from certain phosphorus-rich residues, such as manure, meat, bone meal and agricultural waste, as well as sewage sludge (Frossard et al. 2009; Herzel et al. 2016; Meng et al. 2019).

In the process of municipal wastewater treatment, wastewater treatment plants use enhanced biological phosphorus removal (EBPR), which makes activated sludge rich in a large number of phosphorus accumulating organisms (PAOs) and denitrifying phosphorus accumulating bacteria (DPAOs). These two types of microorganisms accumulate much more phosphorus than physiologically needed for their own metabolism. The excess phosphorus accumulates in their cells in the form of polyphosphates and further on is removed at the final stage of the aerobic treatment process. The process generates residue which contains approximately 25% of the total phosphorus content in sewage. The phosphorus contained in sewage sludge occurs in the inorganic form. The organic form of phosphorus can be found in sewage sludge in relatively scarce amount, which makes it a little less difficult for P to be recovered.

Except for sewage sludge, numerous waste materials contain significant amounts of phosphorus. Table 8.1 presents the concentration of phosphorus in different types of waste.

According to Herzel et al. (2016), the sludge produced by developed countries exceeds 30 million tons per year. In Europe, approximately 11.6 million tons of dry mass of sewage sludge is produced (Environmental, Economic and Social Impacts 2010). Greece stores up over 90% of sewage sludge, while France, Spain and Great Britain use 65% of the produced sludge in agriculture. The Netherlands and Switzerland incinerate all generated sewage sludge. In the recent years sludge incineration has been limited in Great Britain because of its high costs, while pyrolysis has been considered to be potentially

Table 8.1. Concentration of phosphorus in different types of waste (Meng et al. 2019).

Material	P (% P by weight)
Human urine	0.02–0.07
Cow dung	0.04
Human excreta	0.35
Human feces	0.52
Sludge (from biogas digester)	0.48–0.77
Vermicompost	0.65
Farm yard manure (FYM)	0.07–0.88
Meat meal	1.09
Poultry manure	1.27
Bone meal	8.73–10.91

more cost-effective due to higher recovery of energy and other components, e.g. phosphorus (Mills et al. 2014).

8.3 SEWAGE SLUDGE TREATMENT

Except for such valuable nutrients as nitrogen, phosphorus and potassium, sewage sludge can contain pathogens and heavy metals (Anttila et al. 2008). If not disposed of or managed properly, sludge can pose environmental hazards, among others, contamination of aquifers, causing a real threat for the human health and life. The disposal of waste can be divided into three categories: landfill (less and less popular), agricultural use and incineration (Fonts et al. 2012). Further reuse of sewage sludge constitutes the principal objective of wastewater treatment. In Europe, land use is the primary disposal method for sludge, while in the United States approx. 60% of sludge is treated to form bio-solids for the use as farmland fertilizer.

Incinerated sewage sludge (solid waste fly ash) can also be used in the construction industry (Lin et al. 2005; Monzó et al. 2003; Wu et. 2016) and this method of sludge reuse found a large group of followers in Japan. However, the results of endurance tests of the produced building materials should still be analyzed in detail (Baeza-Brotons et al. 2014).

Sewage sludge deposition in landfills is not recommended for various reasons. Rich in organic matter, sludge causes the generation of landfill gas with a high content of methane, which is a much more potent greenhouse gas than carbon dioxide (CO2). Disposal of large amount of sewage to landfills would inevitably induce some of geo-environmental problems, such as: differential settlement, compression or slope instability of the landfill (Lo et al. 2002).

The second disposal method of sewage sludge is agricultural use because of its abundance of organic matter and large amount of nitrogen, phosphorus and potassium, which can be used as fertilizer for crops.

Incineration is the last of the methods of sewage sludge disposal in the effect of which the exhaust gases containing CO_2, water and ash, and due to its exothermal nature – heat is generated. The use of the incineration process decreases the volume of sewage sludge by almost 90% and destroys pathogens. The generated ash can be deposited in landfills or, as it has already been mentioned, may be used for the production of building materials (Lin et al. 2005; Monzó et al. 2003; Wu et. 2016).

In addition to the three major methods of sewage sludge disposal, other sludge treatment methods are considered, namely pyrolysis, gasification and hydrothermal treatment (Frišták et al. 2018; Kleemann et al., 2017; Samolada & Zabaniotou 2014;. Figure 8.1 presents the sources and the disposal of phosphorus.

Pyrolysis is a thermal decomposition process under low-oxygen or anoxic conditions at relatively high temperature (300–1000° C) which provides three basic products such as: pyrolysis gas, bio-oil and solid product, namely biochar (Frišták et al. 2018; Lehmann et al. 2015; Racek et al. 2020).

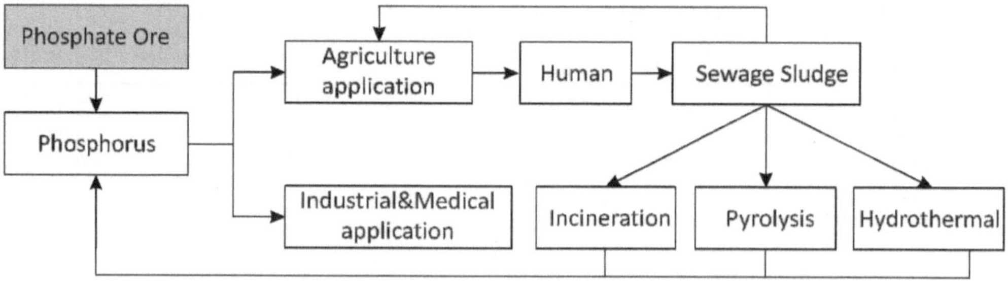

Figure 8.1. Source, use and disposal of phosphorus (Meng et al. 2019).

Participation of particular products in the process, including their qualitative composition, depends mainly on the process conditions, such as temperature, response time, pressure and processed sludge characteristics. However, regardless of the process conditions, the most significant share in the product is taken by the liquid phase (containing 15–30% w/w).

Gasification is a thermal process carried out at a temperature of 400–1400°C. In the effect of this process, the organic matter contained in sewage sludge is converted into flammable gas and synthesis gas with the oxygen used at the rate from 20 – 40% as required for the total burn-off (Samolada & Zabaniotou 2014).

Hydrothermal treatment is a thermal-chemical process carried out in a completely closed system under certain pressure and at relatively lower temperature, as compared to incineration or pyrolysis. The hydrothermal reactions taking place throughout the process are exothermic processes (Huang et al. 2017; Xue et al. 2015).

8.4 IMPACT OF THERMAL TREATMENT ON PHOSPHORUS RECOVERY FROM SEWAGE SLUDGE

Sewage sludge has long been used in agriculture as fertilizer. However, not all types of sludge can be used for that purpose due to the high content of heavy metals and other contaminants (Anttila et al. 2008; Laturnus et al. 2007; Tarayre et al. 2016). The legal regulations in the scope of sewage sludge are becoming more and more restrictive all around the world, especially when it comes to the maximum allowable concentrations of heavy metals introduced into the soil (Tarayre et al. 2016). For this reason, the technologies for sewage sludge treatment and indirect recovery of phosphorus are becoming increasingly more popular. Thermal sewage sludge treatment reduces its volume, causes decomposition of organic contaminants, contributes to the production of valuable by-products and therefore is considered to be the best possible method of sewage sludge utilization (Hossain et al. 2011). Thermal treatment of sewage sludge comprises pyrolysis, incineration and thermal processing. Thermal treatment modifies the physical and chemical structure of sewage sludge impacting the speciation of phosphorus, which determines its mobility and bioavailability.

8.4.1 *Incineration*

In the EU member states, and also partly in Poland, the problem of sewage sludge disposal is solved by its incineration. The incineration process completely eliminates the majority of potentially hazardous organic compounds, parasites and pathogenic microorganisms.

The ashes from sewage sludge contain much higher amounts of phosphorus due to a significant reduction in the volume of incinerated material. The ashes from sewage sludge incineration may contain from 4 to 11% of phosphorus (Franz 2008). The product of incineration is obtained in the powdered form (powdered fly ash) which makes the future processing operations much easier (Weigand et al. 2013).

The main elements of the ash fraction are: Ca, Si, Fe, Al and P. Most of them occur as oxides forming such compounds as CaO, SiO_2, Fe_2O_3 or Al_2O_3. Incineration allows complex management of all solid residues generated in the course of the phosphorus recovery process (Takahashi et al. 2001).

Additionally, it was proven that during incineration at high temperatures, calcium phosphate ($Ca_3(PO_4)_2$), calcium pyrophosphate ($Ca_2P_2O_7$), and hydroxyapatite ($Ca_5(PO_4)_3(OH)$) as well as apatite from Al and Fe phosphates and amorphous Ca phosphate phases, are formed. Phosphorus often occurs as $Fe_4(P_4O_{12})_3$, $Al(PO_3)_3$, mainly because $AlCl_3$, $Al_2(SO_4)_3$, $FeCl_3$, $Fe_2(SO_4)_3$ compounds are used as precipitation agents (Biswas et al. 2009). Using aluminum compounds as precipitation agents poses higher hazard for the environment than iron.

Incineration of sewage sludge is often carried out at the temperature of 850°C, in which phosphorus takes the form of volatile oxides. Upon cooling to a temperature between 40 and 600°C they condense to P_4O_{10} and become a component of the ash retained by filters (Cieślik & Konieczka 2017). Belevi & Langmeier (2000) proved that approximately 80–90% of phosphorus remains in the ash while the remaining part is transported to fly ash. Large amount of fly ash formed during the incineration process considerably decreases the amount of phosphorus in the incineration ash. The study carried out by Thygesen et al. (2011) showed that the incineration temperature should be kept below 700°C, which will ensure the formation of insoluble hydroxyapatite in the ash. Many authors claim that incineration of sewage sludge at the temperature higher than 700°C results in a situation in which the product obtained is useless for the fertilization purposes due to the content of the non-bioavailable form of phosphorus (Yuan et al. 2012). On the other hand, incineration of sewage sludge at the temperature of 1250°C leads to the reduction of the concentration of phosphorus in ash and to its increased concentration in the dust fraction (Zhang & Ninomiya 2007).

In order to increase the available phosphorus in ash, innovative incineration process parameters were proposed. Zhao et al. (2018) claimed that adding cotton stalk to the incineration process improves the bioavailability in the fly ash. Ren & Li (2015) claimed that adding wheat straw to the incinerated sewage sludge leads to the formation of alkali-rich phosphate silicate. The reactions between the phosphorus-rich compounds and alkali metals lead to the formation of species containing K-Ca-P bindings. Beck & Unterberger (2006) indicated that phosphorus tends to bind with Ca during the incineration of sewage sludge with the additive of carbon. Han et al. (2009) investigated that the CaO additive may increase the content of phosphorus and heavy metals during the sewage sludge incineration process while the application of HCl decreased the content of phosphorus and heavy metals in the incineration ash. The efficiency of the heavy metals removal from sewage sludge or ashes exhibited an increasing tendency with the additive of chlorides, especially in respect to Cu, Zn and Pb.

8.4.2 *Pyrolysis*

Among numerous techniques for sewage sludge treatment, pyrolysis has been recognized as one of the most effective alternatives to reduce the volume of sewage sludge, eliminate the pathogens and convert the sludge into valuable materials with multiple applications (Feng et al. 2015).

As a result of pyrolysis, phosphorus remains mainly in solid phase while its share in the liquid and gas phases is rather insignificant. The proportions between particular elements depend mainly on the temperature and pressure in which the process is carried out. In the case of pyrolysis carried out at a temperature 250–600°C almost 100% of phosphorus in the solid phase can be recovered (Azuara et al. 2013; Huang & Tang 2015; Kleemann et al. 2017).

Composition of sewage sludge to a large extent depends on the source of sewage, the method of its treatment and further methods of its management. Organic matter contained in sewage sludge is formed of mainly organic compounds forming the cellular structures such as: proteins, lipids, polysaharydes and DNA. Sewage sludge in the raw sludge is extremely sensitive to pyrolysis and the possibility of its conversion into gas phase increases. Organophosphates contained in sewage sludge may undergo transformations during the pyrolysis process such as: dehydration, decarboxylation, and polymerization (Libra et al. 2011). Such reactions may further on lead to the degradation of organophosphates and formation of orthophosphates, pyrophosphates and/or organophosphates with more condensed functional groups (Meng et al. 2019).

During the slow pyrolysis process, the polyphosphates in sewage sludge degrade into shorter chains. The polyphosphate chain becomes shortened as the pyrolysis temperature increases. The process of the sewage sludge pyrolysis may convert the organic phosphorus into inorganic, reduce the content of the organophosphates and polyphosphates and increase the total phosphorus content by two or three times (Meng et al. 2019). A huge impact of the temperature on the phosphorus transformation in the pyrolysis process was observed. Pyrophosphate forms are formed during pyrolysis at the average temperatures ranging from 300 to 600°C (Xu et al., 2016).

At the temperatures exceeding 700oC, phosphorus occurs almost exclusively in the form of orthophosphate. Higher heating temperatures in the process promote phosphorus migration into the gaseous phase, e.g. phosphorus can volatilize during gasification (Bläsing et al. 2013), and the main gaseous compound in the temperatures of gasification ranging from 900 to 1400°C is PO_2^+.

The process of pyrolysis can affect both the quantity and speciation of phosphorus in biochar; its mobility and bioavailability are closely related to its speciation (Huang & Tang 2015; Qian & Jiang 2014). During the pyrolysis of sewage sludge, decreases of soluble phosphorus were observed (Huang & Tang 2016). In those cases the HCl-extractable phosphorus fractions increased, which is related to the organophosphates decrease and the increase of crystalline Ca phosphates. Such an effect of immobilization increases proportionally to the increase of temperature. Additionally, the increased temperature of pyrolysis would lead to the conversion of the non-apatite inorganic phosphorus to apatite phosphorus (Xu et al. 2016). The content of calcium phosphate and orthophosphate in a sample of sewage sludge increases as the temperature rises. The speed and time of heating have a minimum effect of phosphorus transformation.

The physical and chemical properties of incinerator sewage sludge ash and pyrolysis sewage sludge char were compared for the purposes of phosphorus recovery. Both products obtained from sewage sludge incineration and pyrolysis contain whitlockite. Heavy metals are less soluble in sludge after pyrolysis, because they are more strongly incorporated in particles. The use of the generated biochar to fertilize soil may increase the sequestration of carbon in soil, decrease the concentration of CO2 in the atmosphere as well as the bioavailability of some of the heavy metals. Additionally, it may improve the physicochemical quality of soil as well as change the content and availability of nutrients (Meng et al. 2019).

8.4.3 *Hydrothermal treatment*

Hydrothermal processing is generally divided into thermal and hydrothermal treatment. Depending on the operating conditions, such as temperature or the type of the product obtained, each of the above-mentioned types of treatment can be further divided into the three separate processes of carbonization, liquefaction, and gasification (Bridgwater 2012). The application of hydrothermal techniques can cover a wide range of wastes – from dry to wet and from homogenous to non-homogenous waste (Zheng et al. 2020; Zumbühl 2013).

They may convert solid biowaste into multi-purpose products, in this way achieving the actual goal of the resource recovery (Titirici et al. 2012).

Hydrothermal treatment can also be adopted for sewage sludge disposal as its application has some unquestionable advantages. Sewage sludge is characterized with high moisture content. A substantial fraction of water contained in sewage sludge occurs in the intracellular or chemically bound form, which cannot be removed with the use of thermal treatment. Hydrothermal technique may be used for converting the high moisture content of sewage sludge at mild temperature into valuable products, among others, into biochar (biocarbon). In these cases, hydrothermal treatment can be used to avoid the drying processes (Zumbühl 2013).

The thermal treatment processes described above are carried out mainly in closed systems at relatively low temperatures, unlike other thermal processes where higher temperatures are required. Similarly to other treatment processes hydrothermal treatment can be used for the purpose of degradation of pathogens and organic pollutants. The studies carried out by Von Eyser et al. showed that organic pollutants can be degraded after the hydrothermal carbonization. It was also found that the pharmaceuticals contained in sewage sludge were degraded in more than 90% after hydrothermal treatment at the temperature of 210°C (Vom Eyser et al. 2015).

Hydrothermal treatment can also be recommended to release phosphates from sewage sludge. The phosphorus recovery from the hydrothermal carbonization produced residues commonly exceeds 80% (Heilmann et al. 2011; Heilmann et al. 2014). Metal-complexed and mineral-associated phosphorus compounds contained in sewage sludge, under hydrothermal conditions can be transformed into stable compounds. The studies revealed that the application of hydrothermal processes may lead to the transformation of all types of phosphorus, mainly polyphosphate, phosphonate and organic phosphates into inorganic orthophosphate. With the use of hydrothermal processing, more significant changes in phosphorus speciation can be obtained than when the thermal processing is applied. The non-stable inorganic forms of phosphorus may become dissolved or re-crystallized.

Speciation of phosphorus, above all, depends on the presence of P-binding metals. Sewage sludge treatment process carried out at the temperature of 225°C with the use of hydrothermal carbonization can serve as an example here. The process activated an increasing association of phosphate from iron and calcium which resulted from the fact that the concentration of both cations in the sewage sludge is high and they have a strong tendency to form sediments from precipitates and/or surface complexes (Feng et al. 2015). Thus, the transformation of calcium phosphate phases is an extremely important reaction occurring under the conditions of hydrothermal treatment. Additionally, in the course of hydrothermal processes, the majority of heavy metals can undergo immobilization in the solid phase (Shi et al. 2013). Hydrothermal treatment is also considered a method of preliminary sewage sludge processing for magnesium ammonium phosphate ($MgNH_4PO_3 \cdot 6H_2O$) crystallization.

8.5 CONCLUSIONS

The phosphorus contained in sewage sludge and in its ashes is characterized by a high recovery potential. The methods of phosphorus recovery presented in this study show that there are numerous options for its recycling and further reuse mainly in agriculture which may contribute to the protection of the rapidly depleting natural resources of this valuable element. Due to a higher recycling potential, larger amounts of phosphorus could be recovered after thermal treatment of sewage sludge with the use of pyrolysis, incineration and hydrothermal processing. It is expected that the interest in the recovery of phosphorus with the use of pyrolysis will grow because when compared to incineration, pyrolysis is considered to be one of the most cost-effective methods of thermal sewage sludge treatment).

REFERENCES

Anttila, J., Bergman, R., Horttanainen, M., et al. 2008. Hajautetun energiantuotannon modulaarinen yhdyskunnan sivuainevirtoja hyödyntävä CHP-laitos. http://www.doria.fi/bit-stream/handle/10024/43043/isbn9789522146953.pdf?sequence=1

Azuara, M., Kersten, S.R.A., Kootstra, A.M.J. 2013. Recycling phosphorus by fast pyrolysis of pig manure: concentration and extraction of phosphorus combined with formation of value-added pyrolysis products. *Biomass & Bioenergy* 49: 171–80.

Baeza-Brotons, F., Garces, P., Paya, J., Saval, J.M. 2014. Portland cement systems with addition of sewage sludge ash. Application in concretes for the manufacture of blocks. *Journal of Cleaner Production* 82: 112–124.

Beck J., Unterberger S. 2006. The behaviour of phosphorus in the flue gas during the combustion of high-phosphate fuels. *Fuel* 85(10–11):1541–1549.

Belevi, H., Langmeier, M. 2000. Factors determining the element behavior in municipal solid waste incinerators. 2. Laboratory experiments, *Environmental Science & Technology* 34(12): 2507–12.

Biswas, B.K., Harada, H., Ohto, K., et al. 2009. Leaching of phosphorus from incinerated sewage sludge ash by means of acid extraction followed by adsorption on orange waste gel. *Journal Environmental Sciences* 21(12): 1753–60.

Bläsing, M., Zini, M., Müller, M.J.E. 2013. Influence of feedstock on the release of potassium, sodium, chlorine, sulfur, and phosphorus species during gasification of wood and biomass shell. *Energy & Fuels* 2013(27): 1439–1445.

Bridgwater, A.V. 2012. Review of fast pyrolysis of biomass and product upgrading. *Biomass & Bioenergy* 38: 68–94.

Cieślik B., Konieczka P. 2017. A review of phosphorus recovery methods at various steps of wastewater treatment and sewage sludge management. The concept of "no solid waste generation" and analytical methods. *Journal of Cleaner Production* 142(4): 1728–1740.

Cordell, D., Drangert, J.O., White, S. 2009. The story of phosphorus: global food security and food for thought. *Global Environ Change* 19(2): 292–305.

Cordell, D., Rosemarin, A., Schröder, J.J. Smit, A.L. 2011. Towards global phosphorus security: a systems framework for phosphorus recovery and reuse options. *Chemosphere* 84(6): 747–58.

Dawson, C.J., Hilton, J. 2012. Fertilizer availability in a resource-limited world: production and recycling of nitrogen and phosphorus. *Food Policy*, 36(1), 2012, pp. 14–22.

Desmidt, E., Ghyselbrecht, K., Zhang, Y. et al. 2015. Global phosphorus scarcity and full-scale P-recovery techniques: a review. *Critical Reviews in Environmental Science and Technology* 45(4): 336–384.

Environmental, Economic and Social Impacts of the use of Sewage Sludge on Land. 2010. Milieu Ltd., WRc and RPA, Contract DB ENV.G.4/ETU/2008/0076r, s.l.: s.n.). https://ec.europa.eu/environment/archives/waste/sludge/pdf/part_i_report.pdf

Feng, H., Zheng, M., Dong, H., et al. 2015. Three-dimensional honeycomblike hierarchically structured carbon for high-performance supercapacitors derived from high-ash-content sewage sludge. *Journal of Materials Chemistry A* 3: 15225–15234.

Fonts, I, Gea, G., Azuara, M., et al. 2012. Sewage sludge pyrolysis for liquid production: a review. *Renewable & Sustainable Energy Reviews* 16(5): 2781–2805.

Franz, M. 2008. Phosphate fertilizer from sewage sludge ash (SSA). Waste Manage. 28: 1809–1818.

Frišták, V., Pipíška, M., Soja, G. 2018. Pyrolysis treatment of sewage sludge: a promising way to produce phosphorus fertilizer. *Journal of Cleaner Production*. 172: 1772–1778.

Frossard, E., Bünemann, E., Jansa, J., Oberson, A., Feller, C. 2009. Concepts and practices of nutrient management in agro-ecosystems: can we draw lessons from history to design future sustainable agricultural production systems. *Bodenkultur* 60: 43–60.

Gilber, N. 2009. The disappearing nutrient. *Nature* 461: 716–718.

Han, J., Kanchanapiya, P., Sakano, T., et al. 2009. The behaviour of phosphorus and heavy metals in sewage sludge ashes. *International Journal of Environment and Pollution* 37(4): 357–368.

Heilmann, S.M., Jader, L.R., Sadowsky, M.J., et al. 2011. Hydrothermal carbonization of distiller's grain. *Biomass & Bioenergy* 35(7): 2526–2533.

Heilmann, S.M., Molde, J.S., Timler J.G., et al. 2014. Phosphorus reclamation through hydrothermal carbonization of animal manures. *Environmental Science & Technology*. 48: 10323–10329.

Herzel, H., Krüge, O., Hermann, L., et al. 2016. Sewage sludge ash—a promising secondary phosphorus source for fertilizer production. *Science of the Total Environment* 542(Pt B): 1136–1143.

Hossain, M.K., Strezov, V., Chan K.Y., et al. 2011. Influence of pyrolysis temperature on production and nutrient properties of wastewater sludge biochar. *Journal of Environmental Management* 92(1): 223–228.

Huang, R., Tang, Y. 2016. Evolution of phosphorus complexation and mineralogy during (hydro)thermal treatments of activated and anaerobically digested sludge: Insights from sequential extraction and P K-edge XANES. *Water Research* 100: 439–447.

Huang, R., Fang, C., Lu, X., et al. 2017. Transformation of phosphorus during (Hydro)thermal treatments of solid biowastes: reaction mechanisms and implications for P reclamation and recycling. *Environmental Science & Technology*. 51(18): 10284–10298.

Huang, R., Tang, Y. 2015. Speciation dynamics of phosphorus during (hydro)thermal treatments of sewage sludge. *Environmental Science & Technology* 49: 14466–1446674.

IFDC. 2010. World Phosphate rock reserves and resources. https://pdf.usaid.gov/pdf_docs/Pnadw835.PDF

Kahiluoto, H., Kuisma, M., Koukkanen, A., Mikkilä, M., Linnanen, L. 2014. Taking planetary nutrient boundaries seriously: can we feed the people? *Global Food Security* 3: 16–21.

Kleemann, R., Chenoweth, J., Clift, R., Morse, S., Saroj, D. 2017. Comparison of phosphorus recovery from incinerated sewage sludge ash (ISSA) and pyrolysed sewage sludge char (PSSC). Waste Manage. 60: 201–210.

Koppelaar, R.H.E.M., Weikard, H.P. 2013. Assessing phosphate rock depletion and phosphorus recycling options. *Global Environ Change* 23: 1454–1466.

Laturnus, F., von Arnold, K., Grøn, C. 2007. Organic contaminants from sewage sludge applied to agricultural soils. False alarm regarding possible problems for food safety? *Environmental Science and Pollution Research* 14: 53–60.

Lehmann, J., Joseph, S. 2015. Biochar for environmental management science *Technol. Implement*. 25: 15801–15811.

Libra, J.A., Ro, K.S., Kammann, C., et al. 2011. Hydrothermal carbonization of biomass residuals: a comparative review of the chemistry, processes and applications of wet and dry pyrolysis. *Biofuels* 2: 71–106.

Lin, K., Chiang, K., Lin, C. 2005. Hydration characteristics of waste sludge ash that is reused in eco-cement clinkers *Cement and Concrete Research* 35: 1074–1081.

Lo, I.M.C., Zhou, W.W., Lee, K.M. 2002. Geotechnical characterization of dewatered sewage sludge for landfll. *Revue Canadienne De Géotechnique* 39(5): 1139–1149.

Meng, X., Huang, Q., Xu J., Gao, H., Yan, J. 2019. A review of phosphorus recovery from different thermal treatment products of sewage sludg. *Waste Disposal & Sustainable Energy* 1: 99–115.

Mills, N., Pearce, P., Farrow, J., Thorpe, R.B., Kirkby, N.F. 2014. Environmental & economic life cycle assessment of current & future sewage sludge to energy technologies. *Waste Management* 34: 185–195.

Monzó, J., Payá, J., Borrachero, M.V., Girbés, I. 2003. Reuse of sewage sludge ashes (SSAs) in cement mixtures: the effect of SSA on the workability of cement mortars. *Waste Manage* 23: 373–381.

Qian, T.T., Jiang, H., 2014. Migration of phosphorus in sewage sludge during different thermal treatment processes. *ACS Sustainable Chemistry & Engineering* 2(6): 1411–1419.

Racek, J., Sevcik, J., Chorazy, T., Kucerik, J. Hlavinek, P. 2020. Biochar – Recovery Material from Pyrolysis of Sewage Sludge: A Review. *Waste and Biomass Valorization* 11: 3677–3709.

Ren, Q., Li, L. 2015. Co-combustion of agricultural straw with municipal sewage sludge in a fluidized bed: role of phosphorus in potassium behavior. *Energy & Fuels* 29(7): 4321–4327.

Samolada, M. C., Zabaniotou, A. 2014. Comparative assessment of municipal sewage sludge incineration, gasification and pyrolysis for a sustainable sludge-to-energy management in Greece. Waste Manage. 34: 411–420.

Shi, W., Liu, C., Ding, D., et al. 2013. Immobilization of heavy metals in sewage sludge by using subcritical water technology. *Bioresource Technology* 137: 18–24.

SøRensen, B.L., Dall, O.L., Habib, K. 2015. Environmental and resource implications of phosphorus recovery from waste activated sludge. *Waste Manage.* 45: 391–399.

Suh, S., Yee, S. 2011. Phosphorus use-efficiency of agriculture and the food system in the US. *Chemosphere* 84: 806–813.

Takahashi, M., Kato, S., Shima, H. et al., 2001. Technology for recovering phosphorus from incinerated wastewater treatment sludge. *Chemosphere* 44(1): 23–29.

Tarayre, C., Clercq, L.D., Charlier, R., et al. 2016. New perspectives for the design of sustainable bioprocesses for phosphorus recovery from waste. *Bioresource Technology* 206: 264–74.

Thygesen, A.M., Wernberg, O., Skou, E., et al. 2011. Effect of incineration temperature on phosphorus availability in bio-ash from manure. *Environmental Technology* 32(6): 633–638.

Titirici, M.M., White, R.J., Falco, C, et al. 2012. Black perspectives for a green future: hydrothermal carbons for environment protection and energy storage. *Energy & Environmental Science* 5(5): 6796.

Vom Eyser, C., Palmu, K., Schmidt, T.C., et al. 2015. Pharmaceutical load in sewage sludge and biochar produced by hydrothermal carbonization. *Science of the Total Environment* 537: 180–186.

Weigand, H., Bertau, M., Hübner, W., et al. 2013. RecoPhos: full-scale fertilizer production from sewage sludge ash. Waste Manage. 33(3): 540–544.

Wu, M.H., Lin, C.L. , Huang, W.C., et al. 2016. Characteristics of pervious concrete using incineration bottom ash in place of san sandstone graded material. *Construction and Building Materials* 111: 618–624.

Xu, G., Zhang, Y., Shao, H., et al. 2016. Pyrolysis temperature affects phosphorus transformation in biochar: chemical fractionation and 31P NMR analysis. *Science of the Total Environment* 569–570: 65–720.

Xue, X., Chen, D., Song X., et al. 2015. Hydrothermal and pyrolysis treatment for sewage sludge: choice from product and from energy benefit. *Energy Procedia.* 66: 301–304.

Yuan, Z., Pratt, S., Batstone, D.J. 2012. Phosphorus recovery from wastewater through microbial processes, *Current Opinion in Biotechnology* 23(6): 878–83.

Zhang, L., Ninomiya, Y. 2007. Transformation of phosphorus during combustion of coal and sewage sludge and its contributions to PM10, *Proceedings of the Combustion Institute* 31(2): 2847–54.

Zhao, Y., Ren, Q., Na, Y. 2018. Promotion of cotton stalk on bioavailability of phosphorus in municipal sewage sludge incineration ash. Fuel 214: 351–355.

Zheng, X., Jiang, Z., Ying, Z., Ye, Y., Chen, W., Wang, B., Dou, B. 2020. Migration and Transformation of Phosphorus during Hydrothermal Carbonization of Sewage Sludge: Focusing on the Role of pH and Calcium Additive and the Transformation Mechanism, *ACS Sustainable Chemistry & Engineering* 8: 7806–7814.

Zumbühl, K. 2013. Hydrothermal carbonization as an energy-efficient alternative to established drying technologies for sewage sludge: a feasibility study on a laboratory scale. *Fuels* 27: 454–460.

CHAPTER 9

The use of solubilization for phosphorus recovery

E. Bezak-Mazur & J. Ciopińska
Faculty of Environmental, Geomatic and Energy Engineering, Kielce University of Technology, Kielce, Poland

ABSTRACT: The economy conducted in accordance with the principles of sustainable development imposes the obligation of rational use of environmental resources and recovery of potential raw materials from waste. Considering the scale of formation and phosphorus content, the most important waste materials from which phosphorus can be recovered are sewage sludge and biochar, ashes generated after their thermal treatment or slaughterhouse waste, including animal bones. The article discusses the mechanism of solubilization, considering the ways of carrying it out. The factors influencing the efficiency of solubilization, such as: pH value, organic carbon content, temperature, oxygen availability, and the form of phosphorus, were characterized. The paper presents the examples of the possible implementation of microbial solubilization to recover the phosphorus from various materials such as sewage sludge, ashes, biochar, bones, farm waste, paying attention to the effectiveness of the process and the necessity to use additives containing organic carbon or carbonates and nitrogen in the case of waste containing only mineral ingredients. The role of solubilization in the release of phosphorus from bio-fertilizers and thus providing the bioavailable forms of this element was also discussed.

9.1 INTRODUCTION

In recent years, phosphorus has been included on the EU list mentioning non-renewable environmental resources undergoing depletion (Report on Critical Raw Material for the UE, 2015). In accordance with the principles of sustainable development, particular attention should be paid to the possibilities of recovering raw materials from waste materials, thus limiting the use of the environmental resources to a minimum. The previous analyses indicate that there are many forms of waste materials from which phosphorus can be recovered. The most important is sewage sludge. Around 116 tons of sewage sludge are produced annually in the EU. This corresponds to a content of at least 300,000 tons of phosphorus, which covers 20% of the demand for mineral phosphorus fertilization (Bezak-Mazur 2019). Other waste materials are slaughter waste, animal bones, farmed waste (manure) as well as biochar and ashes obtained after the thermal treatment of sewage sludge (Schroder et al. 2000). It is estimated that the production of meat products in the EU generates an average of 3,500,000 tons of bone per year (Wyciszkiewicz et al. 2016).

The phosphorus present in these waste materials occurs in various forms. The simplest division is non-reactive (NPR) and reactive (PR) phosphorus (Dijk et al. 2016). The reactive phosphorus is inorganic phosphates, orthophosphates, while the non-reactive phosphates contain polymerized inorganic phosphates (metaphosphates, di-tetra phosphates) and organic phosphorus compounds. According to Venkiteshwaran (Venkiteshwaran et al. 2018), only the reactive phosphorus can be transformed and recovered in various processes using the physical, chemical or biological methods. In addition to such a division, speciation analysis is commonly used in the literature to describe phosphorus forms, which allows the presentation of phosphorus forms through the fractions secreted under different conditions.

Among the phosphorus recovery methods, much attention is paid to the biological methods, because they are low-cost, do not burden the environment with waste chemical reagents and do not require large amounts of energy to implement them. The properly selected microorganisms transform non-reactive,

DOI 10.1201/9781003171669-9

insoluble forms of phosphorus into reactive, soluble forms. The process of such a phosphorus transformation is called solubilization, and the microorganisms carrying it out are solubilizing microorganisms (PSM) (Bezak-Mazur et al. 2020).

This article reviews the microorganisms capable of solubilization, analyzes the mechanism of solubilization and its conditioning, and the phosphorus fractions after solubilization. The next part discusses the examples of phosphorus recovery through microbiological solubilization and the use of solubilization in the production of fertilizers (bio-fertilizers), effects on plants as well as soil remediation.

9.2 PHOSPHORUS SOLUBILIZING MICROORGANISMS

PSM are microorganisms that have the ability to dissolve both the organic and inorganic phosphorus. Among them are PSB bacteria, PSF fungi and algae (Sharma et al. 2013). Alori (Alori et al. 2017) in the review from 2017 included the phosphorus solubilizing bacteria: Pseudomonas, Agrobacterium, Bacillus, Burkholderia, Enterobacter, Ervwinia, Rhizobium, and Thiobacillus. In addition, this list should include: Achromobacter, Micrococcus Aerobacter, Flavobacterium, Acinetobacter, Alcaligenes, Rahnella (Hayat et al. 2010; Li et al. 2019). Basically, a particular bacterial strain solubilizes either phosphorus in the inorganic or organic bonds. Kushneria YCWA18 exhibits exceptional solubilization capabilities of both types of phosphorus compounds (Zhou et al. 2011). The fungi capable of solubilizing phosphorus include Alternaria, Aspergillus, Cladosporium, Penicylium, Rhizoctonia, Sacchoromyces, Tonula, Curvularia, Trichoderm (Alori et al. 2017; Hayat et al. 2010; Li et al. 2019; Sharma et al. 2013). Fungi, owing to the secreted enzymes, i.e. phosphatase and phytates, are particularly predestined for the solubilization of the phosphorus in organic connections, e.g. in the form of phospholipid-lecithin (Li et al. 2019). Fungi can use mycorrhiza and cooperate with other microorganisms that promote plant growth, occurring on the roots. The fungi from the Curvularia geniculata strain located inside the roots of plants solubilize phosphorus (Priyadharsini et al. 2017).

The ability to solubilize can be increased by genetic modification (Rodriquez et al. 2006). Gluconic acid is the most commonly produced acid by microorganisms. Glucose is oxidized to gluconic acid by glucose dehydrogenase (GDH) bound to the quinoprotein membrane, the active side of which is in the periplasmic space. Pyrroloquinoline quinone (PQQ) is a cofactor used by GDH for direct (non-phosphorylated) oxidation. Many PSM microorganisms require external pyrroloquinoline quinone (PQQ) for strong phosphorus solubilization. Many bacterial species can produce apoquinoprotein (apoGDH), but not the PQQ cofactor. Therefore, the phosphorus solubilizing microorganisms that are independent of PQQ, are sought. These include Acinetobacter (Ogut et al. 2010). This gene can be isolated, subcloned and the recombinant PQQ gene thus transferred by conjugation to the bacteria lacking it (Rodriquez et al. 2006).

According to Adnan (Alam et al. 2002), bacteria have a higher solubilization potential than fungi. For bacteria, it is determined at a level of 1 to 50%, while for fungi 0.1–0.5% of total phosphorus. Fungi, unlike bacteria, better solubilize at very low pH (Ogbo 2010). Li (Li et al. 2019) compared the solubilizing capabilities of fungi and bacteria. The solubilization capacity was assessed by determining the phosphatase enzyme activity after a 5-day incubation.

It was higher for the Acinetobacter sp. bacteria than for the Aspergillus niger fungi. An additional advantage of bacteria was greater reproductive capacity, compared to fungi. They are also very common in the soil and water environment, from which they can be isolated.

9.3 THE MECHANISM OF SOLUBILIZATION

The main role in the course of solubilization is played by the organic acids secreted as a result of metabolic changes by microorganisms. Bacteria secrete acids such as: gluconic, formic, 2-ketogluconic, citric, oxalic, lactic, isovaleric, succinic or glycolic (hydroxyacetic), aspartic, propionic,

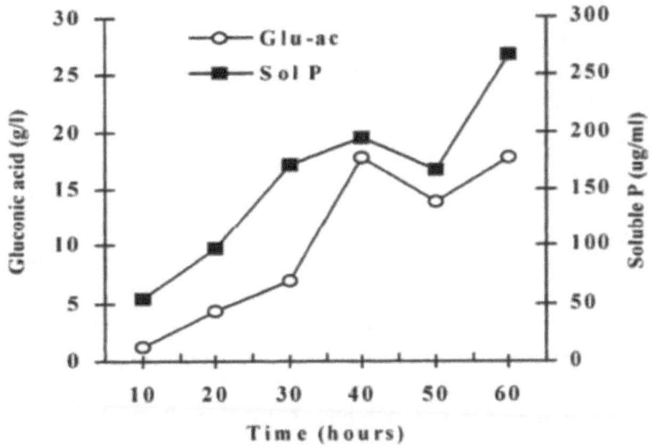

Figure 9.1. Gluconic acid secretion and phosphorus solubilization from phosphate rock (Vassilev et al. 2001).

glutamic, ketobutyric, tartaric, isobutyric. The production of organic acids is considered the basic mechanism for solubilization (Rodriquez et al. 1999). Organic acids are produced in the periplasmic space by direct oxidation. According to Rashid, enzymes (gluconate and 2-cetogluconate) participate in the acid production process, the presence of which increases the production of acids. Gram-negative bacteria are more efficient in the production of enzymes and acids (Rashid et al. 2004). According to Föllmi (Föllmi 1996), in the microbiological solubilization, in addition to the production of organic acids that change pH and dissolve solid phosphates, the so-called chelate mechanism can be identified. Chelates are acidic residues of organic acids that chelate iron and aluminum ions associated with insoluble phosphates, which causes their dissolution, and depending on the reaction of the environment, the release of $H_2PO_4^-$, HPO_4^{2-} and PO_4^{3-}. The carboxyl groups present in weak organic acids can also strongly chelate the calcium ions (Yu et al. 2012).

Enzymes secreted by some microorganisms also play an important role in solubilization. The bacterial enzymes such as phosphatase, phytase, nucleoidase, phosphohydrolase, C-P lyase (Föllmi 1996; Othman et al. 2014) act as specific catalysts. Out of these enzymes, the most popular is phosphatase, which eliminates phosphorus by hydrolyzing phosphoric acid monoesters, which produces PO_4^{3-} ion and free OH^- groups. The second most popular enzyme is phytase, which facilitates the hydrolysis of sodium phytase, resulting in the formation of inorganic phosphorus capable of solubilization (Othman et al. 2014). Enzymes such as glucose dehydrogenase, citrate synthase, and lactate dehydrogenase facilitate the oxidation of glucose, a reaction that is one of the stages in the production of organic acids (Chen et al. 2006). For example, the organic phosphorus compounds present in soil, such as the dominant inositol phosphate and the phosphonoesterase, phosphodiesterase and phospholipid esters, should be transformed into simpler inorganic connections before they are solubilized. Phytohormones, e.g. auxins, gibberellins also play a role in solubilization (Fibach-Paldi et al. 2012). Environmental factors such as reaction, oxygen concentration, humidity, and temperature affect the microbiological solubilization (Chen et al. 2006). The changes in the pH of the solubilization environment depend primarily on the strength of the organic acid produced and the scale of its release. An example of the relationship between the concentration of secreted acid and the amount of solubilized phosphorus is shown in Figure 9.1.

The secretion of organic acids by PSB leads to a significant acidification of the environment, which has been repeatedly presented in the literature (Bezak-Mazur 2019; Bezak-Mazur et al. 2015; Bezak-Mazur et al. 2019; Labuda et al. 2012). While acidification of the environment promotes solubilization, its alkalizatio hinders this process. Adnan (Adnan et al. 2017) showed that the addition of calcium carbonate to the soil samples containing various sources of phosphorus (superphosphate, rock phosphate, poultry manure, farm yard manure) and inoculated PSB (a mixture of several PSBs containing mainly Bacillus (15%) and Pseudomonas (12.7%)) inhibit solubilization. As the carbonate content increased from 4.78% to 20%, the phosphorus solubilization decreased. When using chicken manure (poultry manure and farm yard manure), which allowed soil alkalinity, the effect of lowering solubilization was smaller.

The effectiveness of solubilization also depends on the type of carbon and nitrogen sources available to the bacteria conducting the process. Xiao showed that the bacterial population and the amount of gluconic acid secreted by them is most preferably influenced by nitrogen in the form of ammonium ion (Xiao et al. 2009), while Balan (Balan 2003) confirmed that the best sources of glucose needed for acid production are the organic carbon sources present in the process.

The activity of the PSB microbial community in soil can significantly depend on many environmental factors including soil pH (Ragot et al. 2015), productivity and nutrient content measured by total organic carbon (TOC), total nitrogen (TN), total phosphorus (TP) (Ragot et al. 2017) and fertilization scale (Frasse et al. 2015).

Rhizospheric carbon has been found to have a positive effect on the PSB activity of soil (Long et al. 2018). The effect of the nitrogen-binding bacteria on the effect of solubilization by Bacillus megaterium and Pseudomonas chlororaphis was studied (Yu et al. 2012). The presence of the nitrogen-binding bacteria increases the solubilization effect measured by the so-called available phosphorus determined according to Calwell in a bicarbonate soil extract (Calwell 1965).

9.4 FRACTIONATION OF PHOSPHORUS AND ITS SOLUBILIZATION FROM VARIOUS WASTE MATERIALS

The forms of phosphorus can be more accurately described using the term speciation, and actually using the manner of separation - fractionation. The essence of the phosphorus fractionation mechanism in microbiological solubilization is to distinguish with which element phosphorus is bound. The known forms in the speciation analysis include, the phosphorus fraction associated with Fe-P iron, with Al-P aluminum, with Ca-P calcium and the phosphorus in organic connections. When considering the possibility of migration in the environment and availability for plants, the so-called bioavailable phosphorus fraction can be mentioned. The fractions extracted with water or very mild extractants are considered as bioavailable fractions (Bezak-Mazur 2019).

Wollmann used sequential extraction according to the Hedley's methodology in order to assess the amount of bioavailable fraction secreted during phosphorus solubilization from several bio-fertilizers tested (Wollmann et al. 2018). In the Hedley's method used by Wollman, the fraction extracted with sodium bicarbonate is considered the bioavailable fraction. Nilanjan performed an assessment of the natural mobilization of the Ca-P fraction from fresh sediments (Nilanjan et al. 2015). The fraction changes were assessed according to the Jackson's methodology (Bezak-Mazur 2019). The Ca-P phosphorus fraction was found to show seasonal variability and correlates with pH. Kucey (Kucey et al. 1989) and Fankem (Fankem et al. 2006) using microbiological solubilization to release the phosphorus from sediments, showed that the Ca-P calcium-related phosphorus is better solubilized than the iron, aluminum and manganese phosphorus that is, the Fe-P, Al-P and Mn-P fractions.

Garcia-Lopez analyzed the changes in the phosphorus fraction in soil and the rhizosphere zone in the presence of Bacillus and Trichoderma asperellum (Garcia-Lopez et al. 2018).

The biological control of the bio-fertilizer effect on plants was demonstrated. It was shown that the Bacillus subtilis bacteria were more effective in solubilizing the phosphorus contained in soil, and thus allowed to increase its uptake by plants. The research conducted by the E. Bezak-Mazur's team regarding the possibility of solubilization of the phosphorus in sewage sludge included an analysis of the changes in the phosphorus speciation forms according to the Golterman's methodology (Bezak-Mazur et al. 2015; Bezak-Mazur et al. 2019). During incubation, the changes in pH, population growth of microorganisms, and changes in phosphorus speciation forms were analyzed. The exemplary changes in phosphorus fraction concentrations during 3-day solubilization for the hygienized sludge are shown in Figure 9.2.

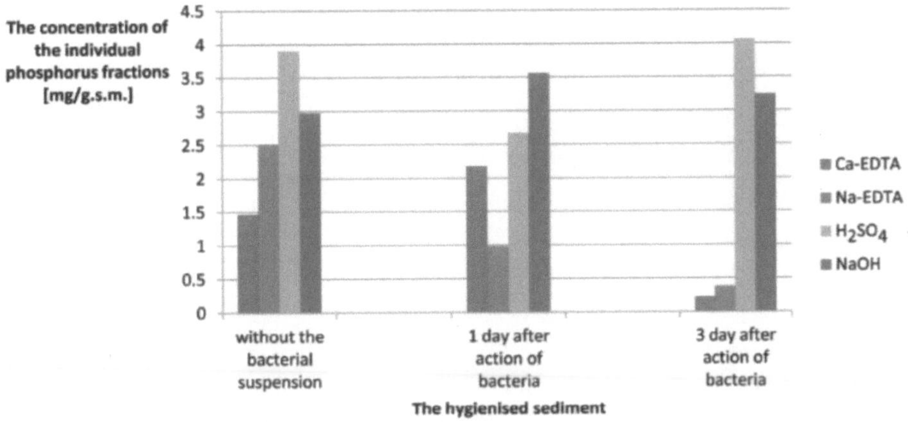

Figure 9.2. The speciation analysis of phosphorus in the hygienised sediments subjected to the operation of the Bacillus megaterium bacteria (Bezak-Mazur et al. 2015).

9.5 APPLICATION OF MICROBIAL SOLUBILIZATION FOR PHOSPHORUS RECOVERY - REVIEW OF EXAMPLES

Selected examples of phosphorus recovery and its use in bio fertilizer production are discussed below.

9.5.1 *Solubilization of* $Ca_3(PO_4)_2$

Avdalovic et al. (Avdalovic et al. 2015) conducted a study of solubilization by Acidithiobacillus sp. bacteria from the phosphate minerals containing phosphate calcium. Mineral apatites were solubilized by the bacteria isolated from copper sulfide ores. It was shown that 34.5% of inorganic phosphorus was solubilized. For comparison, only 3.8% of the bacterial-free sample was leached out (Corbridge 2013).

The studies conducted by Yu (Yu et al. 2012) regarding the solubilization of mineral phosphates after inoculation with solubilizing and nitrogen-binding bacteria showed a synergistic solubilization effect when both types of microorganisms are present. Pseudomonas chlororaphis and Bacillus megaterium were used as the phosphorus solubilizing bacteria, and Arthrobacter pasceus and Burkholderia chloraphis as the nitrogen-binding bacteria. The results of the Yu tests are presented in the figures below (Figure 9.3 and Figure 9.4).

A correlation was observed between the pH changing under the influence of secreted organic acids and the amount of solubilized phosphorus (Figure 9.4). On the 4th day of research, the largest number of microbial populations was observed (Figure 9.3), which was reflected in a decrease in pH by more than two units. The highest concentration of solubilized phosphorus was observed after the 4th day of the experiment. The highest solubilization and lowest pH were recorded for mixed bacterial cultures. According to the authors, the PSB solubilizing and NFB phosphorus binding bacteria can stimulate the plant growth (Yu et al. 2012).

9.5.2 *Phosphate minerals*

The phosphorus solubilization studies with rock phosphate using medium thermophilic Acidithiobacillus caldus and mesophilic Acidithiobacillus thiooxidans were carried out by Xiao (Xiao et al. 2011). He showed the effect of pH and temperature on the process, which was more effective in the case of A.caldus.

Interesting quantitative comparisons of the solubilization abilities with reference to mineral phosphates and those present in animal bones were carried out by Labuda (Labuda et al. 2012). The highest degree of solubilization of 80% was obtained for the phosphates contained in bones, for the $Ca_3(PO_4)_2$

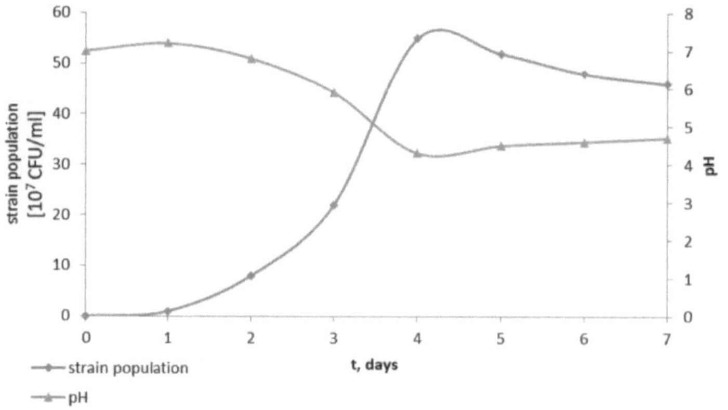

Figure 9.3. Population of S. maltophilia YC in culture medium during 7 days of solubilizing experiment in relation to pH value.

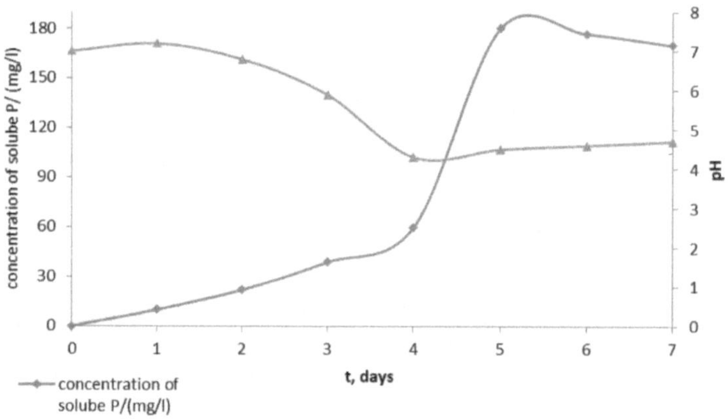

Figure 9.4. Concentration of soluble P in culture medium during 7 days of solubilizing experiment in relation to pH value.

phosphate it was 70%, and for the mineral phosphates – only 45%. Similarly, Saeid observed a higher solubilization for biogenic apatites present in bone than for mineral calcium phosphates (Saeid et al. 2014). The content of the bioavailable phosphorus expressed as the fraction extracted with citric acid after the solubilization of biogenic apatite with Bacillus megaterium was 17%, and after solubilization of phosphate from mineral – 6.5%. These results indicate the necessity of interest in the solubilization of phosphorus in the waste materials containing organic carbon, i.e. sewage sludge, animal bones.

9.5.3 *Bones*

A series of papers devoted to the solubilization of phosphorus from bones with or without various additives has been published by Saeid and colleagues (Labuda et al. 2012; Saeid et al. 2018; Wyciszkiewicz et al., 2017; Wyciszkiewicz et al. 2017). The animal bone phosphates (poultry and fish) were solubilized. Comparing the phosphorus solubilization coefficients (SF) with those for mineral phosphates (Wyciszkiewicz et al. 2017), it was shown that SFmax for bones exceeded 90% (95.7% for raw material dose 2 g/l, and for fish bones 93.9 %), whereas in the case of phosphate rock, SFmax was 90.8% when using twice the dose (4g/l) of this material. Bone phosphorus was also solubilized in the presence of ashes obtained after the thermal treatment of sewage sludge (Saeid et al. 2018). In these studies, the highest solubilization coefficients were obtained for the samples solubilized in the presence of organic carbon. The Bacillus bacteria, i.e. B. megaterium, B. cereus, B. subtilis, and Acidithiobacillus ferroxidans were used as the solubilizing bacteria. The maximum concentration of P_2O_5 (CP_2O_5) for solubilization with Bacillus megaterium for a 5g/l dose of the tested material was, respectively: for bones 270 mg/l, bones 483 mg/l, and for ashes 85.4 mg/l. In the case of phosphate rock this result was only 33.4 mg/l. The conditions of phosphorus solubilization from the bones subjected to composting with spent media after cultivation of fungus (mushrooms) were also analyzed. The used medium contained organic carbon in the form of lignocellulose inoculated with the Bacillus megaterium bacteria (Wyciszkiewicz et al. 2017). In this way, the possibility of utilizing post-production waste, i.e. used substrates after fungus cultivation, was indicated.

9.5.4 *Sewage sludge*

Sewage sludge contains from 2 to 10% phosphorus calculated as P_2O_5 (Blocher et al. 2012; Bień et al., 2011). The phosphorus contained in them occurs in different fractions with different susceptibility to solubilization. In addition, the phosphorus fractions present in the sewage sludge change depending on the method of wastewater treatment and the chemical reagents used. Huang (Huang et al. 2008), using the Hedley's methodology, analyzed the phosphorus fractions in sewage sludge. When using the coagulants based on aluminum compounds, the phosphorus fractions WSP-P, membrane-P, $NaHCO_3$-P, and HCL-P are converted into the Al-P fractions. The coagulants based on the iron compounds and the iron hydroxides present in such wastewater promote the adsorption of the WSP-P, membrane-P fraction on them and an increase in the Fe-P fraction. The use of calcium compounds, e.g. limewash, increases the Ca-P fraction. The above-mentioned changes in phosphorus fractionation in sediments should be taken into account when considering the possibility of phosphorus recovery from sediments. It will also minimize the losses of phosphorus introduced into the soil. The studies on the phosphorus solubilization of lime-sanitized sludge (Bezak-Mazur et al. 2015; Bezak-Mazur et al. 2019) have shown that after a few hours of the process, the Ca-EDTA fraction increases (the fraction most available to plants). It should be noted that the increase in the Ca-EDTA fraction content correlated with the number of multiplied microorganisms (Bezak-Mazur et al. 2019), and with them the increased amount of secreted organic acids.

9.5.5 *Ashes after thermal neutralization of sewage sludge*

The ashes after the thermal treatment under aerobic conditions (combustion) are an interesting alternative source of phosphorus. Their direct use is limited by the very low availability of phosphorus present in the form of insoluble complex phosphate compounds. In laboratory studies, Raymond (Raymond et al. 2018) showed very high solubilization capacity of phosphorus in ashes in relation to Penicillium bilaiae fungi. These abilities were confirmed in field studies (Raymond et al. 2019), where the soil analyses showed an increase in the concentration of the bioavailable phosphorus.

However, the tests of the photosynthesis activity of barley sown on the tested soil did not show an increase in the concentration of carbon dioxide in the MicroResp TM respiration test. Such research results clearly highlight that the impact of the introduced PSM on the native flora of microorganisms is not fully recognized.

9.5.6 *Biochar*

Biochar, obtained as a result of pyrolysis of sewage sludge, is another remarkable waste material from which phosphorus can be recovered. Depending on the pyrolysis conditions, and above all the temperature, the degree of phosphate polymerization and their solubility varies. The research using NMR and XPS (X-ray photoelectron spectroscopy), carried out by Qian and colleagues (Qian et al. 2019), showed that in the biocarbon formed at 400°C, the phosphorus compounds are less polymerized than after pyrolysis at 700°C. Phosphorus was solubilized with both biochar using Pseudomonas putida. Greater phosphorus release from the biochar pyrolyzed at 400°C was observed. The phosphorus compounds with lower polymerization (shorter chains) are more easily solubilized by P. putida than the phosphorus compounds with longer chains present in the pyrolyzed biocarbon at 700°C. The research results showed that the effect of solubilization using P. putida increases along with the greater presence of simple phosphates in the biocarbon.

9.5.7 *Biofertilizers*

Modern agriculture does not use soil supplementation with phosphorus fertilizers, but primarily assumes the use of soil phosphorus reservoirs. The phosphorus compounds present in soil are for the most part organic connections, only 5% of which is available to plants (Maksimov et al. 2011). Therefore, in order to activate the soil phosphorus, the so-called biofertilizers, i.e. substances containing live microorganisms that, when applied to plants, colonize their rhizosphere on soil, promoting the plant growth by providing the nutrients available to plants, have to be used (Vessey et al. 2003). Nutrients, including phosphorus, appear as a result of solubilization of soil phosphorus or with biofertilizer. Some examples of biofertilizer preparation using phosphorus solubilization are presented below.

Animal bones carbonized at 550°C give a porous product called animal bone charcoal. Its main components are phosphorus and calcium. Postma (Postma et al. 2013) used it as a fertilizer with the addition of PSB. The addition of Pseudomonas chlororaphis, Bacillus pumilus, Paenibacillus polymyxa, Streptomyces pseudovenezuelae was considered. Biofertilizer was used to fertilize tomatoes. Studies have shown that the most beneficial effect on tomato harvest was obtained using the fertilizer with Pseudomonas chlororaphis.

The Rolewicz team (Rolewicz et al. 2018) used the ashes from sewage sludge with the addition of bones associated with a granulation facilitating agent (sodium ligninosulphate, dried blood, bentonite, calcium sulfate hemihydrate) to obtain high-strength granules. At 85% ash and 15% dried blood, a biofertilizer containing 15 to 14% P2O5 was obtained. Bacillus megaterium bacteria solubilized phosphorus. The same bacteria and Acidithiobacillus ferrooxidans were used to prepare the fertilizer granules based on animal bones and sewage sludge ashes (Jastrzębska et al. 2018). Their positive effect on the wheat growth was observed and, additionally, the released phosphates modified the phytoavailability of cadmium and lead, limiting it.

Further suggestions for preparing biofertilizer concern composting agricultural waste (grain straw, rice straw) with the addition of mineral phosphates. It should be emphasized that no fractions of bioavailable phosphorus are generated during composting of agricultural waste. The organic phosphates are transformed into inorganic ones, which in the presence of calcium, aluminum, iron bind phosphates into more stable fractions. The addition of solubilizing microorganisms producing organic acids and enzymes promotes their solubilization. According to Vassiliev, agricultural waste composted with 10% mineral phosphate in the presence of Bacillus polymyxa and Pseudomonas increases the phosphorus solubilization by 18% (Vassilev et al. 2003). Vasiliev also analyzed the possibilities of obtaining a biofertilizer after fermentation of agricultural waste with the addition of mineral phosphates.

Solubilization in the presence of A. Niger fungus allowed to convert as much as 76% phosphorus into the bioavailable forms. According to Wang (Wang et al. 2019), the most favorable C/N ratio during composting and solubilization is 15–27. Usmani (Usmani et al. 2019) presented another suggestion to obtain a biofertilizer, which proposes composting coal ash with bovine manure in a ratio of 1:3 in the presence of earthworms (vermicompost). Earthworms produce an enzyme – alkaline phosphatase – which promotes the hydrolysis of organic phosphorus.

9.6 CONCLUSIONS

The paper highlights the possibilities of using microbial solubilization for recovery, as well as increasing the efficiency of phosphorus recovery from various materials, such as sewage sludge, ashes, biochar, bones, and agricultural waste. This is very important in the context of sustainable soil management and constitutes an alternative way to expensive chemical methods for obtaining the phosphate fertilizers. The biological production of fertilizers, especially with the use of waste materials, can be considered more accessible than the current technological methods.

ACKNOWLEDGEMENTS

The article was financed from the Program of the Minister of Science and Higher Education – Regional Initiative of Excellence, financed by the Ministry of Science and Higher Education under contract No. 025 / RID / 2018/19 of 28/12/2018, in the amount of PLN 12 million.

REFERENCES

Adnan, M., Shah, Z., Fahad, S. et al., (2017) Phosphate-Solubilizing Bacteria Nullify the Antagonistic Effect of Soil Calcification on Bioavailability of Phosphorus in Alkaline Soils. Sci Rep.7(16131), 1–13.

Alam S., Khalik S., Ayub N., Rasid M., (2002), In vitro solubilization of inorganic phosphate by phosphate solubilizing microorganism (PSM) from maize rhizosphere, Int. J. Agric. Biol., 4, 454–458.

Alori ET, Glick BR., Babalola OO., (2017) Microbial phosphorus stabilization and its potential for use in sustainable agriculture, Frontiers in Microbiology, 8, (971), 1–8.

Avdalovic J., Beskoski V., Gojgic-Cvijovic, G., et al., (2015) Microbial solubilization of phosphorus from phosphate rock by iron-oxidizing Acidithiobacillus sp. B2, Mineral Engineering, 72, 17–22.

Balan SN., (2003), Progress in selected areas of rhizosphere research on P acquisition, Austr. J. Soil Res., 41, 471–499.

Bezak-Mazur E., Stoińska R., Szeląg B., (2015) The influence of the Bacillus megaterium bacteria on speciation of phosphorus in sewage sludge, Architecture Civil Engineering Environment, 8,(4)81–87.

Bezak-Mazur E., Ciopińska J., Stoińska R., Szeląg B., (2019), The impact of Bacillus megaterium on the solubilisation of phosphorus from sewage sludge 3S Web of Conferences 86, 00032, 1–8.

Bezak-Mazur E., (2019) Frakcjonowanie fosforu w matrycach środowiskowych, Monografie, Studia, Rozprawy nr M124, Politechnika Świętokrzyska Kielce.

Bezak-Mazur E., Ciopińska J., (2020) The application of sequential extraction in phosphorus fractionation in environmental samples, Journal AOAC International, 103, 1–13.

Bień J., Neczaj E., Worwąg M., Grosser A., (2011) Kierunki zagospodarowywania osadów ściekowych w Polsce, Inżynieria Ochrony Środowiska, 14, 375–384.

Blocher C., Niewarsch C., Melin T., (2012) Phosphorus recovery from sewage sludge with a hybrid process of low pressure wet oxidation and nanofiltration, Water Research, 46, 2009–2019.

Calwell JD., (1965), An automatic procedure for the determination of phosphorus in sodium hydrogen carbonate extracts of soil, Chem. Ind., 22, 893–895.

Chen YP., Rekta PD., Arun AB., et al., (2006), Phosphate solubilizing bacteria from subtropical soil and their tricalcium phosphate solubilizing abilities, Applied Soil Ecology, 34, 33–41.

Corbridge, DEC., (2013) Phosphorus: Chemistry, Biochemistry and technology, sixth ed. CRC Press Taylor & Francis Group,

Dijk KC., Lesschen JP., Oenema O., (2016), Phosphorus flows and balances of the European Union Member States, Science of The Total Environment, Vol. 542, 1078–1093.

Fankem H., Nwage D., A. Deubel A., et al., (2006), Occurrence and functioning of phosphate solubilizing microorganism from oil palm tree rhizosphere in Cameroon, African Journal of Biotechnology, 5, 2450–2460.

Fibach-Paldi S., Burdman S., Okan Y., (2012), Key physiological properties contributing to rhizosphere adaptation and plant growth promotion abilities of Azospirillum brasilence, FEMS Microbiol. Lett., 326, 99–108.

Frasse TD., Lynch DH., Bent E., (2015) Soil bacterial phoSgene abudance and expression in response to applied phosphorus and long term management, Soil Biol. Biochem., 88, 137–147.

Föllmi K.B., (1996), The phosphorus cycle, phosphogenesis and marine phosphate-rich deposits, Earth-Science Reviews,40, (1–2), 55–124.

Garcia–Lopez AM., Recena R., Aviles M., A. Delgado A., (2018), Effect of Bacillus subtilis QST713 and Trichoderma Asperellum T34 on Phosphorus uptake wheat and how it is modulated by soil properties, J. Soil Sediments, 18, 727–738.

Hayat R., Ali S., Asmara U., et al., (2010), Soil beneficial bacteria and their role in plant growth promotion, a review, Ann. Microbiol., 60, 579–598.

Huang XL., Chen Y., Shenker M., (2008), Chemical fractionation of phosphorus in stabilized biosolids. J Environ Qual., 37(5):1949–58.

Jastrzębska M., Saeid A., et al., (2018), New phosphorus biofertilizers from renewable raw materials in the aspect of cadmium and lead contents in soil and plants, Open Chem., 16, 35–49.

Kucey RN., Jansen HH., Legett ME., (1989), Microbially mediated increase in plant-available phosphorus, Advances in Agronomy, 42, 198–228.

Labuda M., Saeid A., Chojnacka K., Górecki H., (2012), Zastosowanie Bacillus megaterium w solubilizacji fosforu, Use of Bacillus megaterium in solubilization of phosphorus, Przem. Chem., 91, 5, 837–840.
Li Ch., Li O., Wang Z., et al., (2019) Environmental fungi and bacteria facilitate lecithin decomposition and transformation of phosphorus to apatite, Science Reports, 9, (15291), 1–8.
Long X., Yao H., Huang Y., Wei W., et al., (2018), Phosphate levels influence the utilization of rice rhizodeposition carbon and the phosphate -solubilizing microbial community in a paddy soil, Soil Biotechnology and Biochemistry, 118, 103–114.
Maksimov IV., Abizgildina RR., Pusenkova LI., (2011), Plant growth promoting rhizobacteria as alternative to chemical crop protectors from pathogens, Applied Biochemistry and Microbiology, 47, 4, 333–345.
Nilanjan M., Sanjib MK., Srikanta S., et al., (2015), Ecological significance and phosphorus release potential of phosphate solubilizing bacteria in freshwater ecosystems, Hydrobiologia, 745, 69–83.
Ogbo F. C., (2010), Conversion of cassava wasters for biofertilizer production using phosphate solubilizing fungi, Bioresource Technology, 101, 4120–4124.
Ogut M., Er F., Kandemir N., (2010), Phosphate solubilization potentials of soil Acinetobacter strains, Biol. Fertil. Soils., 46,707–715.
Othman RO., Pankwar QA., (2014), Phosphate solubilizing bacteria improves nutrient uptake in aerobic rise, In: Khan MS, Zaodi A., Musarrat J., (eds) Phosphate Solubilizing Microorganism: Principles and applications of Microphos Technology, Springer International Publishing, 207–224.
Priyadharsini P., Muthukumar T., (2017), The root endophytic fungus Curvularia geniculata from Parthenium hysterophorus roats improves plant growth through phosphate solubilization and phytohormone phosphate, Fungal Ecology, 27, 69–77.
Postma J., Clematis F.,Nijhuis EH., Someus E., (2013), Efficacy of four phosphate-mobilizing bacteria applied with an animal bone, charcoal formulation in controlling Pythium aphanidermatum and Fusarium oxysporum f.sp. radicis lycopersici in tomato, Biological control, 67, 284–291.
Qian T., Yang Q., Jun DCHF., Dang F., Zhou Y., (2019), Transformation of phosphorus in sewage sludge biochar mediated by a phosphate-solubilizing microorganism, Chem. Eng. Journal, 359, 1573–1580.
Ragot SA., Kertesz MA., Bunemann EK., (2015), phoD alkaline phosphate gene diversity in soil, Applied and Environmental Microbiology, 81, 7281–7289.
Ragot SA., Kertesz MA., Meszaros E., et al., (2017), Soil phoD and phoX alkaline phosphatase gene diversity responds to multiple environmental factors, FEMS Microbiology Ecology, 93,1, 1–15.
Rashid M., Khalid S., Ayub N., et al., (2004), Organic acids production and phosphate solubilizing by Phosphate solubilizing microorganism under in vitro conditions, Pakistan J. Biological Sciences,7, 2, 187–196.
Raymond NS., Jensen LS., Muller-Stoven D., (2018), Enhancing the phosphorus bioavailability of thermal converted sewage sludge by phosphate-solubilizing fungi, Ecol. Eng.,120, 11–53.
Raymond NS., Jensen LS., Vander Born F., et al., (2019), Fertilizing effect of sewage sludge ash inoculated with the phosphate-solubilizing fungus Penicillium bilaiae under semi-field conditions, Biology and Fertility of Soils, 55, 43–51.
Rodriquez H., Fraga R., (1999), Phosphate solubilizing bacteria and their role in plant growth promotion, Biotechnol, Adv.17, 319–339.
Rodriquez H., Fraga R., Gonzalez T., Bashan T., (2006), Genetics of phosphate solubilizing and its potential applications for improving plant growth-promoting bacteria, Plant soil, 287, 15–21.
Rolewicz M., Klusek P., Borowik K., (2018), Obtaining of granular fertilizers based on ashes from combustion of waste residues and ground bones using phosphorus solubilization by bacteria Bacillus megaterium, J. Environ. Management, 216, 128–132.
Report on Critical Raw Material for the UE; Report of the Hoc Working Group on Defining Critical Raw Materials; Ref. Ares (2015) 1819503 European Commision, Luxemburg, 2014.
Saeid A., Prochownik E., Dobrowolska-Iwanek J., (2018), Phosphorus solubilization by Bacillus megaterium, Molecules, 23, 2897–2915.
Saeid A., Labuda M., Chojnacka K., (2014), Valorization of bones to liquid phosphorus fertilizes by microbial solubilization, Waste Biomass Valor., 5, 265–272.
Schroder HC., Kurz L., Miller WFG, Lorens B., (2000), Polyphosphate in Bone, Biochemistry, 65, 353–361.
Sharma S.B., Sayyed R.Z., Trivedi M.H., and Gobi T. A., (2013), Phosphate solubilizing microbes: sustainable approach for managing phosphorus deficiency in agricultural soils, SpringerPlus, 2, 587–600.
Usmani Z., Kumar V., Gupta P., Gupta G., Rani R., Chandra A., (2019), Enhanced soil fertility, plant growth promotion and microbial enzymatic activities of vermicomposted fly ash, Sci. Reports, 9, (10455), 1–16.
Vassilev N., Vassilieva M., Fenice M., Federici F., (2001), Immobilized cell technology applied in solubilization of insoluble inorganic (rock) phosphates and phosphate plant acquisition, Bioresource Technology, 79, 3, 263–271.

Vassilev N., Vassileva M., (2003), Biotechnological solubilization of rock phosphate on media containing agroindustrial waste, Appl. Microbiol. Biotechnol., 61, 435–440.

Venkiteshwaran K., McNamara PJ., Mayer BK., (2018), Meta-analysis of non-reactive phosphors in water, wastewater, and sludge, and strategies to convert it for enhanced phosphorus removal and recovery, Sci Total Environ., 644, 661–674.

Vessey JK., (2003), Plant growth promoting rhizobacteria as biofertilizers, Plant and Soil, 255, 571–586.

Wang L., Li Y., Prasher S.O, Yan, B., et al., (2019), Organic matter, a critical factor to immobilize phosphorus, copper, and zinc during composting under various initial C/N ratios, Bioresource Technology, 289, 121745–121751.

Wollmann I., Gauro A., Müller T., Möller K., (2018), Phosphorus bioavailability of sewage sludge-based recycled fertilizers, J. Plant. Nutri. Soil. Sci., 181, 158–166.

Wyciszkiewicz M., Saeid A., Dobrowolska-Iwanek J., (2016), Utilization of microorganism in the solubilization of low-quality phosphorus material, Ecol. Engin., 89, 109–113.

Wyciszkiewicz M., Saeid A., Malinowski P., Chojnacka K., (2017), Valorization of phosphorus secondary raw materials by Acidithiobacillus ferroxidans, Molecules, 22, 473–486.

Wyciszkiewicz M., Saeid A., Samuraj M., Chojnacka K., (2017), Solid-state solubilization of bones by B. megaterium in spent mushroom substrate as a medium for phosphate enriched substrates, J. Chem. Technol.Biotechnol, 92, 1397–1405.

Xiao ChQ., Ruan, Ch., Huan He., et al., (2009), Characterization of tricalcium phosphate solubilization by Stenotrophomonas maltophilia YC isolated from phosphate mines, J. Cent. South Univ. Technol., 16, 581–587.

Yu X., Liu X., Zhu, TH., Liu GH., Cui Mao, (2012), Co-inoculation with phosphate-solubilizing and nitrogen-fixing bacteria on solubilization of rock phosphate and their effect on growth promotion and nutrient uptake by walnut, European J. Soil Biology, 50, 112–117.

Zhou F., Qu L., Hang X., Sun X., (2011), Isolation and characterization of a phosphate solubilizing halophilic bacterium Kushneria sp. yCWA18 from Daqiao Saltern on the coast of Yellow Sea of China, Evid. based Complement. Alternat. Med., 615032, 1–7.

CHAPTER 10

Assimilation of phosphorus contained in the sewage sludge by the *Bacillus megaterium* bacteria

E. Bezak-Mazur, R. Stoińska & B. Szeląg
Faculty of Environmental Geomatic and Energy Engineering, Kielce University of Technology, Kielce, Poland

ABSTRACT: The objective of the research paper was to investigate the possibility of *Bacillus megaterium* bacteria assimilating the phosphorus from hygienized sludge. The research involved the use of sludge hygienized with lime, since this type of sludge is most commonly applied for the agricultural purposes. The phosphorus assimilation by the phosphorus bacteria was confirmed by the tests examining, i.a. the dynamics of the change in: the number of mesophilic bacteria, the amount of phosphorus assimilated by microorganisms, the content of bioavailable phosphorus forms in the sewage sludge, the pH value and the FTIR spectrum of the hygienized sludge. The analysis of the obtained research results enables to draw the conclusion that the *Bacillus megaterium* bacteria can effectively solubilize and assimilate the phosphorus compounds contained in the sewage sludge subjected to hygienization with lime.

10.1 INTRODUCTION

Phosphorus is an element indispensable for the functioning of every living organism. This element is absorbed by plants exclusively in the form of the phosphoric acid (V) ions. Unfortunately, the amount of the bioavailable phosphorus forms, in comparison to the amount of total phosphorus contained in soil, is relatively low (Khan et al. 2014). In Poland, the share of soils with a low or very low assimilable phosphorus content is as high as 40% (Lipiński 2000). Therefore, the role of the bacteria solubilizing hardly soluble phosphorus, i.e. the Phosphate Solubilizing Bacteria (PSB), cannot be overestimated. They form a group of beneficial bacteria capable of producing the acids that hydrolyze organic and inorganic phosphorus compounds (Bishop et al. 1994; Toro 2007; Wani et al. 2004).

Two methods of converting the hardly available phosphorus forms into the bioavailable ones can be distinguished. The organic phosphorus forms are transformed into the bioavailable ones in the process called mineralization (Alori et al. 2017); the non-organic forms undergo the process known as solubilization (Grafe et al. 2017; Khan et al. 2014). The mechanism of solubilizing mineralized phosphate is connected with the bacteria releasing organic acids into the soil, which causes the decrease in the soil pH (Pradhan & Shukla 2005) and thus the dissolution of hardly soluble phosphate salts such as calcium phosphate (V), aluminum phosphate (V), iron(III) phosphate(V) (Gupta et al. 2007; Sharma et al. 2013; Song et al. 2008; Sundara et al. 2002).

Certain actinomycetes and fungi are also capable of producing organic acids. Together with the phosphate solubilizing bacteria, those species of fungi and actinomycetes are included into the group of Phosphate Solubilizing Microorganisms (PSM) (Girmay 2013). Due to the beneficial effect of PSMs, in recent years the fertilizers containing strains of phosphoric microorganisms have been regarded as the so-called phosphoric bio-preparations. PSMs are an important agent in the process of optimizing the agricultural productivity (Gyaneshwar et al. 2002). The application of the mixed cultures of PSMs in practice is due to the fact that different PSMs produce different substances. The most effective mixed cultures include the microorganisms of the following genera: *Bacillus, Streptomyces* an *Pseudomonas* (Molla et al. 1984). Very often, the bio-preparations available on the world market include the *Bacillus megaterium* strain. This strain produces i.a. citric acid, lactic acid and propionic acid, thereby effectively solubilizing phosphorus (Khan et al. 2013).

DOI 10.1201/9781003171669-10

Therefore, the aim of the study was to investigate the possibility of the *Bacillus megaterium* bacteria assimilating the phosphorus from hygienized sludge. The sludge hygienized with lime was selected for the study, because this type of sludge is most commonly used for nature-related purposes. However, it should be noted, that the addition of lime to sewage sludge can also affect the share of mobile biogenic forms such as phosphorus. The hygienization of sewage sludge by the addition of CaO results in the reduction of the phosphorus share in the soluble forms (Stoińska 2013), which may lead to a decreased availability of this element to plants in the soil.

Thus, the practical assessment of sludge as a secondary source of biogenes should involve estimating the content of the bio-available phosphorus included therein, defined as the sum of the readily available phosphorus and the phosphorus transformed into available forms through the natural processes occurring in the sludge (Bezak-Mazur and Stoińska 2013). The share of the bioavailable phosphorus forms can be determined by speciation analysis, i.e. the analytical procedure that permits the quantitative determination of the chemical forms in which the particular analyte is present. The Golterman's method (Golterman 1996) is one of the procedures of speciation analysis. This method involves using chelating reagents (Na-EDTA and Ca-EDTA) as well as sulfuric acid and sodium hydroxide solutions in the analysis.

The inventor of the discussed method noted that the phosphorus adsorbed on the surface of the sludge particles, i.e. the Ca-EDTA and the Na-EDTA fractions, represents the most bioavailable speciation form. The examination of the changes in the bioavailable phosphorus forms is also important in the case of applying the bacteria solubilizing phosphorus compounds onto the hygienized sludge (Golterman 1996).

10.2 MATERIALS AND METHODS

The experiment was conducted to confirm the possibility of the *Bacillus megaterium* bacteria assimilating phosphorus from the sludge hygienized with lime. The sewage sludge was hygienized with a dose of 0.15 kg of CaO/kg dw. According to the reference sources, the dose of lime in the hygienization process usually ranges from 0.15–0.25 kg of CaO/kg dw of sewage sludge (Stachowicz & Wójcik 2017).

For this purpose 0.5-g samples of sewage sludge were placed in 300 cm^3 Erlenmeyer flasks. The sludge was collected at a mechanical-biological treatment plant. The liquid medium was then prepared and sterilized following the procedure of the medium preparation proposed by Labuda et al. (2012). The experiment was aimed at confirming the possibility of phosphorus assimilation by phosphorus bacteria therefore, the prepared medium contained no phosphorus compounds with the hygienized sludge being the sole source of this element. The liquid medium prepared accordingly was cultured with *Bacillus megaterium* bacteria and mixed on the rotary shaker. During the preparation procedure, 150 ml of the liquid, bacteria-containing medium was placed in each flask containing hygienized sludge (0.5 g of the sludge in each flask). Then, the number of mesophilic bacteria was determined in the resulting suspension (the composition of suspension being: the medium, the *Bacillus megaterium* bacteria and the hygienized sewage sludge) using the serial dilution method. Each suspension contained an average of 3×10^3 cfu/ml of mesophilic bacteria. At the same time a control trial was performed, which consisted in completing all the steps outlined above, but without inserting the *Bacillus megaterium* bacteria into the medium. The experiment was conducted at room temperature for a week in order to determine the dynamics of phosphorus assimilation in time.

During the first two days, the samples of suspension were collected for analysis every few hours and then once a day.

The following analyzes were performed:

- the amount of phosphorus assimilated by microorganisms;

The content of phosphorus assimilated by bacteria was determined following the procedure developed by the authors of this article.

Sequential filtration was used to determine the amount of phosphorus assimilated by the microorganisms. In the first stage, the liquid-medium microorganisms were separated from the sewage sludge

Table 10.1. Phosphorus sequential extraction scheme according to Golterman (1996).

Stage	The extraction condition	Fraction
1	0.05 M Ca-EDTA, 4 h	phosphorus associated with oxides and hydroxy oxides of iron, aluminum and manganese
2	0.1 M Na-EDTA, 18 h	phosphorus associated with carbons
3	0.5 M H_2SO_4, 2 h	phosphorus is presented in the soluble organic matter bonds
4	2 M NaOH, 2 h	the remaining phosphorus, bonded with aluminosilicates and organic matter contained in the form of connections, which is not subject to the action of sulfuric acid in stage 3

residue by filtration through glass wool. The resulting microbial suspension was filtered through a cellulose filter with 0.45 μm-diameter pores (using a vacuum filtration kit) to separate the microorganisms from the liquid medium. The bacteria deposited on the cellulose filter were mineralized with aqua regia. The total phosphorus concentration was determined in the resultant post-mineralization liquid samples using phosphate-molybdenum blue in accordance with the PN-EN ISO 6878 (2006) standard on total phosphorus determination.

- the number of mesophilic bacteria in suspension during the experiment;

For this purpose 1 ml of the solution, freshly mixed on the rotary shaker, was collected from each sample and fed into a Petri dish either directly or after dilution (dilutions 10^{-1}–10^{-6}). All the dilutions of the investigated samples were previously intensively mixed using the Sky Line shaker. Subsequently, the suspension on the Petri dish was flooded with liquid nutrient broth and incubated at 37°C for 24 h. The final step involved counting the colonies using a counter and converting the result according to the dilution of the suspension. All the determinations were performed in triplicate. The number of mesophilic bacteria depending on the duration of the experiment was also determined for the control sample, i.e. for the suspension without the introduced *Bacillus megaterium* bacteria.

- changes in the suspension pH;

For this purpose, the pH values of the suspensions were determined using the METLER TOLEDO pH Meter.

- the amount of bioavailable phosphorus forms in sewage sludge;

In the sewage sludge separated from the suspension, the amounts of bioavailable phosphorus forms were determined using the Golterman's method (1996) described in detail in the literature (Bezak-Mazur et al. 2017a,b).

- physicochemical changes in sewage sludge using infrared spectroscopy

The physicochemical changes in sewage sludge during the time of the experiment were confirmed by spectroscopic examination. These analyses were performed using the Perkin Elmer FTIR spectrometer equipped with the ATR accessory. To this end, the samples with the suspension of bacteria and hygienized sludge were collected daily during the time of the experiment. The hygienized sludge was separated from the suspension by filtration through a cellulose filter. The residue deposited on the filter was brought to an air-dry condition and then transferred to ATR accessory using a technique based on the phenomenon of weakened total infrared radiation reflection. The FTIR spectra of sewage sludge residues were performed in the range of 2000–400 cm^{-1}.

10.3 RESULTS AND DISCUSSION

In order to confirm that the mesophilic bacteria determined in the solution included primarily the introduced phosphorus bacteria, the number of mesophilic bacteria was defined depending on the

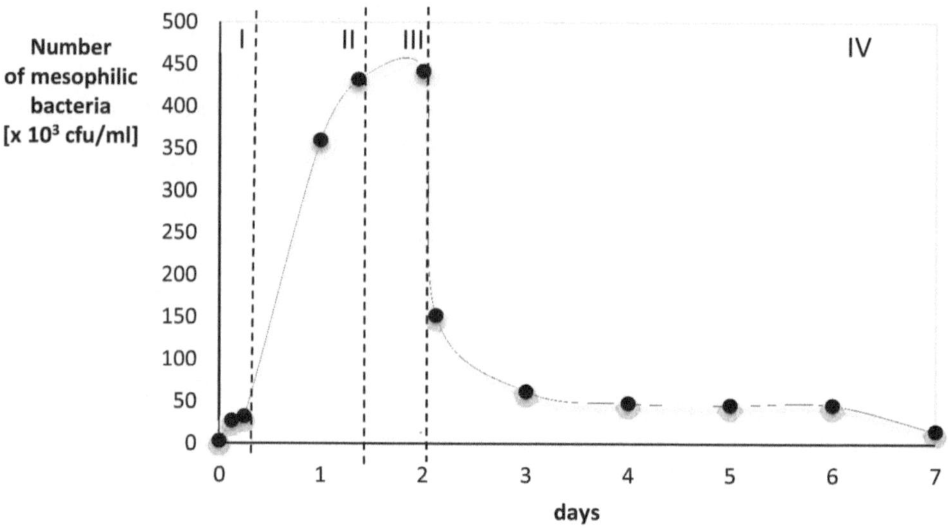

Figure 10.1. Growth curve of mesophilic microorganisms determined in the suspension during the experiment.

duration of the experiment for the tested suspension samples and the control sample, i.e. the suspension without the introduced previously *Bacillus megaterium* bacteria. Comparing the results of the growth of mesophilic bacteria in the samples including the introduced phosphoric bacteria with the control samples it can be stated that the amount of bacteria in the control samples is significantly lower than in the sample with *Bacillus megaterium*. The highest determined number of mesophilic bacteria in the examined solutions with the previously introduced phosphorous bacteria was $4.42 \cdot 10^5$ cfu/ml, whereas in the control samples it was merely $1.4 \cdot 10^3$ cfu/ml. The large difference in the mesophilic bacterial counts may indicate that the introduced *Bacillus megaterium* bacteria are rapidly multiplying.

Determining the amount of the mesophilic bacteria related to the phosphorus bacteria application time in the sewage sludge permitted the preparation of a growth curve for these microorganisms (Figure 10.1). The growth curve is characteristic of a given culture and consists of four phases: primary inhibition (phase I), logarithmic phase (phase II), equilibrium phase (III) and decline phase (IV).

Phase I i.e. the lag phase is very short - 6h for the solution samples with phosphorus bacteria culture. This may be related to the rapid adaptation of microorganisms to the new environmental conditions. In the studies performed by Ciopińska et al. (2019) the duration of the first phase lasted about 9 days. These tests were performed under similar conditions, with a dose of the applied lime as the differentiating factor. In the case of the studies carried out by Ciopińska et al. (2019), a larger dose of lime was used for hygienization, i.e. 0.2 kg CaO/kg dw of sewage sludge. While comparing the duration time of the first phase presented in this paper with the results obtained by Ciopińska et al. (2019) it can be concluded that the initial pH of the sewage sludge has a significant impact on the course of the process. Petrova et al. (2014) studied the effect of the CaO dose in sewage sludge on the decontamination process. Their research showed that as the dose of lime increased, the number of microorganisms in the sewage sludge decreased. A high pH value is a limiting factor for the growth of many microorganisms. By producing organic acids, the Bacillus megaterium bacteria reduce the alkalinity of the suspension and thus moving from the pessimum phase, where the organisms exist on the edge of survival, create more favorable conditions, where the organisms have the opportunity to grow (pejus phase) (Kratochwil 1996).

The logarithmic phase was also very short with the duration of only 24 h. In this phase there is an intensive division of microbial cells.

The logarithmic phase is followed by the equilibrium phase. During this phase, the number of newly emerging cells balances the number of cells dying off at a particular moment.

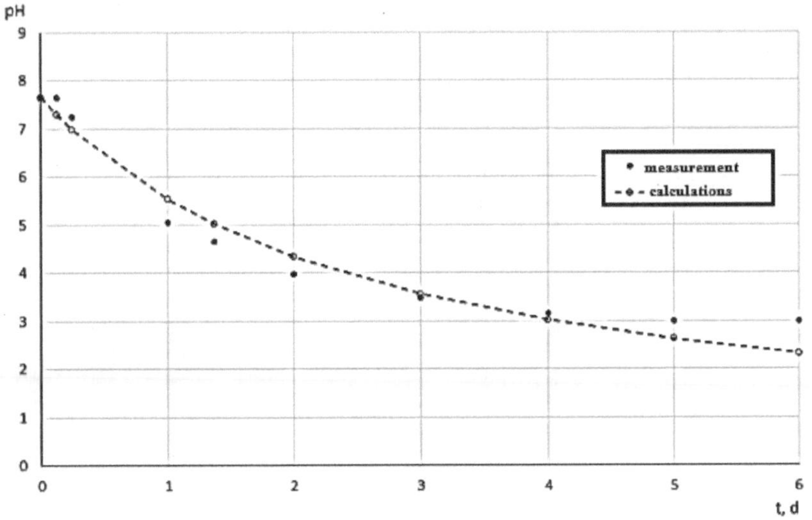

Figure 10.2. Change in the pH value during the process.

Phase III in this experiment lasted one day. This stage occurs when the food sources begin to deplete or the concentration of metabolic products increases to a level that is harmful to the bacteria themselves. The last phase (decline) started rapidly after only two days of the experiment. This phase is dominated by the cells dying off and spores being formed.

The microorganisms applied on the sewage sludge grew in a very rapid manner. Within the first three days all the growth phases can already be distinguished. The decline of microorganisms may result from the depletion of nutrients such as phosphorus in the batch culture. The rapid colonization of the sludge by phosphorous bacteria can also involve the intensive production of organic acids into the environment. The *Bacillus megaterium* bacteria synthesize and release weak organic acids such as citric acid, lactic acid or propionic acid into the environment (Rasid et al. 2004). The acid production causes a significant reduction in the pH of the suspension (Figure 10.2). During the experiment, the pH value was reduced from 7.5 to 3 indicating that the environment was strongly acidified.

On the basis of Figure 10.2, it was found that the pH in the analyzed system is decreasing and its dynamics can be expressed by the second order kinetics equation in the following form:

$$pH = \frac{pH_0}{1 + \alpha \cdot pH_0 \cdot t} \quad (R = 0.9804) \tag{10.1}$$

where: pH_0 – initial pH in the analyzed system, the value of $pH_0 = 7.642$; α – kinetics parameter determined by Levenberg-Maquardt's method (Moré 1997), the value of $\alpha = 0.05$.

Strong acidification of the environment can affect the solubilization of phosphorus to bioavailable forms of this element. The Golterman's analysis was performed in order to confirm the occurrence of the process.

Using this method, the bioavailable forms (Ca-EDTA and Na-EDTA) and organic phosphorus forms (H_2SO_4 and NaOH) can be distinguished. In this article, based on the Golterman's method, the first two phosphorus fractions (Ca-EDTA and Na-EDTA) were analyzed, which are identified with bioavailable phosphorus. The Ca-EDTA fraction is the most mobile fraction, identified with readily assimilable phosphorus, associated with iron, aluminum and manganese oxides and hydroxides.

The subsequently extracted Na-EDTA fraction is identified with phosphorus compounds that can be converted into easily assimilable forms by natural processes in the environment. The phosphorus extracted in the Na-EDTA fraction is of the kind mostly associated with carbonates.

The analysis pertaining to the amount of bioavailable phosphorus forms indicates the ongoing solubilization process, as after the first day of the experiment there was an increase in the most assimilable

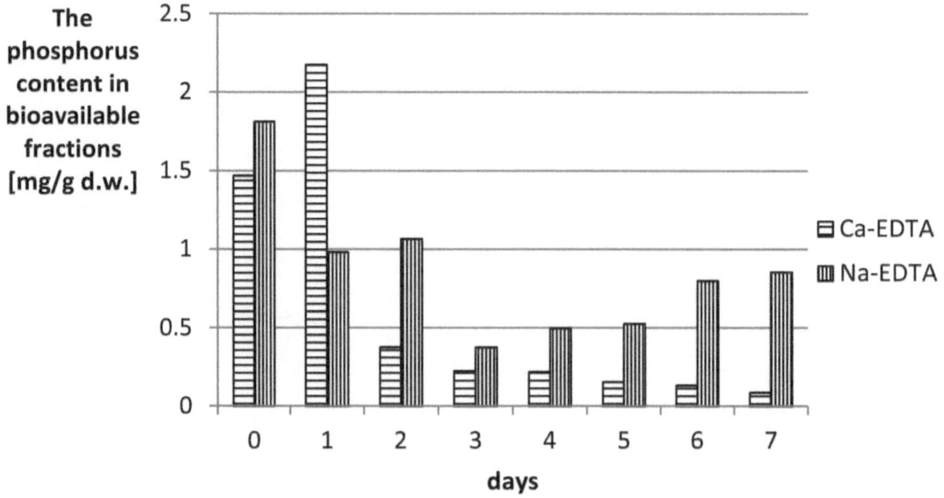

Figure 10.3. Change in phosphorus content in bioavailable fractions in hygienized sludge during the experiment.

forms (Ca-EDTA fraction) and a decrease in the amount of phosphorus with lower bioavailability (Na-EDTA fraction), which may be the effect of acid produced by microorganisms and the conversion of phosphorus forms from Na-EDTA to Ca-EDTA fraction. The increase in the amount of Na-EDTA fraction after three days can be explained by the phosphorus transformations occurring in the phase of decline by the release of this element into the environment. On the third day, the pH value of the suspension is stabilized as well.

The effect of pH on the amount of the bioavailable phosphorus was determined (Figure 10.4) taking into account the changing reaction of the suspension and the decrease in the bioavailable forms (total Ca-EDTA and Na-EDTA fractions). Linear regression relationship between the change in the suspension pH (pH) and the amount of bioavailable forms (bioav.f.) in the sewage sludge was demonstrated:

$$bioav.f. = 7.618\,(\pm 0.976)\cdot pH - 10.785(\pm 0.215) \quad \mathrm{R}^2 = 0.87 \qquad (10.2)$$

The statistically significant relationship presented above confirms the effect of the environment acidification (probably due to the synthesis of organic acids by microorganisms) on the phosphorus solubilization processes. The decrease in the amount of bioavailable phosphorus forms in the sludge during the experiment is closely related to the assimilation of this element by the microorganisms present in the investigated suspension.

The analysis of the data presented in the graph (Figure 10.5) shows that phosphorus is rapidly absorbed immediately after its addition into the suspension. Phosphorus assimilation (P. assim) occurs exponentially for up to three days, which may be related to the logarithmic growth phase of microorganisms. After the third day, a decrease in the amount of phosphorus in the microbial cells occurs, which may result from the inhibition of bacterial division and the formation of spores.

The presented curve of changes in the amount of assimilated phosphorus (P.assim) by microorganisms extends exponentially and is described by equation (3):

$$P.assim = P.assim_0 + (P.assim_{max} - P.assim_0)\cdot t\cdot \exp(-\lambda\cdot t) \quad (\mathrm{R} = 0.994) \qquad (10.3)$$

where: $P.assim_0$ – the initial amount of phosphorus contained in the bacteria, [$P.\ assim_0 = 0.03$]; $P.assim_{max}$ – maximum amount of phosphorus assimilated by phosphorus bacteria depending on the initial amount of bacteria, [$P.\ assim_{max} = 0.36$]; λ – the empirical parameter estimated by the Levenberg-Marquardt's method, $\lambda = 0.354\ (\pm 0.005)$.

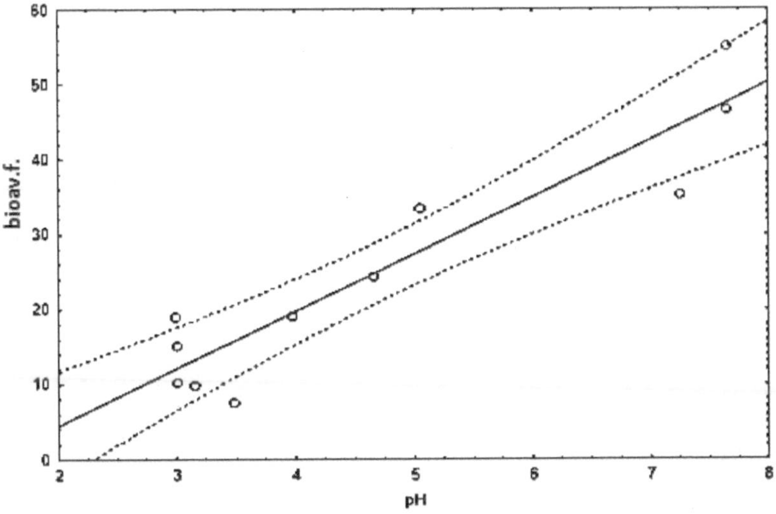

Figure 10.4. Effect of the suspension pH value on the amount of bioavailable phosphorus forms in the hygienized sludge.

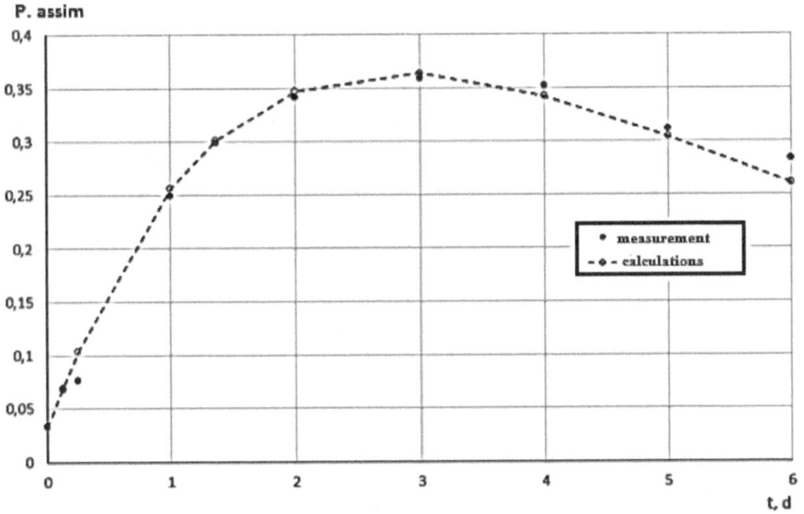

Figure 10.5. Change in the amount of the phosphorous assimilated by microorganisms during the experiment.

After the application of the bacterial suspension into the hygienized sludge (after app. 5 minutes of the experiment), the phosphorus content in the microbial cells amounted to merely 0.036 mg/g.d.w., whereas on the third day of the experiment it was 0.367 mg/g. d.w. (Figure 10.5). Thus, the tenfold phosphorus content increase in the microbial cells during the first three days, and the use of hygienized sludge as the only source of this element, confirms that the *Bacillus megaterium* bacteria can assimilate phosphorus also from sewage sludge previously treated with lime.

A spectroscopic analysis also confirms the possibility of the *Bacillus megaterium* bacteria assimilating phosphorus from sewage sludge. The diagrams below show successive FTIR spectra of hygienized sludge (separated from the suspension) collected on the first three days of the experiment (Figure 10.6) and after the third day of the experiment (Figure 10.7).

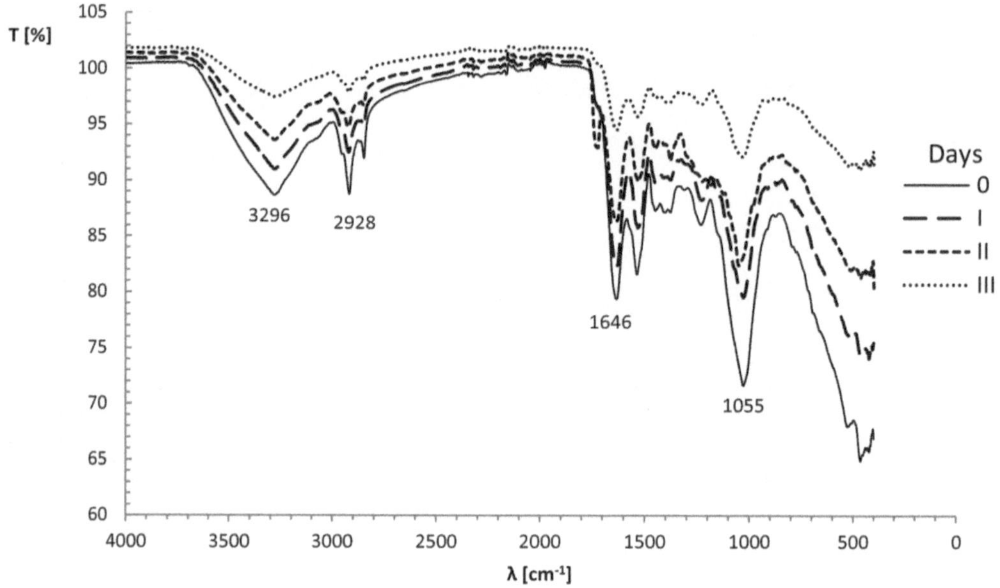

Figure 10.6. FTIR spectra of hygienized sludge separated from the suspension during the first three days of the experiment.

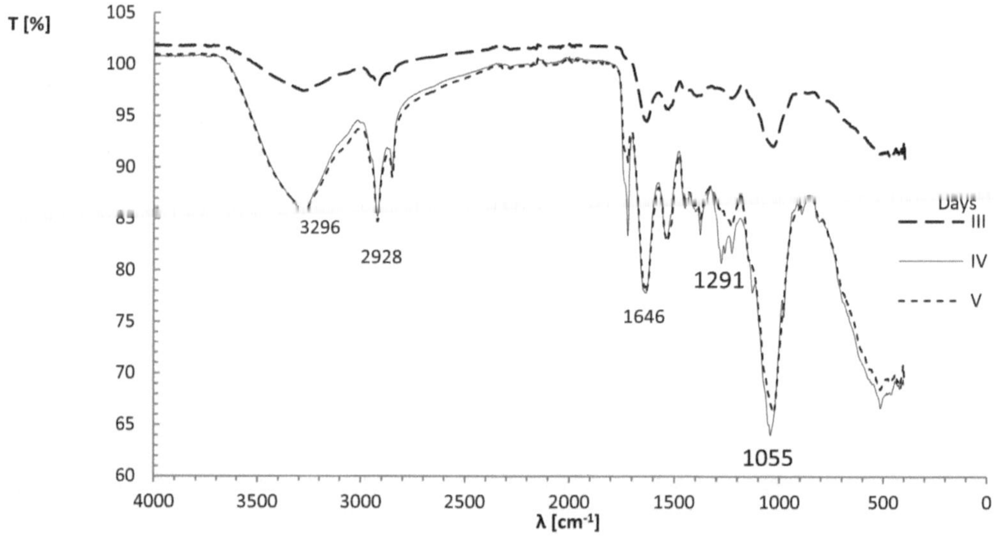

Figure 10.7. FTIR spectra of hygienized sludge separated from the suspension on day three, four and five the experiment.

The analysis of the FTIR spectra (Figures 10.6 and 10.7) shows the change in the transmittance values at individual characteristic wavenumbers for the oscillators of chemical groups. Up to the third day of the experiment (Figure 10.6), the transmittance values of the chemical oscillator bands increased sharply, which indicates a decrease in the concentration of chemical compounds present in the hygienized sludge. The decrease in the concentrations of individual groups may confirm the microbial assimilation of the chemical compounds present in the sludge.

The dactyloscopic area with a band system characteristic of a given particle starts at the 1500 cm^{-1} wavenumber, however, the vibration peaks of oscillators overlap due to the fact that the sludge is a mixture of various chemical compounds. The greatest changes in the transmittance values in the dactyloscopic area are observed at the band within the 1055 cm^{-1} wavenumber, corresponding to the vibrations of the C-O, P-O-C and S-O groups that occur i.a. in organic matter. Within this wavenumber, the vibrations of the P-O group occurring in inorganic phosphorus compounds can also be identified. Therefore, it can be assumed that the increase in the band transmittance within this wavenumber during the first three days may be closely associated with the assimilation of phosphorus from the sludge hygienized by the phosphorus bacteria.

It should be noted that it is not possible to identify clearly the peak growth at the wavenumber of 1055 cm^{-1} with the POC and PO vibrations alone; however, by analyzing the values of the bioavailable phosphorus content (Figure 10.3), the amount of which declined in the sludge up to the third day of the experiment, these assumptions seem justified. In addition, Figure 10.7 shows the FTIR spectra of the hygienized sludge separated from the suspension on day three, four and five of the experiment. On day four and five, a noticeable decrease in the transmittance value is observed, i.e. the concentration of all chemical oscillators increases.

The increase in concentration of chemical groups may be closely related to the ongoing phase of microbial decline, which is characterized by the reduced microbial metabolism, spore formation, and thus the release of substances from cells. The increase in the concentration of chemical oscillator vibrations within the wavenumber of 1055 cm^{-1} after the third day of the experiment coincides with the significant phosphorus content decrease in the sludge assimilated by the microorganisms (Figure 10.5) and the phosphorus content increase in the Na-EDTA fraction in the hygienized sludge (Figure 10.3).

10.4 CONCLUSION

When analyzing the research results presented in the paper, it can be concluded that:

- the application of *Bacillus megaterium* causes the dissolution of the less available Na-EDTA fraction to mobile, easily absorbable Ca-EDTA, which may be of particular importance when using such a suspension for agricultural purposes.
- after the first day of the experiment, phosphorus solubilization already took place because there was an increase in the amount of the most absorbable forms (Ca-EDTA fraction) and a decrease in the amount of less bioavailable phosphorus forms (Na-EDTA fraction), which may be due to the activity of acids produced by microorganisms.
- a tenfold increase in the phosphorus content in the cells of microorganisms during the first three days and the use of hygienized sludge as the only source of this element confirms that the *Bacillus megaterium* bacteria can also assimilate phosphorus from the sewage sludge previously treated with lime.

ACKNOWLEDGEMENTS

The article was financed from the Program of the Minister of Science and Higher Education – Regional Initiative of Excellence, financed by the Ministry of Science and Higher Education under contract No. 025/RID/2018/19 of 28/12/2018, in the amount of PLN 12 million.

REFERENCES

Alori, E., Glick, B., Babalola, O. 2017. Microbial Phosphorus Solubilization and Its Potential for Use in Sustainable *Agriculture, Front Microbiol.* 8: 971.

Bishop, M., Chang, A., Lee, R. Enzymatic mineralization of organic phosphorus in a volcanic soil in Chile. 1994. *Soil Sci.* 157: 238–243.

Bezak-Mazur E., Stoińska R. 2013. Speciation of Phosphorus in Wastewater Sediments from Selected wastewater treatment Plant. *Ecological Chemistry and Engineering. A.* 20: 503–514.

Bezak-Mazur, E, Stoińska, R, Szeląg B. 2017 (a). Analysis of the Effect of Temperature Cycling on Phosphorus Fractionat ion in Activated Sludge. *Annual Set The Environment Protection.* 19, 288–301.

Bezak-Mazur, E, Stoińska, R, Szeląg B. 2017 (b). Analysis of phosphorus speciation in primary sludge in the annual cycle. *Environmental problems.* 2: 83–85.

Ciopińska, J, Bezak-Mazur, E, Stoińska, R, Szeląg, B. 2019. The impact of Bacillus megaterium on the solubilisation of phosphorus from sewage sludge. *First International Scientific Conference On Ecological And Environmental Engineering 2018.* 86: 1–8.

Girmay, K. 2019. Phosphate Solubilizing Microorganisms: Promising Approach as Biofertilizers. *International Journal of Agronomy*. 2019: 1–7.

Golterman, H. 1996. Fractionation of sediment phosphate with chelating compounds: a simplification, and comparison with other methods. *Hydrobiologia* 335: 87–95.

Grafe, M, Goers, M, von Tucher, S, Baum, C, Zimmer, D, Leinweber, P, Vestergaard, G, Kublik, S, Schloter, M, Schulz, S. 2018. Bacterial potentials for uptake, solubilization and mineralization of extracellular phosphorus in agricultural soils are highly stable under different fertilization regimes. *Environ Microbiol Rep.* 10(3): 320–327.

Gupta, N., Sabat, J., Parida, R., Kerkatta, D. 2007. Solubilization of tricalcium phosphate and rock phosphate by microbes isolated from chromite, iron and manganese mines. *Acta Bot Croat.* 66: 197–204.

Gyaneshwar, P, Kumar, G, Parekh L, Poole, P. 2002. Role of soil microorganisms in improving P nutrition of Plants. *Plant and Soil.* 245(1): 83–93.

Khan, M., Ahmad, E., Zaidi, A., Oves, M. 2013. Functional aspect of phosphate-solubilizing bacteria: importance in crop production. in: Maheshwari, D. (eds). Bacteria in agrobiology: crop productivity. *Springer* Berlin 237–265.

Khan, M., Zaidi, A., Ahmad, E. 2014. Mechanism of Phosphate Solubilization and Physiological Functions of Phosphate-Solubilizing Microorganisms in: Khan, M.., Zaidi, A., Mussarat, J. *Phosphate Solubilizing Microorganisms. Springer International Publishing Switzerland* 2014: 31–62.

Kratochwil, A. 1999. Biodiversity in Ecosystems: Some Principles. Kuwer Academic Publishers 5–38.

Labuda, M., Saeid, A., Chojnacka, K., Górecki, H. 2012. Use of Bacillus megaterium in solubilization of phosphorus, *Przemysł Chemiczny* 91: 837–840.

Lipiński, W. 2000. Soil acidity and nutrients content on the basis of chemical analysis provided by agrochemical stations. *Fertlizers Fertilization* 3a: 89–105.

Molla, M. A. Z., A. A. Chowdhury, A. Islam and S. Hoque. 1984. Microbial mineralization of organic phosphate in soil. *Plant Soil* 78: 393–399.

Moré, J. 1977. The Levenberg-Marquardt algorithm: implementation and theory. Argonne National Lab, United States.

Marking of phosphorus. 2004. Spectrophotometric method with ammonium molybdate. PN-EN ISO 6878:2004.

Petrova, T, Marinova-Garvanska, S, Kaleva, M, Zaharinov, B, Gencheva, A, Bayko D. 2014. Decontamination of sewage sludge by treatment with calcium oxide. *Int.J.Curr.Microbiol.App.Sci*, 3: 184–192.

Pradhan, N., Shukla L.B. 2005. Solubilization of inorganic phosphates by fungi isolated from agriculture soil. *Afr J Biotechnol.* 5: 850–854.

Rasid, M, Khalid, S, Ayub, N, Alam,S, Latif, F. 2004. Organic acids production and phosphate solubilizing by phosphate solubilizing microorganism under in vitro conditions, *Pakistan J. Biological Sciences* 7: 187–196.

Sharma, S.B., Sayyed, R.Z., Trivedi, M.H., Gobi, T.A. 2013. Phosphate solubilizing microbes: sustainable approach for managing phosphorus deficiency in agricultural soils. *Springerplus* 2: 587.

Song, O., Lee, S.J., Lee, Y.S., Lee, S.C., Kim, K.K., Choi, Y.L. 2008. Solubilization of insoluble inorganic phosphate by Burkholderia cepacia DA23 isolated from cultivated soil. *Baraz J Microbiol.* 39:151–156.

Stachowicz, F, Wójcik, M. 2017. Ecological and economic benefits from sewage sludge hygienisation with the use of lime in a medium-size treatment plant. *Physics for economy* 1: 49–63.

Stoińska, R. 2013. The Influence of the Process of Hygienisation on the Speciation of Phosphorus in Wastewater Sediments. *Transcom Proceedings* 7: 307–310.

Sundara, B., Natarayan, V., Hari, K. 2002. Influence of phosphorus solubilizing bacteria on the changes in soil available phosphorus and sugarcane and sugar yields. *Field Crops Research* 77: 43–49.

Toro, M. 2007. Phosphate solubilizing microorganisms in the rhizosphere of native plants from tropical savannas: an adaptive strategy to acid soils? In: Velazquez, C., Rodriguez-Barrueco, E. (eds) Developments in plant and soil sciences. *Springer, The Netherlands*. 249–252.

Wani, P.A., Khan, M.S., Zaidi, A. 2007. Synergistic effects of the inoculation with nitrogen fixing and phosphate-solubilizing rhizobacteria on the performance of field grown chickpea. *J Plant Nutr Soil Sci.* 170: 283–287.

CHAPTER 11

Variability of nutrients content during co-digestion process of sewage sludge and brewery spent grain

M. Lebiocka
Faculty of Environmental Engineering, Lublin University of Technology, Poland

ABSTRACT: Digestates are often used as an organic soil fertilizers therefore content of nutrients is important from practical point of view. Assessment of the variability of N and P concentrations during co-digestion process of sewage sludge (SS) and brewery spent grain (BSG) was the aim of the study. The study was carried out in the reactors operating at temperature of 35°C in a semi-flow mode. Two experiments for the co-digestion process were performed at different hydraulic retention times (HRT). As a result, a slight decrease (of about 12%) in the concentration of total nitrogen was observed, to the average values measured in digestates ranged from 2765 to 3806 g m^{-3}. An analogous tendency was observed in the case of total phosphorus. The addition of BSG as a co-substrate in the anaerobic digestion of SS and shortening of HRT did not cause changes in the nutrients concentration in digestate and its supernatant.

11.1 INTRODUCTION

Anaerobic digestion (AD) is one of the most widely used processes for stabilizing sewage sludge. The widespread use of this technique is a result of its potential advantages, which include a reduction of the volume of sludge ultimately requiring disposal by 30–50%, and the production of energy from methane, which is obtained in excess of the amount required to operate the process. Sewage sludge is usually stabilized under the mesophilic or thermophilic conditions (with an optimum temperature of 35 or 55°C, respectively). The thermophilic anaerobic digestion has some advantages over the mesophilic digestion, including faster reaction and a higher organic loading rate; as a result, the former exhibits higher biogas and methane productivity than the latter. However, the thermophilic conditions also have some drawbacks, such as decreased stability and quality on the effluent; accumulation of NH_3, and volatile fatty acids (VFA); susceptibility to the environmental conditions; and increased net energy requirements relative to the mesophilic conditions. Although the mesophilic systems exhibit better process stability and higher microbial richness, they obtain lower both methane yields and substrate biodegradability (Villamil et al. 2020). Anaerobic co-digestion (AcoD), i.e. the simultaneous digestion of two or more substrates, is a well-established option to overcome the drawbacks of mono-digestion and improve the economic feasibility of biogas plants. The latter is a result of the higher methane production and the treatment of several types of wastes in a single facility (Mata-Alvarez et al. 2014; Solé-Bundó et al. 2019). In order to overcome the mono-digestion problems, AcoD has been implemented with co-substrates such as animal manure (Hubin & Zelić, 2013; Labatut et al. 2011; Vivekanand et al. 2018), food waste (El-Mashed et al. 2008; Maragkaki et al. 2018) and sewage sludge (Fernández et al. 2014; Zielewicz et al. 2012). Co-digestion has already been used for many applications. The absence of the inhibition phenomena for the co-digestion of sewage sludge as well as fruit and vegetable wastes, was reported by Di Maria et al. (2015) at high organic loading rates (OLR) >3 kgVS/m^3· day and decreased hydraulic retention times (HRT), i.e. close to 10 days.

An increase in the biogas production, from 0.34 to 0.49 m^3/kg VS, was reported by Cavinato et al. (2013) for the co-digestion of the organic fraction of solid waste and sludge at OLR>2.2 kgVS/m^3·day. Furthermore, co-digestion could be an opportunity for extending the anaerobic treatment also to other sectors at reduced investment costs (Salehiyoun et al. 2020). Among these, brewery spent grain

DOI 10.1201/9781003171669-11

(BSG) generated globally in significant quantities should be considered as an additional substrate in the anaerobic digestion process of sewage sludge. Employing an additional co-substrate, capable of eliminating the shortcomings of a mono-component system, constitutes a viable solution for a successful co-digestion of the residues. Adding brewery spent grain (BSG) may possibly increase the buffer capacity as well as the process stability and the carbon/nitrogen (C/N) ratio. Substantial amounts of wastes and by-products are produced in the brewing industry, including BSG, spent hops and yeast. BSG comprises approximately 16.8–25.4% of cellulose, 21.8 to 28.4% of hemicellulose (predominantly arabinoxylans) as well as 11.9–27.8% of lignin (Mussatto et al. 2006; Santos et al. 2003). The literature contains numerous studies on the anaerobic digestion of BSG. It mainly comprises cellulose, hemicellulose and lignin. It is difficult to conduct anaerobic degradation of these substances, as a result of the occurrence of degradation products, including phenolic compounds that inhibit the process (Panjičko et al. 2017). Sežun et al. (2011), reported that BSG constitutes a mono substrate which – regardless of pretreatment – cannot be subjected to anaerobic digestion under the mesophilic conditions in a semi-continuous bioreactor. Depending on the employed substrate pretreatment, the production of biogas was significantly inhibited. The inhibition of biogas production from BSG – being a prospective lignocellulosic substrate – occurred due to the presence of intermediate lignocellulosic degradation products, especially p-cresol. The difficulties in the biodegradation of lignin stem from the complexity of its structure, insolubility, as well as chemical stability and high molecular weight. The anaerobic digestion process may be potentially inhibited as a result of the inappropriate nutrient structure, low content of nitrogen, lack of diversified microbes and high organic load. The process dynamics were influenced by pesticide contamination, as well as high lignin percentage and C/N content. In this paper the results of the effect of BSG addition to the municipal sewage sludge eon the variability of nutrients in the anaerobic digestion of sewage sludge.

11.2 MATERIALS AND METHODS

11.2.1 *Material characteristics*

In the experiment, sewage sludge constituted the main substrate, and BSG employed as co-substrates were co-digested. The SS (primary and excess sludge) were collected weekly in the municipal wastewater treatment plant in Puławy, Poland. The laboratory experiment involved mixing primary and excess sludge at the volume ratio of 60:40. Afterwards, it was homogenized, passed through a sieve with 3 mm mesh size and portioned. The prepared samples of sludge were stored for up to a week in a laboratory refrigerator at a temperature of 4°C. Table 11.1 presents the sewage sludge composition.

Table 11.1. Composition of the SS used in the experiments (average value and 95% confidence limits are given).

Parameter	Unit	Average value
COD	mg dm^{-3}	53 240 51 400/55 090
SCOD	mg dm^{-3}	3 147 2 927/3 368
VFA	mg dm^{-3}	1 747 1 585/1 909
pH	–	5.83 5.78/5.88
Alkalinity	mg dm^{-3}	875 788/961
TS	g kg^{-1}	34.8 33.9/35.7
VS	g kg^{-1}	26.1 25.3/26.9
TN	mg dm^{-3}	2920 2 866/2 974
TP	mg dm^{-3}	1038 709/1367
NH_4^+-N	mg dm^{-3}	96.7 52.7/140.7
PO_4^{3-}-P	mg dm^{-3}	123.1 88.2/158.1

BSG was obtained from the Grodzka 15 brewery (Lublin, Poland), which employs barley for the beer brewing process. The samples of BSG were collected twice in the form of a warm, raw material, which was promptly taken to the laboratory. In order to achieve stable composition of the substrate, BSG was subjected to drying at 60°C for 2 h in a laboratory dryer. Prior to weighing and portioning, it was ground by means of a laboratory ball mill (for three minutes, at 200 r min^{-1}) to a particulate size of 2.0 mm. Table 11.2 presents the composition of BSG. The BSG for feedstock preparation involved storage in the indoor air until reaching 20°C and subsequent homogenization with a low-speed mixer.

11.2.2 *Experimental set-up*

The investigations were carried out in 40L semi-flow anaerobic reactors (Figure 11.1) at a temperature of 35°C. The laboratory setup comprised three completely mixed digesters (operating in parallel) with a heating jacket at a stable temperature, as well as a gaseous installation (consisting of pipelines, a

Table 11.2. Composition of the BSG used in the experiments (average value and standard deviation are given)

Parameter	Unit	BSG composition
COD	mg dm^{-3}	48 400 ± 427
SCOD	mg dm^{-3}	-
VFA	mg dm^{-3}	3 448 ± 237
pH	-	6.94 ±0.52
Alkalinity	mg dm^{-3}	2 383 ±298
TS	g kg^{-1}	224.9 ± 2.1
VS	g kg^{-1}	215.9 ± 1.9
TN	mg dm^{-3}	615 ± 11.3
TP	mg dm^{-3}	149.5 ± 65.7
NH_4^+-N	mg dm^{-3}	26.4 ± 1.13
PO_4^{3-}-P	mg dm^{-3}	48.8 ± 7.74

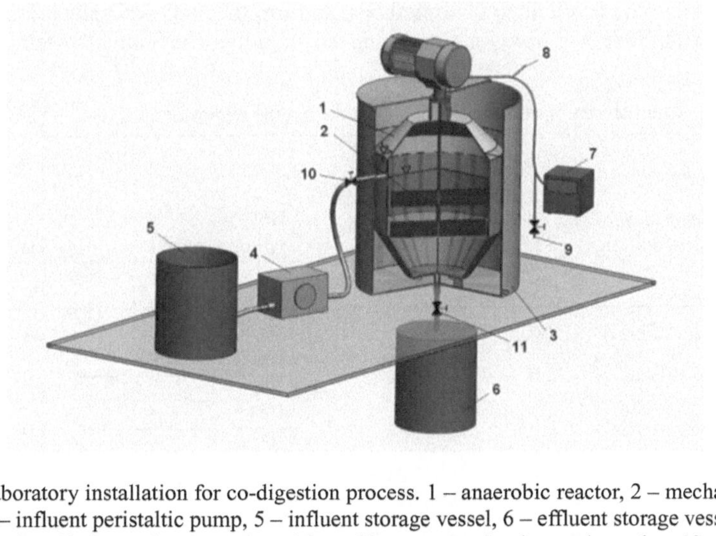

Figure 11.1. Laboratory installation for co-digestion process. 1 – anaerobic reactor, 2 – mechanical stirrer, 3 – heating jacket, 4 – influent peristaltic pump, 5 – influent storage vessel, 6 – effluent storage vessel, 7 – drum gas meter, 8 – gaseous installation and gas sampler with a rubber septum, 9 – dewatering valve, 10 – inlet valve, 11 – outlet valve

gas sampler with a rubber septum, a pressure equalization unit, gas valves, a dewatering valve and a mass flow meter). The reactors were fed once per day.

Additionally, the setup comprised an influent peristaltic pump as well as the vessels for the influent and effluent storage. A mechanical stirrer characterized by a rotational speed of 50 min^{-1} was used for mixing.

The inoculum used in the laboratory reactors was obtained from a wastewater treatment plant in Puławy. It was collected from the mesophilic anaerobic digester which operated with a volume of 2500 m^3 at HRT of 25 d. The digestate was adapted in reactors for 30 d. Two experiments with different HRT and feedstock compositions were conducted. The experiments consisted in performing several runs (including the control ones) which lasted for 90 d. The schedule comprised 30 d devoted for acclimatization as well as 60 d for the main stage in which the measurements were conducted. Two experiments for the co-digestion process were performed at different hydraulic retention times (HRT) of 20 (run 1.2) and 18 d (run 1.4). In each one, the separate control run supplied daily with 2 L of SS and operated at HRT of 20 d was provided (runs 1.1 and 1.3). Table 11.3 shows the experimental settings. It was observed that the characteristics of SS varied among the conducted experiments. The contributing factors include the inconsistent wastewater composition due to the seasonal discharge of the agricultural and industrial wastewater (from the companies processing fruit and vegetables), fluctuations of temperature, as well as the varying characteristics of wastewater. Hence, it was necessary to perform separate control runs, i.e. R 1.1, R 1.3 (Table 11.3).

11.2.3 *Analytical methods*

The sewage sludge was subjected to analysis on a weekly basis, following its delivery to the laboratory. The measured parameters included: total chemical oxygen demand (COD), total solids (TS), volatile solids (VS), total nitrogen (TN) and total phosphorus (TP). On the other hand, the supernatant was analyzed for soluble chemical oxygen demand (SCOD), VFA, alkalinity, pH level, ammonia nitrogen ($N\text{-}NH_4^+$) and orthophosphate phosphorus ($P\text{-}PO_4^{3-}$). Sewage sludge was centrifuged at 4000 r min^{-1} for 30 min and filtered in order to obtain the supernatant samples. Once delivered to the laboratory, the characteristics of SS, ACW and BSG, i.e. COD, TS, VS, VFA, alkalinity, and pH, were determined. The analysis of feedstock composition was performed on a weekly basis. In turn, the parameters of the digestate were measured twice a week, following the same pattern as in the case of the feedstock. The majority of analyses were carried out using Hach Lange UV–VIS DR 5000 (Hach, Loveland, CO, USA) and the Hach analytical methods. In turn, of a HQ 40D Hach-Lange multimeter (Hach, Loveland, CO, USA) was used for monitoring the pH value. The Standard Methods for the Examination

Table 11.3. Characteristics of operational set-up during the experiment.

Run	Feedstock composition	Component volume	Additive mass	BSG mass: feedstock volume ratio*	HRT	OLR	
		SS	BSG			Avg.	Upp./low. 95% mean
		dm^3	g	$g\ dm^{-3}$*	d	$kg\ VS\ m^{-3}d^{-1}$	
R 1.1	SS (control)	2.0	-	100	20	1.35	1.23/1.46
R 1.2	SS + BSG	2.0	20	100:0 / 100	20	1.73	1.68/1.78
R 1.3	SS (control)	2.0	-	10:1* / 100:0	20	1.49	1.41/1.58
R 1.4	SS + BSG	2.2	21.7	100:0 / 10:1*	18	1.98	1.84/2.13

*ratio of the BSG weight to feedstock volume ($g\ dm^{-3}$)

of Water and Wastewater (APHA 2005) were employed for determining the value of total and volatile solids.

11.3 RESULTS AND DISCUSSION

In order to evaluate the effect of co-digestion on the variability of the nutrients content the concentration of TN, N-NH_4^+, TP, and P-PO_4^{3-} in the reactor feedstock and digest were analyzed. The ratio between the concentration of organic matter and structural elements such as nitrogen and phosphorus is an important parameter characterizing the reactor feedstock. The ratio of COD /TN and COD/ P-PO_4^{3-} was determined (Table 11.4).

The value of the COD/TN ratio in reactors feedstock was evolving at an unfavorable level, below the optimal value (20). After the addition of BSG, it increased to 13.2 and 14.8 for the run 1.2 and 1.4, respectively. In the control runs, it reached the values from 9.9 (R 1.1) to 13.0 (R 1.3). The introduction of BSG as a co-substrate to mixture feeding the reactor slightly affected the value of the COD/P-PO_4^{3-} ratio, causing it to increase to 370.4 in the R 1.2 reactor and 192.3 in R 1.4. The value of this proportion in the control samples was 357.1 in R 1.1 and 172.4 in R 1.3, respectively. The effect of co-substrate addition on the concentration of total nitrogen and total phosphorus in the reactor feeding and digest was also investigated (Figure 11.2).

In the presence of BSG, a decrease in the average total nitrogen and total phosphorous concentrations in the feed was noted compared to the control runs (Figure 11.2). As a result of anaerobic digestion, a slight decrease in the concentration of total nitrogen in digest in co-digested runs was observed. The highest reduction rate of 26.5% was obtained in run 1.3, where the reactor was fed only with SS. For two-component feedstock mixtures (R 1.2 and R 1.4), the average concentration of this parameter was 3812.5 g m^{-3} and 3641 g m^{-3}, respectively. For sewage sludge, the average concentrations reached 3975 g m^{-3} (R 1.1) and 3741 g m^{-3} (R 1.3). As a result of using BSG as a co-substrate, the TN concentrations in digestate were at a comparable level to control runs.

In sewage sludge, the average concentration was 3658 g m^{-3} in R 1.1 and 2765 g m^{-3} in R 1.3, and for co-digestion runs - 3806 g m^{-3} (R 1.2) and 3315 g m^{-3} (R 1.4). During the experiment, the concentrations of total phosphorus in the reactor feed and digestate waste were also analyzed (Figure 11.2b). The introduction of BSG caused differences in the value of this parameter in the mixture feeding the reactor. As a result of the brewery spent grain addition to sewage sludge, the concentration dropped from 972 g m^{-3} (R 1.1) to 921 g m^{-3} (R 1.2).

However, in the next stage a decrease in concentration from 1407 g m^{-3} (R1.3) to 760 g m^{-3} (R1.4) was noted. Total phosphorus concentrations in the digestate averaged 669 g m^{-3} and 590 g m^{-3} in control runs, and 647 g m^{-3} (R 1.2) and 605 g m^{-3} (R 1.4) in the presence of BSG.

The introduction of BSG to the feed supplying the reactor caused a slight increase in the concentration of ammonium in comparison with the control tests (Figure 11.3a).

The average concentration of this parameter reached 70 g m^{-3} in series 1.2 and 73 g m^{-3} in series 1.4. In the case of sewage sludge, the ammonium nitrogen concentration was 51 and 62 g m^{-3}, in series 1.1 and 1.3, respectively. As a result of co-fermentation in all runs, a statistically significant increase in the ammonium nitrogen concentration in supernatant was noted compared to the values of this parameter measured in feedstocks, which was the result of ammonification. The highest concentrations

Table 11.4. Ratio of COD /TN and COD/N-NH_4^+/ P-PO_4^{3-} in reactors feedstock.

Parameter	R 1.1	R 1.2	R 1.3	R 1.4
COD /TN	49.3:5 (9.86)	66.1:5 (13.2)	65.2:5 (13.04)	73.9:5 (14.8)
COD/P-PO_4^{3-}	100:0.28 (357.1)	100:0.27 (370.4)	100:0.58 (172.4)	100:0.52 (192.3)

of ammonium nitrogen were observed in the case of sewage sludge fermentation, on average 452 g m^{-3} in run 1.1 and 441 g m^{-3} in run 1.3.

In the presence of brewery spent grain in the feedstock, the average values reached 440 g m^{-3} and 433 g m^{-3}, respectively.

To evaluate the difference between concentrations of the inorganic forms of N and P both in the feed and digest, the release degree was calculated considering N-NH^{4+} and P-PO_4^{3-}. This was defined as a ratio of the reactor effluent load to influent load and indicated as f_{NH4} and f_{PO4}, respectively. Compared to the control samples, the value of the ammonia nitrogen release degree for two-component mixtures decreased from 8.8 (R 1.1) to 6.2 (R 1.2), and from 7.1 (R 1.3) to 5.9 (R 1.4). After adding the BSG, the average orthophosphate concentration in the reactor feedstock was 138 g m^{-3} (run 1.2) and 201 g m^{-3} (run 1.4), while for sewage sludge it was 109 g m^{-3} and 278 g m^{-3} (Figure 11.3b). The release degree values for orthophosphate were 1.2 in R 1.2 and 1.4 in reactor R 1.4, respectively, which means that phosphates were not released into supernatant as much as ammonium nitrogen. In the case of sewage sludge fermentation, only in R 1.1 there was an increase in orthophosphate concentration, the value of f_{PO4} was 1.58 and 0.64 in R 1.1 and R 1.3, respectively.

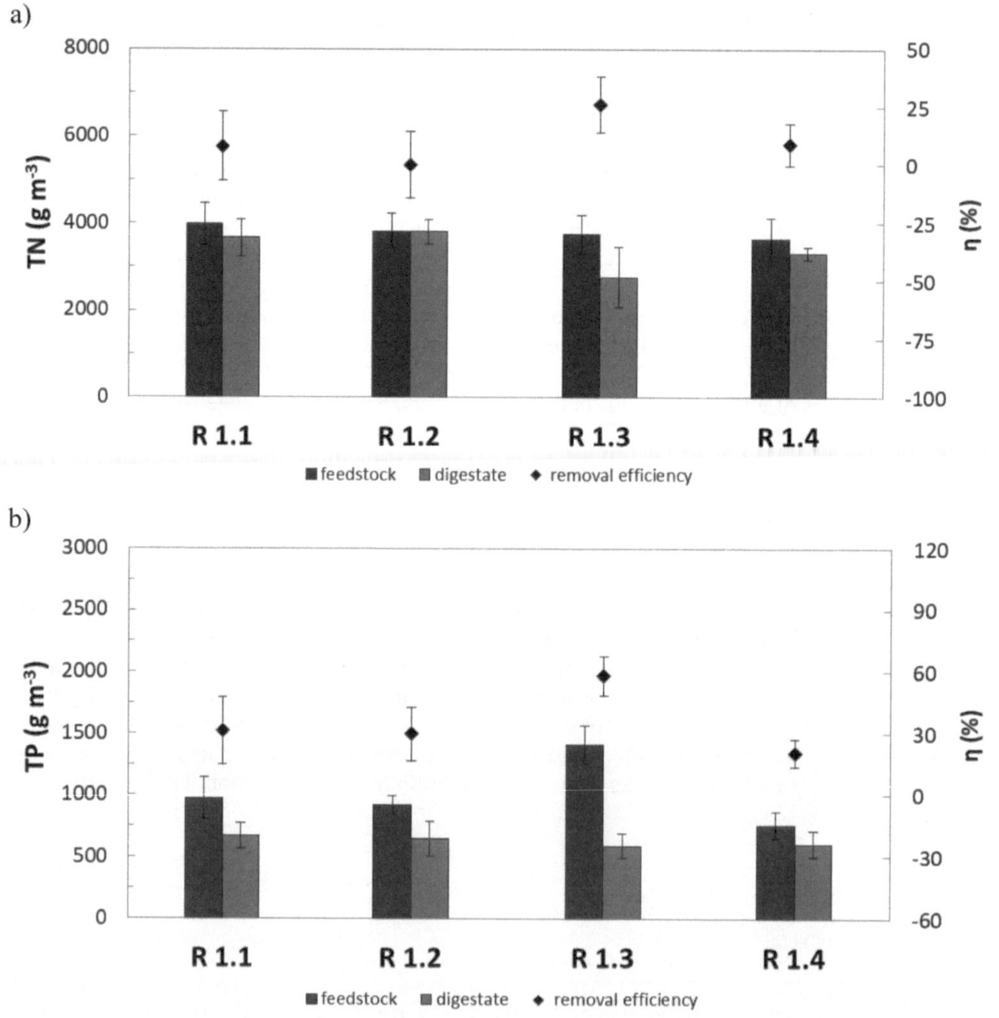

Figure 11.2. Concentration of a) total nitrogen and b) total phosphorus in the feedstock and digestate (average values), and removal efficiency (error bars represent confidence levels, $\alpha = 0,05$)

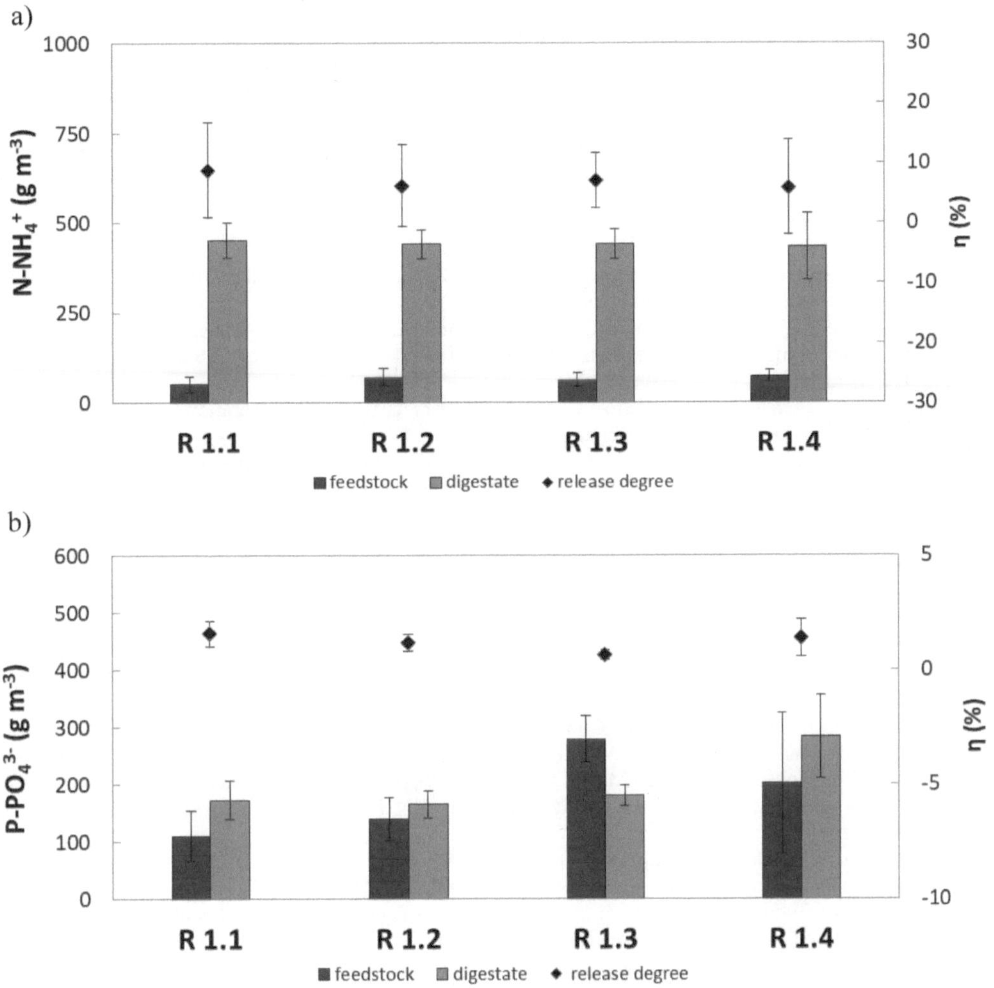

Figure 11.3. Concentration of a) ammonium nitrogen and b) orthophosphate in the feedstock and digestate supernatant (average values), as well as the release degree (error bars represent confidence interval, $\alpha = 0,05$)

11.4 CONCLUSION

In summary, the addition of BSG to sewage sludge did not cause significant differences in the concentration of nutrients in the reactor feed. There was also no significant increase in the concentration of these elements in digestate supernatant. It can be stated that the addition of BSG as a co-substrate in the anaerobic digestion of SS and shortening of HRT did not cause changes in the nutrients concentration in digestate and its supernatant.

REFERENCES

Cavinato C., Bolzonella D., Pavan P, Fatone F., Cecchi F. 2013. Mesophilic and thermophilic anaerobic co-digestion of waste activated sludge and source sorted biowaste in pilot- and full-scale reactors. Renewable Energy 55: 260–265.

Di Maria F., Sordi A., Cirulli G., Micale C. 2015. Amount of energy recoverable from an existing sludge digester with the co-digestion with fruit and vegetable waste at reduced retention time. Applied Energy 150: 9–14.

El-Mashad, H.M., McGarvey, J.A., Zhang, R. 2008. Performance and microbial analysis of anaerobic digesters treating food waste and dairy manure. Biological Engineering Transactions 1 (3): 233–242.

Fernández, C., Blanco, D., Fierro, J., Martínez, E.J., Gómez, X. 2014. Anaerobic co-digestion of sewage sludge with cheese whey under thermophilic and mesophilic conditions. International Journal of Energy Engineering 4: 26–31.

Hublin, A., Zelić, B. 2013. Modelling of the whey and cow manure co-digestion process. Waste Management and Research 31 (4): 353–360.

Labatut, R.A., Angenent, L.T., Scott, N.R. 2011. Biochemical methane potential and biodegradability of complex organic substrates. Bioresource Technology 102 (3): 2255–2264.

Maragkaki, A.E., Vasileiadis, I., Fountoulakis, M., Kyriakou, A., Manios, T. 2018. Improving biogas production from anaerobic co-digestion of sewage sludge with a thermal dried mixture of food waste, cheese whey and olive mill wastewater. Waste Management 71: 644–651.

Mata-Alvarez, J., Dosta, J., Romero-Güiza, M.S., Fonoll, X., Peces, M., Astals, S. 2014. A critical review on anaerobic co-digestion achievements between 2010 and 2013. Renewable and Sustainable Energy Reviews 36: 412–427.

Mussatto, S.I., Dragone G., Roberto I.C., 2006. Brewers' spent grain: generation, characteristics and potential applications. Journal of Cereal Science 43: 1–14.

Panjičko M., Zupančič G., Faned L., Logar R.M., Tišma M., Zelić B., 2017. Biogas production from brewery spent grain as a mono-substrate in a two-stage process com-posed of solid-state anaerobic digestion and granular biomass reactors. Journal of Cleaner Production166: 519–529.

Salehiyoun A.R., Di Maria F., Sharifi M., Norouzi O., Zilouei H., Aghbashlo M. 2020. Anaerobic co-digestion of sewage sludge and slaughterhouse waste in existing wastewater digesters. Renewable Energy 145: 2503–2509.

Santos M., Jimenez J.J., Bartolome B., Gomez-Cordoves C., del Nozal M.J. 2003. Variability of brewers' spent grain within a brewery. Food Chemistry 80: 17–21.

Sežun M., Grilc V., Zupančič G.D., Marinšek-Logar R. 2011. Anaerobic digestion of brewery spent grain in a semi-continuous bioreactor: inhibition by phenolic degradation products. Acta Chimica Slovenica 58(1): 158–166.

Solé-Bundó M., Passos F., Romero-Güiza M.S., Ferrera I, Astals S. 2019. Co-digestion strategies to enhance microalgae anaerobic digestion: A review, Renewable and Sustainable Energy Reviews 112: 471–482.

Villamil J.A., Mohedano A.F., San Martín J. Rodriguez J.J, de la Rubia M.A. 2020. Anaerobic co-digestion of the process water from waste activated sludge hydrothermally treated with primary sewage sludge. A new approach for sewage sludge management. Renewable Energy 146: 435–443.

Vivekanand, V., Mulat, D.G., Eijsink, V.G.H., Horn, S.J. 2018. Synergistic effects of anaerobic co-digestion of whey, manure and fish ensilage. Bioresource Technology 249: 35–41.

Zielewicz, E., Tytła, M., Liszczyk, G. 2013. Possibility of sewage sludge and acid whey co-digestion process. Architecture Civil Engineering Environment 5: 87–92.

CHAPTER 12

The evaluation of the synthesis of zeolites from sewage sludge ash

J. Latosińska
Kielce University of Technology, Kielce, Poland

ABSTRACT: The usage of ash for the production of zeolites is the activity consistent with the closed economy principles. In the literature, there is no research demonstrating the optimal method of the sewage sludge ash conversion into a material with useful properties. This article shows the evaluation of the influence of the conversion method and its parameters on the formation of zeolites from the obtained material. After the conversion with the hydrothermal method, (hydroxy)sodalite and zeolite X were formed in some of the samples. The introduction of fusion resulted in the formation of zeolite Y, zeolite P and (hydroxy)cancrinite apart from the (hydroxy)sodalite and zeolite X. Zeolites were found in more samples after the conversion with the fusion method than with the hydrothermal method. It was determined that mainly the prolongation of the crystallization time has the impact on the formation of zeolites.

12.1 INTRODUCTION

Among the methods of the utilization of municipal sewage sludge, which is a by-product of wastewater treatment plants, there has been an increase in the usage of the thermal methods (Eurostat 2020). The dominant technology of the incineration of sewage sludge is a fluidized bed, which is considered Best Available Technology. The occurring ashes resulting from the incineration of sewage sludge, in comparison to the energy wastes of coal, are characterized by the composition similar to slag with simultaneously higher loss on ignition and the concentration of phosphorus (Zabielska–Adamska 2015). There are various known methods of utilizing the ashes obtained from the incineration of municipal sewage sludge and coals. Some of the methods of such waste management are similar, for instance the utilization in road construction (Filipiak 2011; Lind et al. 2008). Sewage sludge ashes could be the source of zeolites, which from the 1980s had been obtained from coal fly ashes (Querol et al. 2002).

Zeolites form a developed inner structure, which makes them very good sorbents and ion exchange materials. The properties of zeolites allowed for their broad usage in chemical industry, agriculture and environmental engineering (Franus 2012; Zhang et al. 2019). Synthetic zeolites are most frequently obtained as a result of the synthesis with the hydrothermal or fusion method (Deng et al. 2016).

The synthesis of zeolites depends on such parameters as the temperature, the reaction time, the alkalinity of a solution (Molina & Poole 2004). The use of different parameters of the conversion process of ash into a zeolite, even within one synthesis method, allows for obtaining different forms of crystalline zeolites from the same raw material (Pimraksa et al. 2010).

The literature proves the possibility of modifying the properties of municipal sewage sludge ash by using the conversion with the hydrothermal and fusion methods (Latosińska 2019; Latosińska et al. 2019; Je–Seung et al. 2008; Lee et al. 2007; Zang et al. 2018). So far, the authors have not demonstrated which method of conversion and what process parameters modify the properties of sewage sludge ash favorably, in the aspect of its usage in the environmental engineering.

The aim of the conducted research was to determine the conditions of the chemical conversion and the temperature of the thermal treatment of sewage ash, for which the obtained material would contain zeolitic structures.

DOI 10.1201/9781003171669-12

12.2 EXPERIMENT

12.2.1 *Materials and methods*

The tested material was the sewage sludge ash obtained as a result of the thermal treatment of sludge under laboratory conditions. The sewage sludge was taken from a wastewater treatment plant located in the central part of Poland.

In order to evaluate the influence of temperature of the thermal treatment of sludge on the properties of ash, which is important in the aspect of the formation of zeolites, the research included the ash samples obtained at three different temperatures: 600°C (S–600), 790°C (S–790) and 980°C (S–980). The differentiation of the incineration temperatures was applied taking into consideration the fact that under the industrial conditions, the temperature declared by the operator of the installation can fluctuate in a certain interval depending on the characteristics of sewage sludge and the furnace.

The analysis of the oxide composition of sewage ash was performed with XRF. The analysis of the phase composition of samples after the process of zeolitization was conducted with the X-ray diffraction method.

The modification of the properties of ash after the thermal treatment of sludge was performed with the hydrothermal and fusion methods with the use of NaOH (Figure 12.1).

The research involved conducting 96 independent experiments. The variable parameters of the conversion process were the ratio of the mass of sewage sludge ash to the mass of sodium hydroxide (SSA:NaOH), the temperature of activation (t_a), the temperature of crystallization (t_c) and the time of crystallization (T_c; Tables 12.1–12.2).

In order to conduct the statistical analysis, the phase composition of samples was transcoded to a binary system. Every phase element was attributed weight within the range from 0 to 1. Assuming that the desired phases are the zeolite structures, they were attributed higher weights. The values of weights for the structures that did not belong to zeolites were not differentiated because they were found undesirable. The sum of arithmetical products of weights of the phase composition of samples was described as the zeolitization effect. Due to the application of a two–level type plan, three pairs

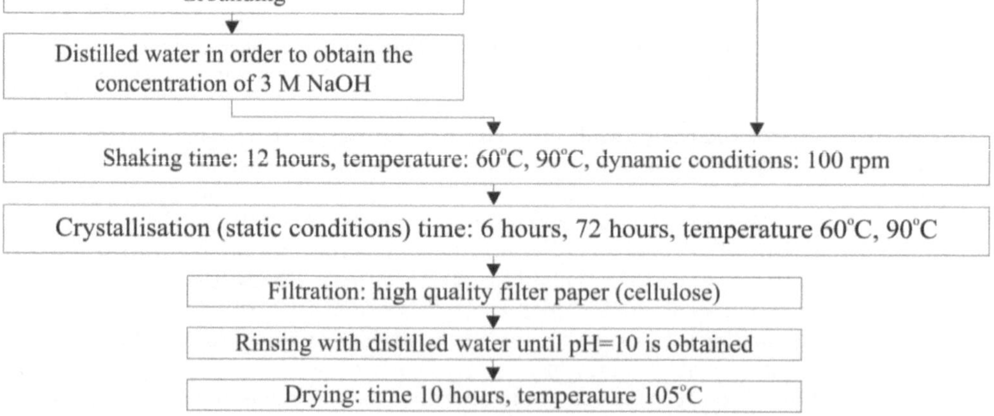

Figure 12.1. The conversion procedures of the sewage sludge ash: S-600, S-790, S-980.

Table 12.1. The parameters of the chemical conversion of SSA–hydrothermal method.

Samples			t_a, °C	T_c, hours	t_c, °C	SSA: NaOH
S–600	S–790	S–980				
S1	S17	S33	60	6	60	1.0:1.4
S2	S18	S34	60	6	90	1.0:1.4
S3	S19	S35	90	6	60	1.0:1.4
S4	S20	S36	90	6	90	1.0:1.4
S5	S21	S37	60	6	60	1.0:1.8
S6	S22	S38	60	6	90	1.0:1.8
S7	S23	S39	90	6	60	1.0:1.8
S8	S24	S40	90	6	90	1.0:1.8
S9	S25	S41	60	72	60	1.0:1.4
S10	S26	S42	60	72	90	1.0:1.4
S11	S27	S43	90	72	60	1.0:1.4
S12	S28	S44	90	72	90	1.0:1.4
S13	S29	S45	60	72	60	1.0:1.8
S14	S30	S46	60	72	90	1.0:1.8
S15	S31	S47	90	72	60	1.0:1.8
S16	S32	S48	90	72	90	1.0:1.8

Table 12.2. The parameters of the chemical conversion of SSA–fusion method.

Samples			t_a, °C	T_c, hours	t_c, °C	SSA: NaOH
S–600	S–790	S–980				
S49	S65	S81	60	6	60	1.0:1.4
S50	S66	S82	60	6	90	1.0:1.4
S51	S67	S83	90	6	60	1.0:1.4
S52	S68	S84	90	6	90	1.0:1.4
S53	S69	S85	60	6	60	1.0:1.8
S54	S70	S86	60	6	90	1.0:1.8
S55	S71	S87	90	6	60	1.0:1.8
S56	S72	S88	90	6	90	1.0:1.8
S57	S73	S89	60	72	60	1.0:1.4
S58	S74	S90	60	72	90	1.0:1.4
S59	S75	S91	90	72	60	1.0:1.4
S60	S76	S92	90	72	90	1.0:1.4
S61	S77	S93	60	72	60	1.0:1.8
S62	S78	S94	60	72	90	1.0:1.8
S63	S79	S95	90	72	60	1.0:1.8
S64	S80	S96	90	72	90	1.0:1.8

of temperature levels were used in the thermal treatment of sewage sludge: 600°C and 790°C, 790°C and 980°C, 600°C and 980°C.

The coefficients of the regression function were determined with the least squares method, while the selection of elements constituting the final form of the model was conducted with the step selection method. The models of the regression function contained linear elements and the interactions of all possible factor pairs. The significance of the coefficients of the regression function was verified with the F test. The elimination of the statistically insignificant elements of the regression function was performed with the assumption of the limit value for the level of significance at $p = 0.1$. The statistical calculations were performed in the SAS 9.3 program.

The corrected coefficient of determination $\check{R}^2$ including the correction resulting from the different number of elements in the model was introduced in order to evaluate the compliance of the estimated regression model with the experimental data.

The corrected coefficient of determination $\check{R}^2$ applies the values from the range between 0–1. One can speak of the adjustment of the model to the data:

- weak, when $0.0 < \check{R}^2 \leq 0.3$;
- satisfactory, when $0.3 < \check{R}^2 \leq 0.6$;
- good, when $0.6 < \check{R}^2 \leq 0.8$;
- very good, when $0.8 < \check{R}^2 \leq 1.0$.

12.3 RESULTS AND DISCUSSION

The characteristics of the sewage sludge ash are presented in Tables 12.3–12.4. The properties of ash obtained in the laboratory as a result of the thermal treatment of sludge did not differ from the data presented in the literature (Adam et al. 2007; Stempkowska et al. 2015).

12.3.1 *The conversion of ash from sludge with the hydrothermal method*

The presence of monomineral zeolites was not found in any of the samples from the group S1–S48. The results confirmed the presence of ash elements that did not undergo the conversion. The tests proved the formation of (hydroxy)sodalite and zeolite X. The greatest number of samples containing

Table 12.3. Oxide composition of sewage sludge ash, % mas. and ratio of SiO_2:Al_2O_3 (Latosińska 2019).

Oxide composition	S–600	S–790	S–980
SiO_2	20.8±0.5	25.5±0.5	26.1±0.5
Al_2O_3	5.28±0.13	6.68±0.13	6.98±0.13
P_2O_6	19.4±0.5	24.2±0.5	25.3±0.5
CaO	14.3±0.4	17.9±0.4	18.5±0.4
Fe_2O_3	9.45±0.3	11.8±0.3	12.4±0.3
MgO	3.72±0.09	4.29±0.09	4.56±0.09
K_2O	1.51±0.04	1.89±0.04	1.98±0.04
SO_3	1.45±0.04	2.14±0.06	0.573±0.017
TiO_2	0.725±0.025	0.942±0.025	0.975±0.025
Na_2O	0.407±0.023	0.575±0.023	0.602±0.023
SiO_2:Al_2O_3	3.9	3.8	3.7

Table 12.4. Phase quantitative analysis of sewage sludge ash and the contribution of the amorphous substance, % mas. (Latosińska 2019).

Phase	S–600	S–790	S–980
Amorphous substance	76.71±0.62	27.4±1.2	13.61±0.92
Quartz, SiO_2	9.12±0.13	15.27±0.21	15.44±0.17
Calcite, $CaCO_3$	4.06±0.1	1.36±0.13	nd.
Dolomite, $CaMg(CO_3)_2$	3.28±0.11	nd.	nd.
Potassium feldspar, $KAlSi_3O_8$	1.72±0.29	3.25±0.37	4.96±0.34
Muscovite, $KAl_2(Si_3Al)O_{10}(OH,F)_2$	4.42±0.46	5.91±0.58	nd.
Plagioclase, $(Ca,Na)(Si,Al)_4O_8$	0.472±0.092	nd.	nd.
Whitlockite, $Ca_9(Ma,Fe)(PO_4)_6(PO_3OH)$	nd.	28.36±0.50	34.19±0.34
Hematite, Fe_2O_3	nd.	nd.	10.76±0.13
Stanfieldite, $Ca_4(Mg,Fe^{++},Mn)_5(PO_4)_6$	nd.	13.09±0.44	13.05±0.32
Tridymite, SiO_2	nd.	2.88±0.26	7.99±0.33

nd.- not detected

Table 12.5. Crystalline phases obtained as a result of the conversion of sewage sludge after the thermal treatment at 600°C, hydrothermal method.

Sample	Apatite	(Hydroxy) sodalite/ Sodalite	Quartz	Plagioclase	Potassium feldspar	Calcite	Dolomite	Muskovit	Hematite	Magnetite
S1	+	-	+	+	+	+	+	+	-	-
S2	+	+	+	-	+	+	+	-	-	-
S3	+	+	+	-	+	-	+	-	-	-
S4	+	+	+	-	+	-	+	-	-	-
S5	+	-	+	+	+	-	+	+	-	-
S6	+	+	+	-	+	-	+	+	-	-
S7	+	+	+	-	+	+	+	-	-	-
S8	+	+	+	+	+	-	+	-	+	-
S9	+	+	+	+	+	-	+	+	-	-
S10	+	+	+	+	+	-	+	+	-	-
S11	+	+	+	-	-	-	+	-	-	-
S12	+	+	+	+	+	-	+	-	-	+
S13	+	-	+	+	+	+	+	+	-	-
S14	+	+	+	+	+	-	-	+	-	-
S15	+	+	+	-	-	-	+	+	-	-
S16	+	+	+	+	+	-	-	+	-	-

Table 12.6. Crystalline phases obtained as a result of the conversion of sewage sludge after the thermal treatment at 790°C, hydrothermal method.

Sample	Apatite	(Hydroxy) sodalite/ Sodalite	Quartz	Plagioclase	Potassium feldspar	Calcite	Hematite	Muskovite	Whitlockite
S17	+	-	+	+	+	+	+	+	+
S18	+	-	+	-	+	+	+	+	+
S19	+	+	+	-	+	+	+	+	+
S20	+	+	+	-	+	+	+	-	+
S21	+	-	+	-	+	+	-	+	+
S22	+	-	+	-	+	+	+	+	+
S23	+	+	+	-	+	+	+	+	+
S24	+	-	+	+	+	+	+	+	+
S25	+	+	+	+	+	+	+	+	+
S26	+	+	+	-	+	-	+	+	-
S27	+	+	+	+	+	-	+	+	+
S28	+	+	+	-	+	+	+	+	-
S29	+	-	+	+	+	-	+	-	+
S30	+	+	+	+	+	+	+	+	-
S31	+	+	+	+	+	+	+	+	+
S32	+	+	+	+	+	+	+	+	-

(hydroxy)sodalite was found after the zeolitization of S–600 ash, subsequently for S–980 ash, and the lowest for S–790 ash (Tables 12.5–12.7). The chemical conversion of S–980 ash resulted in the formation of (hydroxy)sodalite structure and also zeolite X.

Table 12.7. Crystalline phases obtained as a result of the conversion of sewage sludge after the thermal treatment at 980°C, hydrothermal method.

Sample	Apatite	(Hydroxy) sodalite/ Sodalite	Zeolite X	Quartz	Plagioclase	Potassium feldspar	Hematite	Whitlockite
S33	+	+	+	+	-	+	+	+
S34	+	+	+	+	-	+	+	+
S35	+	+	-	+	-	+	+	+
S36	+	+	-	+	-	+	+	+
S37	+	-	+	+	-	+	+	+
S38	+	-	+	+	-	+	+	+
S39	+	+	+	+	-	+	+	+
S40	+	+	-	+	-	+	+	+
S41	+	-	+	+	-	+	+	+
S42	+	-	-	+	-	+	+	+
S43	+	+	-	+	-	+	+	+
S44	+	+	-	+	+	+	+	+
S45	+	-	+	+	-	-	+	+
S46	+	+	-	+	-	-	+	+
S47	+	+	-	+	-	-	+	+
S48	+	+	-	+	-	+	+	+

Table 12.8. Crystalline phases obtained as a result of the conversion of sewage sludge after the thermal treatment at 600°C, fusion method.

Sample	Apatite	(Hydroxy) sodalite/ Sodalite	Zeolite Y	Zeolite X	Zeolite P	Magnetite	Quartz
S49	+	-	-	-	-	+	+
S50	+	-	-	-	-	+	+
S51	+	-	-	-	-	+	+
S52	+	[illegible]	-	-	-	+	+
S53	+	-	-	-	-	+	+
S54	+	-	+	-	-	-	+
S55	+	+	-	-	-	+	+
S56	+	+	-	-	-	+	+
S57	+	-	-	-	-	+	+
S58	+	-	-	-	+	+	+
S59	+	-	-	-	-	+	+
S60	+	-	-	-	+	+	+
S61	+	-	-	+	-	+	+
S62	+	+	-	+	-	+	-
S63	+	-	-	-	+	+	+
S64	+	+	-	-	-	+	+
S65	+	-	-	+	+	+	-

12.3.2 *The conversion of ash from sludge with the fusion method*

Regardless of the process parameters and the temperature of the thermal treatment of sludge, monomineral zeolites were not obtained (Table 12.8–12.10). (Hydroxy)sodalite, zeolite X, zeolite Y, zeolite P and hydroxycancrinite were formed in some of the samples. Ash elements, which did not undergo the conversion, were also present in the samples.

Table 12.9. Crystalline phases obtained as a result of the conversion of sewage sludge after the thermal treatment at 790°C, fusion method.

Sample	Apatite	(Hydroxy) sodalite/ Sodalite	Zeolite X	Magnetite	Quartz
S66	+	+	-	-	-
S67	+	+	-	-	-
S68	+	+	-	-	+
S69	+	+	-	-	+
S70	+	+	-	-	-
S71	+	+	-	-	-
S72	+	+	-	-	-
S73	+	+	-	-	-
S74	+	-	+	-	-
S75	+	+	-	-	+
S76	+	+	-	-	+
S77	+	+	-	-	+
S78	+	+	+	-	-
S79	+	+	-	-	+
S80	+	+	-	-	-
S81	+	+	-	+	-

Table 12.10. Crystalline phases obtained as a result of the conversion of sewage sludge after the thermal treatment at 980°C, fusion method.

Sample	Apatite	(Hydroxy) sodalite/ Sodalite	Zeolite X	Hydroxycancrinite
S82	+	-	-	-
S83	+	+	-	-
S84	+	+	-	-
S85	+	+	-	-
S86	+	+	-	-
S87	+	+	-	-
S88	+	+	-	+
S89	+	+	-	-
S90	+	+	+	-
S91	+	+	-	-
S92	+	+	-	-
S93	+	+	-	-
S94	+	+	+	-
S95	+	+	-	-
S96	+	+	-	-

12.3.3 *The influence of the temperature of the thermal treatment of sludge*

The influence of the temperature of the thermal treatment (t_{tt}) of sludge on the properties of ash after the conversion was presented together with the influence of the conversion method. It was the consequence of the applied plan of the experiment.

12.3.4 *The influence of the conversion method – the hydrothermal method*

For the thermal treatment of sewage sludge at temperatures of 600°C and 790°C, the temperature of activation was a statistically significant parameter. Moreover the ratio of ash:NaOH, together with

Table 12.11. Models describing the effect of zeolitization, hydrothermal method.

Independent variable	Parameter evaluation	Standard error	Statistical value F	Pr.>F (value p)
Temperature of the thermal treatment of sludge: 600°C, 790°C				
Intercept	0.61872	0.14724	17.66	0.0002
SSA:NaOH* t_{tt}	−0.00001211	0.00000275	19.33	0.0001
t_a	0.00327	0.00114	8.31	0.0075
t_c * T_c	0.00001215	0.00000658	3.41	0.0752
Coefficient of determination $R^2 = 0.526$; Corrected coefficient of determination $\check{R}^2 = 0.475$				
Temperature of the thermal treatment of sludge: 790°C, 980°C				
Intercept	−0.57848	0.14886	15.10	0.0005
t_{tt}	0.00113	0.00016724	45.48	<0.0001
Coefficient of determination $R^2 = 0.603$; Corrected coefficient of determination $\check{R}^2 = 0.589$				
Temperature of the thermal treatment of sludge: 600°C, 980°C				
Intercept	0.36221	0.05971	36.80	<0.0001
t_a * t_{tt}	0.00000217	9.607516E-7	5.12	0.0310
Coefficient of determination $R^2 = 0.146$; Corrected coefficient of determination $\check{R}^2 = 0.117$				

SSA:NaOH – sewage sludge ash: NaOH ratio, t_a – the temperature of activation; t_c – the temperature of crystallization; T_c – the time of crystallization; t_{tt} – temperature of the thermal treatment.

Table 12.12. Models describing the effect of zeolitization, fusion method.

Independent variable	Parameter evaluation	Standard error	Statistical value F	Pr.>F (value p)
Temperature of the thermal treatment of sludge: 600°C, 790°C				
Intercept	−0.32916	0.09671	11.58	0.0019
SSA:NaOH* t_{tt}	0.00001603	0.00000225	50.96	<0.0001
Coefficient of determination $R^2 = 0.629$; Corrected coefficient of determination $\check{R}^2 = 0.617$				
Temperature of the thermal treatment of sludge: 790°C, 980°C				
Intercept	0.24006	0.09486	6.40	0.0171
t_c * t_{tt}	−0.00000167	8.086588E-7	4.26	0.0482
SSA:NaOH* ttt	0.00000622	0.00000191	10.56	0.0029
Coefficient of determination $R^2 = 0.277$; Corrected coefficient of determination $\check{R}^2 = 0.227$				
Temperature of the thermal treatment of sludge: 600°C, 980°C				
Intercept	−0.13578	0.06401	4.50	0.0426
SSA:NaOH* t_{tt}	0.00000987	0.00000124	63.43	<0.0001
t_c* T_c	0.00001162	0.00000569	4.17	0.0503
Coefficient of determination R^2=0.6998; Corrected coefficient of determination $\check{R}^2 = 0.679$				

SSA:NaOH – sewage sludge ash: NaOH ratio, t_c – the temperature of crystallization; T_c – the time of crystallization; t_{tt} – temperature of the thermal treatment.

the temperature of the thermal treatment, determined the formation of zeolitic forms, whereas such influence was negative. Therefore the above-mentioned factors and their interactions did not favor the synthesis of zeolites, in contrast to the changes in the temperature of crystallization coinciding with the changes of the time of crystallization (Table 12.11). For the samples after the treatment at the temperatures of 790°C and 980°C the temperature of the thermal treatment of sludge was the only statistically significant factor. Poor adjustment of the model describing the interaction of the temperature of activation and the temperature of the thermal treatment of sludge was obtained for the samples after the thermal treatment at the temperature of 600°C and 980°C. It can be considered favorable because the values of the factors are not rigorously limited and at the same time easier to apply. The obtained models confirmed the literature data stating that the chemical conversion is a complex process.

12.3.5 *The influence of the conversion method – the fusion method*

In every analyzed set of the thermal treatment temperatures, the interaction of the SSA:NaOH ratio has a favorable influence on the formation of zeolitic structures. For the set of samples with a higher range of temperatures of the thermal treatment, the interaction between the temperature of crystallization and the temperature of the thermal treatment has a negative influence. The model for the data of a higher pace of the change of the temperature of the thermal treatment includes the interaction between the temperature of crystallization and the time of crystallization. The interaction between these two factors stimulates the formation of zeolitic structures. However, the change of their values results in a very little change of the zeolitization effect (Table 12.12).

12.4 CONCLUSIONS

The conducted research and the statistical analyses enabled to draw the following conclusions:

- no monomineral zeolites were obtained as a result of the experiment,
- the temperature of the thermal treatment of sludge is a factor with an insignificant influence on the phase composition of ash after the conversion, both with the hydrothermal and fusion methods,
- the formation of zeolitic structures are determined mainly by the crystallization time,
- the fusion method is more favorable for the formation of zeolitic structures than the hydrothermal method.

ACKNOWLEDGEMENTS

The Programme of the Polish Ministry of Science and Higher Education – the Regional Initiative of Excellence, financed by the Polish Ministry of Science and Higher Education on the basis of the contract no 025/RID/2018/19 of 28 December 2018, the amount of funding: 12 million PLN.

"No competing financial interests exist"

REFERENCES

Adam, C., Kley, G. & Simon, F.G. 2007. Thermal treatment of municipal sewage sludge aiming at marketable P-fertilisers. *Materials Transactions* 12: 3056–3061.

Deng, L., Xu, Q. & Wu, H. 2016. Synthesis of zeolite-like material by hydrothermal and fusion methods using municipal solid waste fly ash. *Procedia Environmental Sciences* 31: 662–667.

Eurostat. Available at www.ec.europa.eu/eurostat

Filipiak, J. 2011. Fly ash in construction industry. Strength tests of soil stabilized with mixture of ash and cement. *Annual Set The Environment Protection* 13(1): 1043–1054 (in Polish).

Franus, W. 2012. *The usage of zeolites obtained from fly ash for the removal of contamination from water and sewage.* Lublin. Polish Academy of Science (in Polish).

Je-Seung, L., Sook-Nye, C. & Chul-Hwi, P. 2008. Synthesis of Zeolite P1 and Analcime from Sewage Sludge Incinerator Fly Ash. *Journal of Korean Society of Environmental Engineers* 30(6): 659–665.

Latosińska, J, Muszyńska, J. & Gawdzik, J. 2019. Remediation of Landfill Leachates with the Use of Modified Ashes from Municipal Sewage Sludge, In A. Krakowiak–Bal & M. Vaverkova (eds.) *Infrastructure and Environment.* 136–143. Springer.

Latosińska, J. 2016. Zeolitization of sewage sludge ash with a fusion method. *Journal of Ecological Engineering* 17(5): 138–146.

Latosińska, J (ed.) 2019. *Modification of municipal sewage sludge ash by zeolitization. Selected issues.* Kielce: Kielce University of Technology Publishing House (in Polish).

Lee, J.S., Eom, S.W. & Choi, H.Y. 2007. Synthesis of zeolite from sewage sludge incinerator fly ash by hydrothermal reaction in open system. *Korean Journal of Environmental Health Science* 33(4): 317–324.

Lind, B.B., Norrman, J., Larsson, L.B., Ohlsson, S.-A. & Bristav, H. 2008. Geochemicals anomalies from bottom ash in a road construction – Comparison of the leaching potential between an ash road and the surroundings. *Waste Management* 28(1): 170–180.

Molina, A. & Poole, C. 2004. A comparative study using two methods to produce zeolites from fly ash. *Minerals Engineering* 17(2): 167–173.

Pimraksa, K., Chindaprasirt, P. & Setthaya, N. 2010. Synthesis of zeolite phases from combustion by–products. *Waste Management and Research* 28: 1122–1132.

Querol, X., Moreno, N., Umaña, J.C., Alastuey, A., Hernăndez, E., López-Soler, A. & Plana, F. 2002. Synthesis zeolites from coal fly ash: an overview. *International Journal of Coal Geology* 50(1–4): 413–423.

Stempkowska, A., Kępys, W. & Pietrzyk, J. 2015. The influence of incinerated sewage sludge ashes physical and chemical properties in possibility of usage in red ceramic. *Mineral Resources Management* 31(2): 109–121.

Zabielska-Adamska, K. 2015. Combustion product of municipal sewage sludge as anthropogenic soil. *Annual Set The Environment Protection* 17(2): 1286–1305 (in Polish).

Zhang, L., Zhang, S. & Li, R. 2019. Synthesis of hierarchically porous Na-P zeotype composites for ammonium removal. *Environmental Engineering Science* 36(9): 1089–1099.

Zhang, Y., Leng, Z., Zou, F., Wang, L., Chen, S.S. & Tsang, D.C.W. 2018. Synthesis of zeolite A using sewage sludge ash for application in warm mix asphalt. *Journal of Cleaner Production* 172: 686–695.

CHAPTER 13

Assessment of the possibility of using sludge and ash mixtures for nature/reclamation purposes based on the PHYTOTOXKIT test

J. Kiper & A. Głowacka
Department of Sanitary Engineering, West Pomeranian University of Technology in Szczecin, Szczecin, Poland

ABSTRACT: The implementation of innovative technologies for the processing of sewage sludge, improves the energy balance of the entire treatment plant and creates the possibility of using sludge as a valuable raw material in other industries.The aim of the study was to assess the toxicity of substances contained in municipal and industrial sewage sludge. The phytotoxicity degree is particularly important in the case of the waste utilized for natural purposes. The subsoils for growing the test plants were prepared on the basis of sewage sludge and by-products of coal and biomass combustion. The assessment of the germination capacity as well as the growth of plant roots and stems was carried out. The study was conducted on Hordeum vulgare L., Sinapis alba L., Lepidium sativum, using typical lab equipment as well as PHYTOTOXKIT test. The study allowed determining the usefulness of sludges as well as ash and sludge blends for the natural purposes.

13.1 INTRODUCTION

At the end of 2015, the European Commission addressed a communication: Closing the loop - An EU action plan for the Circular Economy, to the European Parliament, the Council, the European Economic and Social Committee and the Committee of the Regions. The biological waste is used in many branches of the industry as a valuable raw material with renewable, biodegradable and compostable properties (Union 2014).

An integral part of the closed-loop economy is the municipal sewage sludge generated in wastewater treatment plants. The implementation of innovative solutions for sewage sludge processing will improve the energy balance of the entire plant and create the opportunities for using the sludge as a valuable raw material in other industries.

The main directions of municipal sewage sludge management include thermal transformation and natural use, which includes the application of sludge in agriculture, land reclamation and cultivation of plants intended for production of compost. Despite the significant increase in the share of thermal sludge processing (from 3% in 2003 to 41% in 2018), most sewage sludge is still being managed in nature (Figure 13.1) (Eurostat 2019).

Traditionally, the sludge intended for thermal disposal is analysed only in terms of the fuel parameters and hydration. However, the natural use of sewage sludge is required to meet the standards set out in the Regulation of the Minister of the Environment of 6 February 2015 on municipal sewage sludge. This Regulation also defines the permissible amounts of heavy metals present in the top layers of land on which the municipal sewage sludge is to be applied. The Regulation of the Minister of Agriculture and Rural Development of 18 June 2008 on the implementation of certain provisions of the Act on fertilizers and fertilization provides the information on the permissible contents of pollutants and the required content of biogenic substances in fertilizers or plant cultivation aids. Both documents regulate the presence of heavy metals, which include: cadmium (Cd), copper (Cu), nickel (Ni), lead (Pb), zinc (Zn), mercury (Hg), chromium (Cr), live eggs of intestinal parasites such as Ascaris sp., Trichuris sp., Toxocara sp., as well as Salmonella bacteria.

The limitations in the use of sewage sludge for natural purposes also consider the presence of toxic compounds that have a direct impact on the plant growth. The study was conducted on higher

DOI 10.1201/9781003171669-13

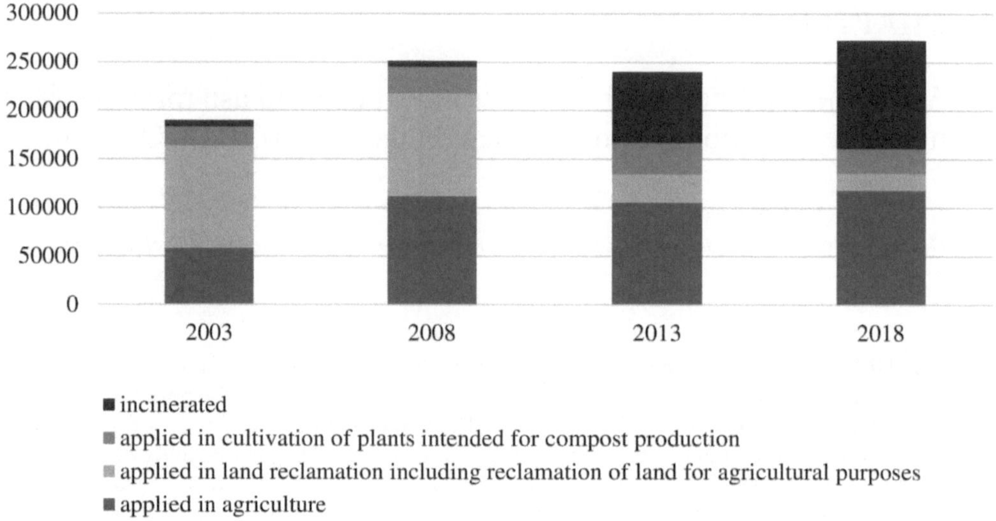

Figure 13.1. The amount and methods of municipal sewage sludge management in Poland (Eurostat 2019).

plants. The determination of the influence of toxicity of the sludge and subsoils prepared on its basis was performed. The tests were conducted in three series in November 2016, April and September 2019.

The use of phytotests allows to easily determine the degree of toxicity of the tested substrate to plant growth, and the result can be obtained after just 3 days. The experiment enables direct observation and measurement of the length of the roots and stems. The preparation of the experiment does not require the use of complicated devices and materials. Determination of the phytotoxicity of the substrate, along with the analysis of the physicochemical composition, allows for a complete assessment of materials intended for natural use, and helps in choosing the way of their management.

Using fast germination seeds can extend the typical range of tests performed for the sludges used for nature.

13.2 MATERIAL AND METHOD

For the purpose of the study, the samples of sludge-ash blends were prepared. The specimens were created based on the municipal and industrial sewage sludges as well as by-products of coal and biomass combustion (Table 13.1).

Sludge-ash mixtures were prepared by mixing organic materials (sewage sludge) and inorganic materials (ashes and slags) in a ratio of 1: 1 in terms of dry mass.

Ashes and slags are characterized by a high content of micro- and macroelements, improving the nutritional status of plants. The management of this waste together with sewage sludge allows, among others to get rid of problems related to combustion waste landfills and the management of sewage sludge. Additionally, the alkaline reaction of the ashes supports the stabilization of sewage sludge (Antonkiewicz 2004).

The first series of tests was performed on the sludge taken from "Pomorzany" Municipal Wastewater Treatment Plant in Szczecin. The second and third series of tests utilized the same sludge with the addition of the sludge generated in the Bosman brewery and by-products of municipal sludge combustion.

The screening of the prepared sludge-ash mixes was performed using PHYTOTOXKITTM. The setup is used to determine the influence of toxic compounds on the growth of plants in different stages. The toxins can either be detected after being absorbed by leaves, root system or regardless

Table 13.1. Components of the sludge-ash blends and their designation.

	organic components		mineral components				
series	municipal sewage sludge	industrial sewage sludge	straw ash	wood ash	fly ash	ash-slag	SSA
I series	P1	–	S	W	F1	A1	SSA1
II series	P2	B2	–	–	F2	A2	–
III series	P3	B3	–	–	F3	A3	SSA3

Table 13.2. Characteristic of materials used in the first series of tests in 2016 (Głowacka 2017; Kiper 2017; Kiper 2018; Tarnowski 2017).

		P1	S	W	F1	A1	SSA1
pH	–	7.54	8.19	12.80	10.99	7.78	8.03
Conductivity	mS/cm	1.275	174.66	32.25	2.78	0.31	2.84
Dry mass	% d.m.	20.34	99.34	73.13	99.88	86.65	93.50
Humidity	% d.m.	79.66	0.66	26.88	0.12	13.35	6.50
Organic matter	% d.m.	67.14	2.00	4.22	3.16	9.52	1.27
Mineral substance	% d.m.	32.86	98.00	95.78	96.84	90.48	98.73
Cd	mg/kg d.m.	1.25	8.64	7.25	0.0	0	0.30
Co		5.89	0.65	4.00	13.1	7.14	13.97
Cu		271.53	29.91	98.46	28.16	22.93	1591.56
Mn		115.50	60.23	9303.43	359.3	323.5	367.96
Ni		18.67	8.24	17.15	52.12	22.23	66.76
Pb		18.62	10.99	8.60	26.56	7.19	17.96
Zn		1548.21	281.60	2843.64	72.17	29.62	4209.92
Fe	g/kg d.m.	60.20	0.76	9.77	17.83	35.03	179.88
K	mg/kg d.m.	0.78	189.57	143.26	1.83	0.67	132.8
Na		30.59	272.59	277.44	0.78	0.31	0.87
Mg		0.13	0.04	60.45	33.66	0.16	47.95
P	% d.m.	4.81	1.40	1.87	0.29	1.5	10.12
N		7.65	–	–	–	–	–

of the absorption method. The substances absorbed by the root system will travel through xylem along with water to the upper parts of the plants. The compounds absorbed by leaves are transported by phloem to other parts of the plant, along with the products of photosynthesis. The toxicity screening is particularly important in the case of substances used in agriculture (Kvesitadze 2006).

13.2.1 *The first series*

The first series of tests was carried out using typical laboratory equipment. A similar methodology was used by Czop et al. (Czop 2016). The general characteristics of sludge and combustion by-products used in the first series of tests are presented in Table 13.2.

The assessment of the germination and growth of plant roots and stems was carried out after 7 days of growth on the prepared subsoil mixes. The tests were conducted on seeds of Hordeum vulgare L. (spring barley), Sinapis alba L. (white mustard) and Lepidium sativum (pepper). The experiment was performed in two sets, by planting 25 of the prepared seeds on 40 grams of the sludge-ash mixes. The subsoils were placed on petri dishes with a diameter of 11.0 and 11.5 cm. The reference specimen was made by placing seeds in redistilled water.

Table 13.3. The classification system for toxicity.

Effect [%]	Toxicity class
<25	non-toxic
25–50	low toxicity
50–75	toxic
>75	high toxicity

The toxicity was determined based on a method provided by the Regulation of the Minister of Environment dated 13 May 2004 on conditions in which it is assumed that waste is not hazardous (Dz.U. 2004 no. 128, item 1347 – Journal of Laws), "Determination of cytotoxic activity in garden cress" (Ordinance of the Minister of the Environment of May 13, 2004)

13.2.2 *The second and third series*

The second and third series of tests were carried out using specialized diagnostic microbiotests Phytotoxkit^{TM}. The kit contains the necessary elements to perform a complete phytotoxicity test with one monoecious plant (Sorghum saccharatum) and two dioecious plants (Lepidium sativum, Sinapis alba). The determination is carried out using special transparent plates which enable to directly observe and perform measurment using image analysis at the end of the test (Phytotoxicity test 2019). For the comparison purposes, the experiment was carried out for 7 days, in three repetitions, without incubating the samples. The control sample was prepared using the soil attached to the test set.

Toxicity was expressed as a percentage of inhibition (or stimulation) of seed germination and root growth in relation to the control:

$$I = \frac{(L_c - L_t)*100}{L_c} \tag{13.1}$$

L_c – average seed germination and root length in the control. [cm]
L_t – average seed germination and root length in the test. [cm].

The results of the tests were compared by means of a 4-step classification system reflecting the strength of the effect observed in samples (Table 13.3) (Antonopoulou 2016; Dudziak 2016; Ricco 2004; Umar 2016).

13.3 RESULTS AND DISCUSSION

The recorded ambient temperature and humidity during the three series of tests is shown in Figure 13.2. The first series of tests was carried out at 38±3% relative humidity and 23.7±1.2°C.

The soils were regularly sprayed with redistilled water and covered with a film to maintain appropriate conditions for the plant growth. The second series of tests was conducted at the lowest RH=25.5±2.5% at 22.6±2.0°C. The lowest average air temperature was recorded during the third series of tests (19.4°C), with an average humidity of 46% – the highest in all series. Due to being enclosed within the plates, the samples in the last series did not require additional watering or covering. Due to the slight differences in temperature and degree of humidity, their effect on the growth of the test plants cannot be determined. During the following series of experiments, efforts were made to maintain constant environmental conditions.

The results of seed germination rate evaluation are presented in Table 13.5. The soils with the best average germination rate were observed in the first series: P1+A1 (73%), P1+SSA1 (81%) and in the third series: B3+SSA3 (90%). The results for the remaining samples ranged from 4% in the case of the cress growth on P1+W to 83% in the case of the sorghum growth on B3+SSA3 (Figure 13.3).

No plant growth was observed in nine of the 15 analysed samples. None of the samples in the second series exhibited germination. A possible reason for the lack of plant growth is the increased ammonia

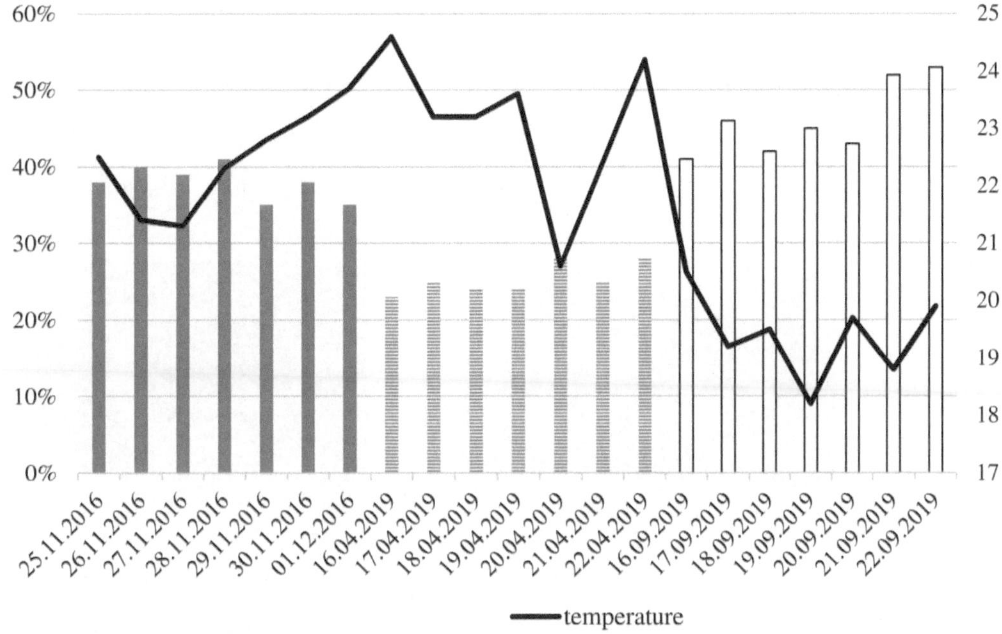

Figure 13.2. Measurements of air temperature and humidity during three series of test.

Table 13.5. Percentage of germination of test plants.

series	component	*Lepidium sativum*	*Sinapis alba L.*	*Hordeum vulgare L. / Sorghum saccharatum*	mean
I	control	92%	88%	52%	77%
	P1 + S	0%	0%	0%	0%
	P1 + W	4%	4%	6%	5%
	P1 + F1	20%	50%	24%	31%
	P1 + A1	94%	86%	38%	73%
	P1 + SSA1	94%	86%	62%	81%
II	control	93%	83%	97%	91%
	P2 + F2	0%	0%	0%	0%
	P2 + A2	0%	0%	0%	0%
	B2 + F2	0%	0%	0%	0%
	B2 + A2	0%	0%	0%	0%
III	conrol	83%	93%	97%	91%
	P3 + F3	0%	0%	0%	0%
	P3 + A3	0%	0%	0%	0%
	P3 + SSA3	0%	0%	0%	0%
	B3 + F3	0%	0%	0%	0%
	B3 + A3	7%	43%	83%	44%
	B3 + SSA3	100%	77%	93%	90%
mean		33%	34%	31%	32%
mean (without control)		21%	23%	20%	22%
SD		0,441	0,404	0,391	0,395

content in the analysed subsoils and the beginning of putrefaction. Apart from heavy metals, ammonia is one of the main toxic substances, as shown by the studies conducted by (El Fels 2014; Nóvoa-Muñoz 2008; Tiquia 2002). The lack of seed germination may also be associated with insufficient sludge stabilisation. The results of the analysis pertaining to the physicochemical composition of the collected samples (Table 13.2) indicate insufficient stabilisation of municipal sewage sludge.

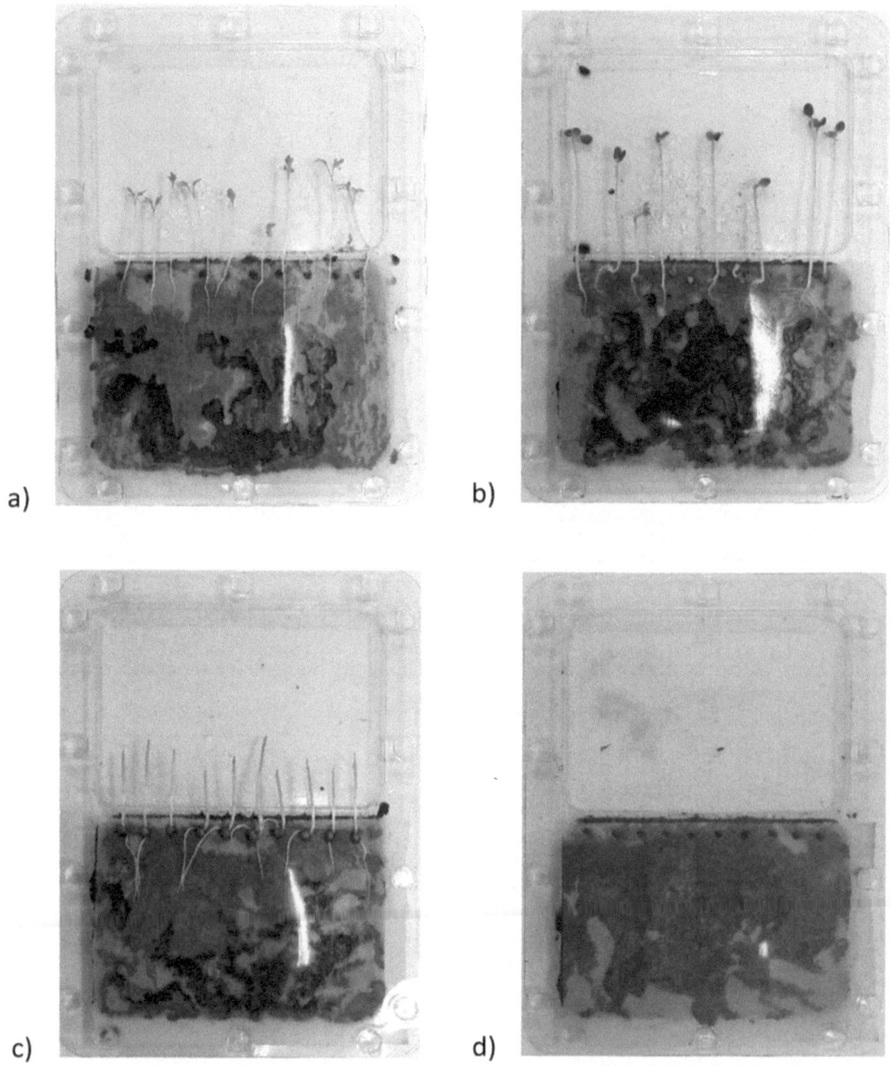

Figure 13.3. The growth of tested seeds on the B3+SSA3 subsoil in third series of tests (a-Lepidium sativum, b-Sinapis alba L., c-Sorghum saccharatum) and inhibition of germination of Lepidium sativum on P3+SSA3 subsoil (d).

The content of organic matter in the second and third series of tests was 70.08 and 77.06% d.m., respectively.

Corrêa Martins et al. (Martins 2016) carried out a study on the effect of sludge on the growth of common onion (A.Cepa). The researchers did not find any significant differences between the concentration of most heavy metals and the toxicity of the analysed materials. However, the results of the study suggest a large influence of the degree of sludge stabilisation, which may have a greater impact on the toxicity than the pollutants level.

Only a miniscule growth and lack of seedling emergence on the P1+S and P1+W soil blends is probably associated with a high sodium and potassium content. Precipitated salt crystals were observed on the specimen. The straw ash contained 189.57 mg K/d.m. and 272.59 mg Na/d.m., while coal ash contained 143.26 mg K/d.m. and 277.44 mg Na/d.m. Increased salt and potassium depositions

Table 13.6. Toxicity of test plant root growth [%].

		Lepidium sativum		*Sinapis alba L.*		*Hordeum vulgare L. / Sorghum saccharatum*	
series	component	inhibition	stimulation	inhibition	stimulation	inhibition	stimulation
	P1 + S	0,0	–	0.0	–	0.0	–
	P1 + W	91.4	–	94.6	—	96.0	–
I	P1 + F1	97.1	–	94.6	–	87.1	–
II	P1 + A1	95.7	–	98.2	–	95.7	–
	P1 + SSA1	95.8	–	93.9	–	66.8	–
	P2 + F2	100	–	100	–	100	–
III	P2 + A2	100	–	100	–	100	–
	B2 + F2	100	–	100	–	100	–
	B2 + A2	100	–	100	–	100	–
	P3 + F3	100	–	100	–	100	–
	P3 + A3	100	–	100	–	100	–
	P3 + SSA3	100	–	100	–	100	–
	B3 + F3	100	–	100	–	100	–
	B3 + A3	65.5	–	9.6	–	91.6	–
	B3 + SSA3	–	7.4	43.4	–	47.9	–

deteriorate the soil structure and limit the water uptake by plants (Lakhdar 2010; Siyal 2002; Widłak 2016).

Table 13.6 shows the results of toxicity screening in the subsoils from the first and third series of tests. The results were expressed as a percentage of growth inhibition or stimulation of the plants in relation to the control sample. Except for the third test series subsoil made from industrial sludge mixed with sewage sludge incineration waste (B3+SSA) and Lepidium sativum seeds (7.4%), all other soil blends exhibited a delayed plant growth. Using the previously mentioned plant toxicity classification system used by many researchers (Antonopoulou 2016; Dudziak 2016; Umar 2016), most subsoils showed high toxicity to the test plants. Lower toxicity was exhibited by:

P1+SSA – toxic effect on Hordeum vulgare L. (inhibition of 66.8%),

B3+A3 – toxic effect on Lepidium sativum (inhibition of 65.5%) and non-toxic to Sinapis alba L. (9.6%),

B3+SSA3 – toxic effect on Sinapis alba L (43.4%) and Sorghum saccharatum (47.9).

Other studies concur the observation (Fuentes 2006; Oleszczuk 2012), proving that the root growth inhibition is a more precise measure of phytotoxicity.

Satisfactory plant growth and low toxicity of subsoils prepared with industrial sludge (B) confirm the applicability of the food industry generated sludges. Those sludges tend to have lower content of heavy metals and much lower number of microorganisms, including the pathogenic ones (Nowak 2010).

The research conducted by Oleszczuk et al. (Oleszczuk 2012) indicated an increase in the phytotoxicity of subsoils with an addition of a sewage sludge. The research indicates how important it is to take other parameters influencing the utilization of sewage sludge into account.

13.4 CONCLUSIONS

The conducted tests proved the necessity of evaluating the sewage sludge (physicochemical, microbiological, phytotoxic properties) before its application for the natural purposes, which concurs with other studies (Adamcová 2016; Oleszczuk 2012). The obtained results do not allow to unequivocally state which of the plants is the most susceptible to the toxic effects of the sludge-ash soil blend. The PhytotoxkitTM microbiotest proved to be an effective tool in the phytotoxicity determination.

REFERENCES

Adamcová, D., Vaverková, M. D., & Břoušková, E. 2016. The toxicity of two types of sewage sludge from wastewater treatment plant for plants. *Journal of Ecological Engineering,* 17(2). DOI: 10.12911/22998993/62283

Antonkiewicz, J., Jasiewicz C., 2004. Effect of various sludge-ash and peat-ash mixtures on the yield and quality of a mixture of grasses with trefoil. *Bulletin of the Plant Breeding and Acclimatization Institute*, 234, 227–236. (in Polish).

Antonopoulou, M., Hela, D., & Konstantinou, I. 2016. Photocatalytic degradation kinetics, mechanism and ecotoxicity assessment of tramadol metabolites in aqueous TiO2 suspensions. *Science of the Total Environment*, 545, 476–485. DOI: 10.1016/j.scitotenv.2015.12.088

Czop, M., Żorawik, K., Grochowska, S., Kulkińska, L., & Januszewska, W. 2016. Tests of phytotoxicity of mining wastes on selected group of plants. *Archives of Journal of Waste Management and Environmental Protection*, 18(4). (in Polish).

Dudziak, M., & Kopańska, D. 2016. Phytotoxicological assessment of selected made grounds. *Proceedings of ECOpole*, 10(1), 121–127. DOI: 10.2429/proc.2016.10(1)014

El Fels, L., Zamama, M., El Asli, A., & Hafidi, M. 2014. Assessment of biotransformation of organic matter during co-composting of sewage sludge-lignocelullosic waste by chemical, FTIR analyses, and phytotoxicity tests. *International Biodeterioration & Biodegradation*, 87, 128–137. DOI: 10.1016/j.ibiod.2013.09.024

European Statistical Office Eurostat (2019) (https://ec.europa.eu/eurostat/ (19.03.2019)).

Fuentes, A., Llorens, M., Saez, J., Aguilar, M. I., Pérez-Marín, A. B., Ortuño, J. F., & Meseguer, V. F. 2006. Ecotoxicity, phytotoxicity and extractability of heavy metals from different stabilised sewage sludges. *Environmental pollution*, 143(2), 355–360. DOI: 10.1016/j.envpol.2005.11.035

Głowacka, A., Rucińska, T., & Kiper, J. 2017. The slag original from the process of sewage sludge incineration selected properties characteristic. In *E3S Web of Conferences* (Vol. 22). EDP Sciences. DOI: 10.1051/e3sconf/20172200054.

Kiper, J. 2017. The possibilities of natural development of ash-sludge blends. *Ecological Engineering*, 18(3), 74–82. DOI: 10.12912/23920629/70260 (in Polish).

Kiper, J., & Głowacka, A. 2018. Analysis and methods of waste management of sludges obtained from exhaust gas treatment plant. *Gas, Water and Sanitary Technology,* 20–24. DOI 10.15199/17.2018.1.5

Kvesitadze, G., Khatisashvili, G., Sadunishvili, T., & Ramsden, J. J. 2006. *Biochemical mechanisms of detoxification in higher plants: basis of phytoremediation*. Berlin: Springer Science & Business Media.

Lakhdar, A., Scelza, R., Scotti, R., Rao, M. A., Jedidi, N., Gianfreda, L., & Abdelly, C. 2010. The effect of compost and sewage sludge on soil biologic activities in salt affected soil. *Revista de la ciencia del suelo y nutrición vegetal*, 10(1), 40–47. DOI: 10.4067/S0718-27912010000100005

Martins, M. N. C., de Souza, V. V., & da Silva Souza, T. 2016. Cytotoxic, genotoxic and mutagenic effects of sewage sludge on Allium cepa. *Chemosphere*, 148, 481–486. DOI: 10.1016/j.chemosphere.2016.01.071

Nowak, M., Kacprzak, M., & Grobelak, A. 2010. Sewage sludge as a substitute of soil in the process of remediation and reclamation of sites contaminated with heavy metals. *Engineering and Protection of Environment*, 13, 121–131. DOI: 10.7862/rb.2014.4 (in Polish).

Nóvoa-Muñoz, J. C., Simal-Gándara, J., Fernández-Calviño, D., López-Periago, E., & Arias-Estévez, M. 2008. Changes in soil properties and in the growth of Lolium multiflorum in an acid soil amended with a solid waste from wineries. *Bioresource technology*, 99(15), 6771–6779. DOI: 10.1016/j.biortech.2008.01.035

Oleszczuk, P., Malara, A., Jośko, I., & Lesiuk, A. 2012. The phytotoxicity changes of sewage sludge-amended soils. *Water, Air, & Soil Pollution*, 223(8), 4937–4948. DOI: 10.1007/s11270-012-1248-8

Ordinance of the Minister of the Environment of May 13, 2004 on conditions in which waste is considered not to be hazardous, Journal of Laws, Dz.U. 2004 no. 128, item 1347 (in Polish).

Phytotoxicity test – phytotoxkit solid samples standard operating procedure https://www.microbiotests.com/wp-content/uploads/2019/07/phytotoxicity-test_phytotoxkit-solid-samples_standard-operating-procedure.pdf

Ricco, G., Tomei, M. C., Ramadori, R., & Laera, G. 2004. Toxicity assessment of common xenobiotic compounds on municipal activated sludge: comparison between respirometry and Microtox®. *Water Research*, 38(8), 2103–2110. DOI: 10.1016/j.watres.2004.01.020.

Siyal, A. A., Siyal, A. G., & Abro, Z. A. 2002. Salt affected soils their identification and reclamation. *Pakistan Journal of Applied Sciences*, 2(5), 537–540. DOI: 10.3923/jas.2002.537.540

Tarnowski, K., Bering, S., Kiper, J., Lendzion-Bielun, Z., Mazur, J., & Glowacka, A. 2017. Studies of phytoremediation of heavy metals-containing sewage sludge-ash mixtures with spring barley. *Przemysl Chemiczny*, 96(8), 1733–1735. (in Polish).

Tiquia, S. M., Wan, H. C., & Tam, N. F. 2002. Microbial population dynamics and enzyme activities during composting. *Compost science & utilization*, 10(2), 150–161. DOI: 10.1080/1065657X.2002.10702075

Umar, M., Roddick, F., & Fan, L. 2016. Impact of coagulation as a pre-treatment for UVC/H2O2-biological activated carbon treatment of a municipal wastewater reverse osmosis concentrate. *Water research*, 88, 12–19. DOI: 10.1016/j.watres.2015.09.047

Union, I. 2014. Communication from the Commission to the European Parliament, the Council, the European Economic and Social Committee and the Committee of the Regions. *A new skills agenda for europe. Brussels*.

Widłak, M. 2016. Natural indicator of soil salinity. *Proceedings of ECOpole*, 10(1), 359–365. DOI: 10.2429/proc.2016.10(1)039 (in Polish).

Environmental pollutants migration and removal

CHAPTER 14

Degradation of selected cytostatic medicines in hospital wastewater by ozonation

J. Czerwiński & S. Skupiński
Faculty of Environmental Engineering, Lublin University of Technology, Poland

ABSTRACT: In recent years, a significant increase in the application of chemotherapeutic drugs is observed, along with their appearance in the environment. For many years, there were no standards regulating the limitations on the introduction of medicines to the environment. Significant amounts of these pollutants were discharged with municipal wastewater to the treatment plants which were not adapted for their removal. It was shown that the standard technologies are characterized by low efficiency in terms of removing the active substances of drugs and their metabolites. Therefore, it is necessary to implement additional steps in the technological system of treatment plants, based on advanced oxidation processes such as ozonation and UV irradiation. Moreover, numerous substances negatively impact the biotechnological processes, inhibiting biochemical reactions and multiplication of utilized microorganisms. Determinations of the analyzed cytostatics were carried out with the HPLC-MS/MS method, using Agilent 1200 series HPLC and Q-trap 4000 MS/MS system. The results showed that cyclophosphamide was present in hospital wastewater in the range 98–841 $ng \cdot L^{-1}$, while iphosphamide was present in some months in the effluents from only one wastewater treatment plant and hospital effluents in the range 45–622 $ng \cdot L^{-1}$. Degradation of the analyzed cytostatics ranged from ~80 till ca.100%.

14.1 INTRODUCTION

Micropollutants, such as pharmaceuticals, constitute a new challenge for wastewater treatment plants and the industry. Medicines are designed to ensure their maximum biological activity. Cytotoxic agents, used in the treatment of cancer are one of the compounds characterized by the strongest effect (Avella 2010).

In recent years, a significant increase in the application of chemotherapeutic drugs is observed, along with their appearance in the environment (Souza 2018; Santos 2018).

For many years, there were no standards regulating the limitations on the introduction of medicines to the environment. Significant amounts of these pollutants were discharged with municipal wastewater to the treatment plants which were not adapted for their removal. It was shown that the standard technologies are characterized by low efficiency in terms of removing the active substances of drugs and their metabolites. Therefore, it is necessary to implement additional steps in the technological system of treatment plants, based on advanced oxidation processes such as ozonation and UV irradiation (Czerwinski 2015). Moreover, numerous substances negatively impact the biotechnological processes, inhibiting biochemical reactions and multiplication of utilized microorganisms. Many studies on micropollutants focus only on the efficiency of their removal, not accounting for the hindrance of the entire process (Avella 2010).

One of the main issues in the studies on the presence and removal of micropollutants involve the problems with inadequacy of the analytic method, resulting from the presence of trace amounts of analytes in the compiled matrices, with composition varying in time.

It is often necessary to carry out separation and enrichment prior to conducting the analysis. These operations may cause errors connected with incomplete recovery and introduction of pollutants, resulting in undesirable chemical reactions and interferences. Thus, a thorough validation of the employed analytic protocol seems necessary. Reference materials and isotope labelled substances play a special role in quality control.

DOI 10.1201/9781003171669-14

It should be mentioned that the physicochemical properties of particles, characterized by high polarity, constitute an additional issue in terms of quantitative determinations (Kovalova 2009). Therefore, it is necessary to employ dedicated sorbents for polar substances in the solid-phase extraction (SPE) method.

The cytotoxic agents drawing the greatest attention among scientists include 5-fluorouracil and platinum compounds (Cisplatin, oxaliplatin and carboplatin) (Kosjek 2013; Kovalova 2009). However, there are few literature reports compared to the papers on antibiotics and nonsteroidal anti-inflammatory drugs. This gap in literature should be explained by the analytical difficulties; however, due to the high toxicity, this topic becomes increasingly popular.

The sources of therapeutic substances include the pharmaceutical industry, application by individual consumers, medicine, veterinary medicine, agriculture and animal husbandry (Kummerer 2009). Pharmaceuticals are discharged to hospital, municipal and industrial wastewater. Then, they spread in the environment, become bio-available and are retained in its elements.

The first documented observations on the occurrence of drugs in the environment appeared in 1981; in that study, clofibric acid was detected in aquatic environment in the concentration of 0.5–2 μg/l. In 1998, Thomas Terens made the first attempt in Europe to identify drugs in water, streams and wastewater in Germany. His monitoring studies confirmed the presence of painkillers, anti-inflammatory and psychotropic drugs as well as hormones (Czerwinski 2015).

This issue was recognized globally, which led to its investigation by numerous scientific centers, also in Poland (Gdańsk University of Technology, Silesian University of Technology, Poznan University of Technology, and Lublin University of Technology). The properties of the analyzed anti-cancer drugs are given in Table 14.1.

Table 14.1. Properties of the analyzed anti-cancer drugs (DrugBank 2020).

Compound CAS	Structure	Molecular mass [g/mol^{-1}]	Water solubility [mg/mL^{-1}]	Log K_{ow}	Biodegra-dability
Cis-platin 15663-27-1	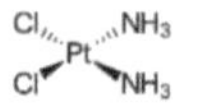	300.05	2.5	−2.19	no
Carboplatin 41575-94-4	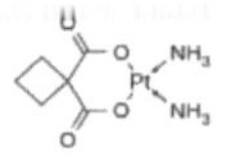	371.26	79.8	0.14	no
Iphosphamide 3778-73-2	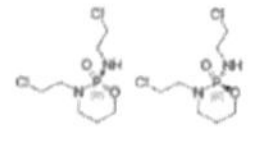	261.08	3.8	0.86	no
Cyclophosphamide 50-18-0	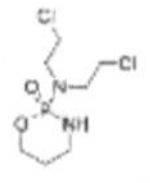	261.08	40.0	0.63	no
Methotrexate 59-05-2	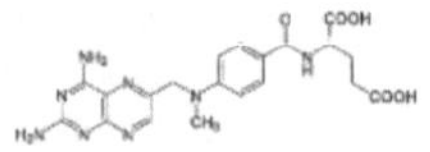	454.45	2.6	−1.85	yes
Doxorubicin 23214-92-8	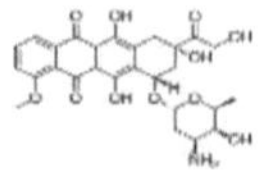	543.52	2.6	1.27	no
Epirubicin 56420-45-2	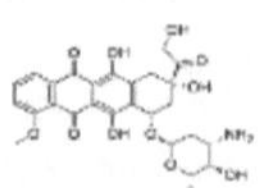	543.52	2.6	1.27	no

Long-term exposure to trace amounts of cytotoxic agents may result in rash, skin irritation, hair loss, reduced number of white blood cells, and increased susceptibility to infections.

Organ toxicity may occur as well. Most importantly, the exposure to chemotherapeutic drugs may damage and mutate DNA as well as induce carcinogenic effect (Gajski 2018; Machnik 2004, 2007; Smerkova 2017).

The first cytotoxic drug analyzed in the hospital wastewater corresponded to 5-fluorouracil (Kovalova 2017). Table 14.2 presented typical concentration of anti-cancer drugs in the analyzed samples.

The data presented in Table 14.1 indicate the presence of considered cytotoxic agents in hospital wastewater at different stages of their treatment. The authors cited in the table mainly employed chromatography coupled with mass spectrometry as their analytic method. In turn, SPE was employed as the sample preparation method. As far as determining the platinum residues is concerned, virtually only the ICP-MS method is used.

The traditional methods of wastewater treatment are insufficient in the case of pharmaceuticals and their metabolites which are being constantly introduced to the environment. Therefore, more efficient

Table 14.2. Concentrations of selected cytotoxic agents from hospital wastewater, treatment plant influent and treated wastewater, as reported in literature.

Cytotoxic agent	Sample type	Analytical method	Result [μg/L]	Reference
Cyclophospamide	Hospital wastewater	GC/MS	0.019- 4.486	Steger-Hartmann 1997
Cyclophospamide	Hospital wastewater	LC/MS-MS	1.080	Isidori et al. 2016
Cyclophospamide	Hospital wastewater	LC/MS-MS	22.100	
Cyclophospamide	Hospital wastewater	LC/MS-MS	0.032	
Cyclophospamide	Treatment plant influent	LC/MS-MS	0.027	
Cyclophospamide	Treatment plant influent	LC/MS-MS	0.027	
Cyclophospamide	Treatment plant influent	LC/MS-MS	0.019	
Cyclophospamide	Treatment plant influent	LC/MS-MS	0.006	
Cyclophospamide	Treated wastewater	LC/MS-MS	0.017	
Methotrexate	Hospital wastewater	LC/MS-MS	<LOD – 4.756	Olalla et al. 2018
Methotrexate	Hospital wastewater	LC/MS-MS	0.019	Isidori et al. 2016
Methotrexate	Hospital wastewater	LC/MS-MS	3.920	
Methotrexate	Hospital wastewater	LC/MS-MS	0.029	
Methotrexate	Treatment plant influent	LC/MS-MS	0.303	
Methotrexate	Treatment plant influent	LC/MS-MS	0.029	
Methotrexate	Treatment plant influent	LC/MS-MS	0.0290,0083	
Iphosphamide	Hospital wastewater	LC/MS-MS	<LOD – 0.031	Olalla et al. 2018
Iphosphamide	Hospital wastewater	LC/MS-MS	0.058-4.761	
Iphosphamide	Hospital wastewater	GC/MS-MS	0.048	Isidori et al. 2016
Doxorubicin	Hospital wastewater	HPLC	0-265	Machnik et al. 2007
Doxorubicin	Hospital wastewater	DLLME HPLC-FLD	4.64	Souza et al. 2018
Doxorubicin	Hospital wastewater	DLLME HPLC-FLD	2.08	
Epirubicin	Hospital wastewater	DLLME HPLC-FLD	6.22	
Epirubicin	Hospital wastewater	DLLME HPLC-FLD	2.67	
Pt (as sum of cisplatin, carboplatin and oxali-platin)	Hospital wastewater	ICP-MS	0.226	Isidori et al. 2016
Pt	Hospital wastewater	ICP-MS	0.352	
Pt	Treatment plant influent	ICP-MS	0.027	
Pt	Treatment plant influent	ICP-MS	0.023	

methods of removing toxic compounds are sought by the academia. The application of deep oxidation processes seems to be a promising direction for future development.

14.2 MATERIALS AND METHODS

14.2.1 *Material characteristics*

The places of wastewater samples collection were hospitals in Lubelskie Voivodship and additionally hospitals in Gdańsk and Szczecin. The samples for the analysis of degradation were obtained by spiking of the filtrated sewage supernatant by known concentration (ca.100 ng·L^{-1}) of cytostatic drug.

14.2.2 *Experimental set-up*

Decomposition of anti-cancer drugs with ozone was run on the laboratory test stand shown in Figure 14.1.

The tests were carried out in a specially designed reactor, which is shown in Figure 14.2. Each time, the reactor was filled with 2 dm^3 of filtrated hospital wastewater supernatant, a sample containing the examined analytes was added to obtain a concentration of ca. 100 ng·L^{-1}. An Akwatech diffuser was installed on the bottom of the reactor, which allowed introducing air and ozone for oxidation. The flow of the gas was possible owing to the water pump included to the system. The concentration of ozone was 20 mg·L^{-1}.

For the needs of the presented paper, several chemical determinations were run. Each time, the following samples were taken: a 0 one, and after 5, 15, 30 and 60 minutes of ozonation. Chromatographic determination was carried out at least three times for the same samples just after collecting

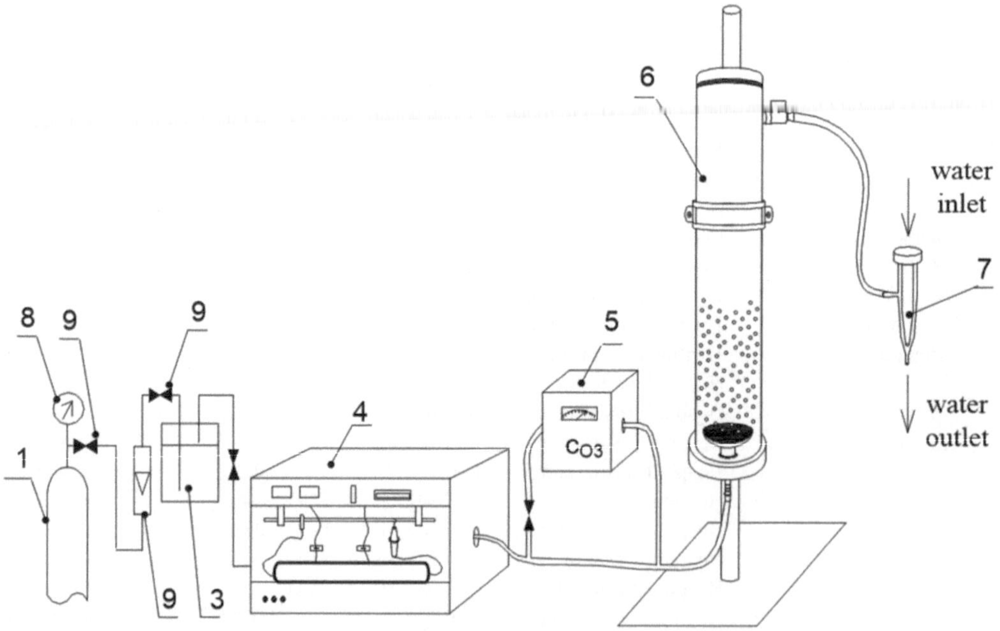

Figure 14.1. The scheme of laboratory test stand. 1 – compressed air cylinder, 2 – rotameter, 3 – gas drier (adsorber), 4 – ozone generator, 5 – meter of ozone concentration, 6 – reactor, 7 – water pump, 8 – manometer, 9 – valves.

Figure 14.2. Photo of the reactor during the study.

the sample. In addition, initial research concerning the bonding of the residual ozone in the samples with the use of anhydrous sodium thiosulphate was performed.

14.2.3 *Analytical methods*

For Pt (sum of cis-platin, carboplatin and oxaliplatin) analysis, the samples were digested in a Microwave 3000 solv (Anton Paar) digester in a HF mineralization vessels. The HCl/HNO_3 mixture was used for the digestion.

All acids were Ultrex grade (JT Baker), and water was obtained from Integral 5 MilliQ system (Millipore). The final analyses of the digested samples were carried out on a Agilent 8900 ICP-MS/MS system. The detection limits for these analyses were estimated at low ppt levels. The HPLC analysis was performed by injecting 50 μL of the sample onto the Lichrospher®100RP-18 column (125 × 4.6 mm) with particle size 5 μm (Merck, Germany). The flow rate was 1000 μL/min. The mobile phase, under gradient conditions, was as follows: mobile phase A – methanol; mobile phase B – 10 mM ammonia and formic acid buffer adjusted to pH 3; time program: 0 min 20%A/80%B; 7 min, 80%A/20%B; 9 min: 80%A/20%B 10 min: 20%A80%B and this mobile phase composition was maintained for 4 minutes. Total run time was no longer than 30 min.

The MS/MS analysis was performed on the 4000 Q-Trap triple quadrupole mass spectrometer equipped with an electrospray ionization source (ESI) operating in positive ion mode. Analyst software (Applied Biosystems/MDS Sciex, v1.5) was used for the instrument control and data collection. The instrument was operated in multiple reaction monitoring (MRM) mode and the following ion transition (precursor/product) were monitored: m/z 261/140 and 261/92 for cyclophosphamide and for iphosphamide 544/361 and 544/397. The electrospray voltage was 5000 V and dwell time was 200 msec. Curtain gas (CUR) 40 psi; collision energy (CE) 31V; declustering potential (DP) 80V; collision cell exist potentials (CEX) 22V were used for the analysis of CP and IP.

14.3 RESULTS AND DISCUSSION

Table 14.3 shows the concentrations of the analyzed cytostatic drugs in hospital effluents.

Table 14.3. Concentrations of the analyzed cytostatic drugs in hospital effluents.

Sample	Sampling point	Concentration Pt [ng·L^{-1}]	Iphosphamide [ng·L^{-1}]	Cyclophosphamide [ng·L^{-1}]
1	PSK-4 Lublin	154.5 ± 5.9 (RSD 3.8%)	231±22	476±21
2	COZL	137.0 ± 19.8 (RSD 14.4%)	216±17	346±24
3	Staszica Lublin	126.2 ± 23.8 (RSD 18.9%)	143±17	98±11
4	Zamość	102.9 ± 14.2 (RSD 13.8%)	74±14	133±17
5	Chełm	85.4 ± 20.0 (RSD 23.4%)	45±13	212±17
6	PSK-4 Lublin II	185.7 ± 17.4 (RSD 9.4%)	219±21	622±34
7	Laborat. COZL	99.8 ± 16.6 (RSD 16.6%)	175±33	716±23
8	Szczecin Dąbie	54.1 ± 16.1 (RSD 29.8%)	427±32	548±48
9	Staszica II Lublin	82.6 ± 20.7 (RSD 25.0%)	143±14	219±43
10	COZL Lublin	92.1 ± 17.4 (RSD 18.9%)	622±29	841±31
11	Chełm II	95.1 ± 12.0 (RSD 12.6%)	78±14	324±21
12	Gdańsk GUM	134.7 ± 13.8 (RSD 10.3%)	411±18	683±26

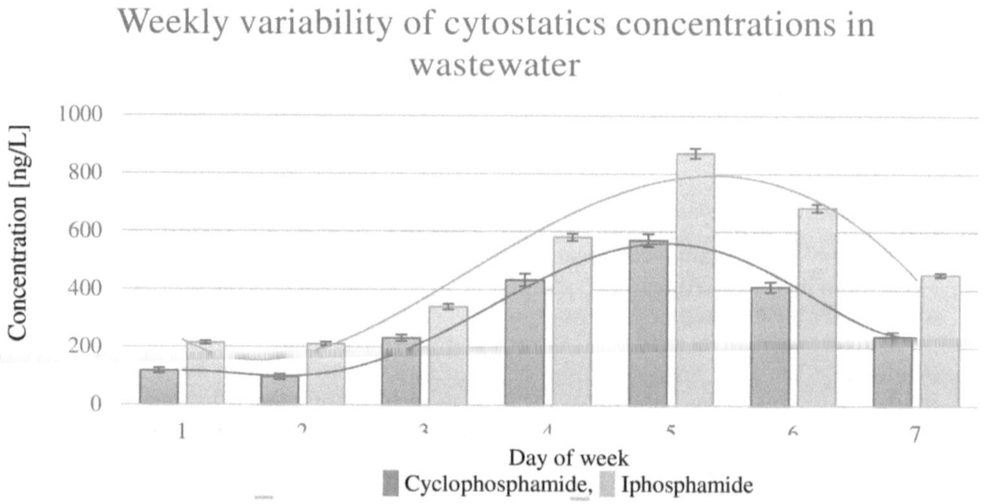

Figure 14.3. Weekly variability of cytostatics concentrations in hospital wastewater

The concentration levels of Pt ranging from 54.1 to 185.7 ng·L^{-1} were obtained as a sum of analyzed compounds (cis-platine, carboplatine and oxaliplatine).

The results showed that cyclophosphamide was present in hospital wastewater in the range 98 - 841 ng·L^{-1}, while iphosphamide was present in some months in hospital effluents in the range 45 - 622 ng·L^{-1}.

Figure 14.3 shows of weekly variability of two cytostatics presented in effluents from hospital in Gdańsk. It shows that the maximum of concentration is observed on Friday. It is connected with cycles of chemotherapy which is applied to patients.

It was observed that during the pandemic period the concentrations of cytostatics in hospital wastewater have been lower than in the corresponding period of 2019. The levels of neoplastic drugs presented in the samples of hospital wastewater of our investigations (12 samples) are similar to the results presented by other investigators (Kovalowa 2015). The degradation efficiency ranged from 80 to ca. 100%, but the products of degradation should be further investigated.

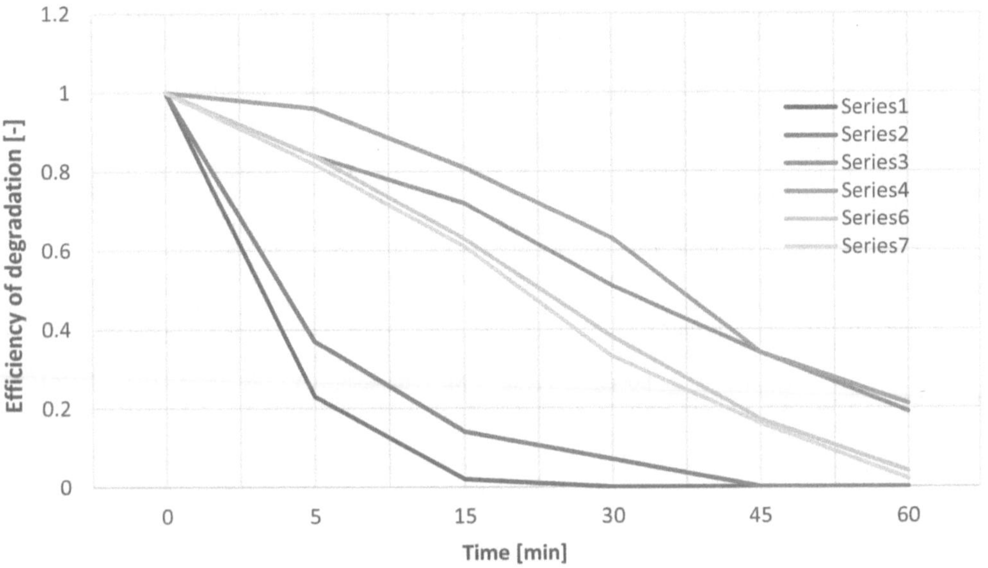

Figure 14.4. Efficiency of degradation selected cytostatics by means of ozone. Series corresponding to: 1 – cis-platin, 2 – carboplatin, 3 – methotrexate, 4 – cyclophosphamide, 5 – iphosphamide, 6 – doxorubicin, and 7 –epirubicin

14.4 CONCLUSION

Hospital effluents are the most important source of cytostatic drugs in the environment. In practice, precautionary mitigation approaches such as drug waste management should be adopted to prevent cytostatic residues from being continuously released into the environment. Further fundamental understanding is required on the trace contaminants removal by various technical alternatives, such as advanced membrane bio-reactor and reverse/forward osmosis filtration, as well as advanced oxidation processes. Degradation of these compounds by means of ozone requires more investigation because there is no information about the degradation products of cytostatics.

ACKNOWLEDGEMENTS

This work was supported by internal grant of LUT. Project No 7514/2019

REFERENCES

Avella, A. C., Delgado, L. F., Görner, T., Albasi, C., Galmiche, M., & de Donato, P. (2010). Effect of cytostatic drug presence on extracellular polymeric substances formation in municipal wastewater treated by membrane bioreactor. *Bioresource Technology,* 101(2), 518–526. doi:10.1016/j.biortech.2009.08.057

Czerwiński J., Kłonica A., Ozonek J. (2015), Pozostałości farmaceutyków w środowisku wodnym i metody ich usuwania", *Czas. Inś. Lądowej śr. Archit.*, t. z. 62, (1) 2015. doi: 10.7862/rb.2015.3

Gajski, G., Ladeira, C., Gerić, M., Garaj-Vrhovac, V., & Viegas, S. (2018). Genotoxicity assessment of a selected cytostatic drug mixture in human lymphocytes: A study based on concentrations relevant for occupational exposure. *Environmental Research*, 161, 26–34. doi:10.1016/j.envres.2017.10.044

Isidori, M., Lavorgna, M., Russo, C., Kundi, M., Žegura, B., Novak, M., …Heath, E. (2016). Chemical and toxicological characterisation of anticancer drugs in hospital and municipal wastewaters from Slovenia and Spain. *Environmental Pollution*, 219, 275–287. doi:10.1016/j.envpol.2016.10.039

Kosjek, T., Perko, S., Žigon, D., & Heath, E. (2013). Fluorouracil in the environment: Analysis, occurrence, degradation and transformation. *Journal of Chromatography A,* 1290, 62–72. doi:10.1016/j.chroma.2013.03.046

Kovalova, L., McArdell, C. S., & Hollender, J. (2009). Challenge of high polarity and low concentrations in analysis of cytostatics and metabolites in wastewater by hydrophilic interaction chromatography/tandem mass spectrometry. *Journal of Chromatography A*, 1216 (7), 1100–1108. doi:10.1016/j.chroma.2008.12.028

Lin, A. Y.-C., Hsueh, J. H.-F., & Hong, P. K. A. (2014). Removal of antineoplastic drugs cyclophosphamide, ifosfamide, and 5-fluorouracil and a vasodilator drug pentoxifylline from wastewaters by ozonation. *Environmental Science and Pollution Research*, 22(1), 508–515. doi:10.1007/s11356-014-3288-7

Mahnik, S., Rizovski, B., Fuerhacker, M., & Mader, R. (2004). Determination of 5-fluorouracil in hospital effluents. *Analytical and Bioanalytical Chemistry,* 380(1). doi:10.1007/s00216-004-2727-6

Mahnik, S. N., Lenz, K., Weissenbacher, N., Mader, R. M., & Fuerhacker, M. (2007). Fate of 5-fluorouracil, doxorubicin, epirubicin, and daunorubicin in hospital wastewater and their elimination by activated sludge and treatment in a membrane-bio-reactor system. *Chemosphere*, 66(1), 30–37. doi:10.1016/j.chemosphere.2006.05.051

Mullot, J.-U., Karolak, S., Fontova, A., Huart, B., & Levi, Y. (2009). Development and validation of a sensitive and selective method using GC/MS-MS for quantification of 5-fluorouracil in hospital wastewater. *Analytical and Bioanalytical Chemistry*, 394(8), 2203–2212. doi:10.1007/s00216-009-2902-x

Olalla, A., Negreira, N., López de Alda, M., Barceló, D., & Valcárcel, Y. (2018). A case study to identify priority cytostatic contaminants in hospital effluents. *Chemosphere*, 190, 417–430. doi:10.1016/j.chemosphere.2017.09.129

Santos, M. S. F., Franquet-Griell, H., Alves, A., & Lacorte, S. (2018). Development of an analytical methodology for the analysis of priority cytostatics in water. *Science of The Total Environment*, 645, 1264–1272. doi:10.1016/j.scitotenv.2018.07.232

Smerkova, K., Vaculovic, T., Vaculovicova, M., Kynicky, J., Brtnicky, M., Eckschlager, T., ...Adam, V. (2017). DNA interaction with platinum-based cytostatics revealed by DNA sequencing. *Analytical Biochemistry*, 539, 22–28. doi:10.1016/j.ab.2017.09.018

Souza, D. M., Reichert, J. F., & Martins, A. F. (2018). A simultaneous determination of anti-cancer drugs in hospital effluent by DLLME HPLC-FLD, together with a risk assessment. *Chemosphere*, 201, 178–188. doi:10.1016/j.chemosphere.2018.02.164

Steger-Hartmann, T., Kümmerer, K., & Hartmann, A. (1997). Biological Degradation of Cyclophosphamide and Its Occurrence in Sewage Water. *Ecotoxicology and Environmental Safety*, 36(2), 174–179. doi:10.1006/eesa.1996.1506

CHAPTER 15

Leaching of pollutants from drilling waste containing water-based muds

M. Chomczyńska & M. Pawłowska
Faculty of Environmental Engineering, Lublin University of Technology, Lublin, Poland

P. Jakubiec
Graduate student, Faculty of Environmental Engineering, Lublin University of Technology, Lublin, Poland

ABSTRACT: The assessment of a potential impact of drilling waste on the water and soil environments was the aim of the study. Two types of drill cuttings different in terms of the chemical and physical parameters, were examined in the batch leaching tests conducted in distilled water and acid solutions (initial pH ~5.0). The mobility of ten heavy metals and six other chemical elements, as well as organic substances (indicated as TOC) were determined. It was found that the tested waste did not pose a threat to the environment due to the mobility of Cu, Co, Mn, Pb, Ni, and Zn but, may be considered as hazardous because of the high content of organic substances and elevated leachability of Ba, Cd, Cr, Na and Tl. Additionally, it was observed that acidification increased the leachability of almost all the examined metals; however, it reduced the leaching of the organic compounds.

15.1 INTRODUCTION

Exploration of oil and natural gas contributes to the generation of different types of waste (Duda et al. 2016). Drilling processes alone generate the second largest amount of the waste, called drilling waste, following the produced water (Onwukwe & Nwakaudu 2012). The drilling waste mainly consists of spent drilling fluids/muds that were used during the drilling of the well, and drill cuttings, which consist of drilled rocks (Śliwka et al. 2012). Due to the large quantities and the peculiar properties, such a high pH, high concentrations of heavy metals, chlorides, sodium, potassium, barium, organic and mineral salts, biocides, occasionally radioactive materials and petroleum substances (Cel et al. 2017; Gaurina-Međimurec et al. 2020; Mikos-Szymańska et al. 2018), drilling waste poses a threat to the environment and creates a serious management problem. The properties of the waste depend mainly on the type of the fluid being used and the kind of rock being drilled. In addition, the drilling technology and other factors associated with the characteristics of the drilled well influence the amount and composition of the waste (Jamrozik et al. 2015).

The drilling waste management in Poland is regulated by national (Act of 10 July 2008 on mining waste) and European (Directive 2006/21/EC) legal provisions. The general rule resulting from these provisions is that the extractive waste must be managed without endangering the human health and without using the processes or methods that could pose a threat to the environment, in particular to water, air, soil, and living organisms.

According to the EU legislation, the waste may be disposed of or recovered (Waste Framework Directive 2008/98/EC). Deep injection of the drilling waste into the geological formations, land treatment, deposition in specially engineered landfills are the drilling waste disposal methods applicable in EU. The recovery of drilling waste is driven by their processing. The physical, chemical, and biological methods are used in the drilling waste disposal or recovery (Jamrozik et al. 2015; Kujawska 2014). Selection of proper methods for drilling waste management depends on the properties of drill cuttings and fluids, especially on their chemical composition. The properties of waste, such as ecotoxicity or carcinogenicity (Waste Framework Directive 2008/98/EU) resulting from the presence of biocides, hydrocarbons or heavy metals, render it hazardous. Thus, the waste should be stored in a hazardous

DOI 10.1201/9781003171669-15

waste landfill. However, the drilling waste which contains fluids produced on the base of water are generally classified to the non-hazardous group of waste. Nevertheless, the current regulations impose restrictions on the drilling waste landfilling. According to the prevailing hierarchy of waste management given in Waste Framework Directive there is a tendency towards enhancing the waste reuse or recycling (Onwukwe & Nwakaudu 2012). The complex and multistage technologies in the drilling waste processing are needed to obtain the good quality waste-derived products, which will be able to replace other materials that would have otherwise been used to fulfill a particular function in various applications. Such a replacement is a mandatory condition for recovery. Additionally, the waste processing and the final products should be cost-effective and environmentally friendly. If the waste or waste-derived products have a direct contact with the environment, the emissivity of pollutants must be determined. High leachability of the pollutants from drilling waste can be a beneficial property when considering the possibility of waste treatment leading to the reduction in their harmfulness or waste recovery, but on the other hand, it is highly undesirable when the waste is placed in soil or stored under improper conditions.

The purpose of the research presented in the paper was to determine the potential impact of drill cuttings on the soil and water environment. Batch tests on the leaching of selected elements (including heavy metals) and organic compounds expressed as TOC from the waste into the aqueous solutions with various initial pH were conducted to obtain the aim of the study.

15.2 MATERIALS AND METHODS

Two kinds of drill cuttings originating from boreholes in the Podkarpackie voivodeship (south-eastern Poland) and the Pomorskie voivodeship (north Poland) were used in the study. The potassium-polymer (K-P) and polymer (P) drilling fluids were contained in these waste, respectively, thus the abbreviations: K-PDCs and PDCs are used for them in the paper. The samples of drill cuttings from both the locations were collected within a few days of extraction from the borehole.

The following physical and chemical properties of the waste were determined: bulk and specific density (Kopecky's cylinder method and La Chatelier's flask method, respectively), pH (potentiometric method), water content (gravimetric method), organic substances content (loss of ignition (LOI) method), total organic carbon (by TOC Analyzer SSM-5000A Shimadzu), total nitrogen content (Kjeldahl's method), calcium carbonate content (Scheibler's method).

The single batch leaching tests were performed to determine the mobility of selected elements and organic substances. For this purpose, (distilled) water suspensions of raw waste were prepared in the solid to liquid phase ratio of 1 : 10 (weight : volume). The flasks with the samples were placed in a rotary mixer (Janke & Kunkel KS 501 D) for 30 minutes and mixed at 150 rpm. After mixing, the samples were allowed to stand for 10 minutes, and then a 0.1 M solution of nitric acid (V) as well as concentrated sulfuric acid (VI) were added to some of them to obtain an initial value of pH=5. After adding the acids and closing the flasks, all the samples were again placed in a rotary mixer for 24 hours and mixed at a speed of 150 rpm.

The way of the waste extract preparation and their initial and final pH are shown in Table 15.1.

In the resultant solutions/eluates, the concentration of heavy metals, defined as metals with the density ≥ 3.5 g/cm^3 (Ba, Cd, Co, Cr, Cu, Ni, Mn, Pb, Tl, Zn) and others elements (Al, Ca, Mg, Sr, K, Na)

Table 15.1. Characteristics of water solutions of drill cuttings.

Drilling cuttings	Water suspension		Initial pH	Added acid [cm^3]		pH after 10 minutes of mixing	Final pH (after 24 hours of mixing)
	Water [cm^3]	Sample [g]		0.1 M HNO_3	Concentrated H_2SO_4		
K-PDCs (depth of 200 m)	400	40	13.40	-	-	-	13.81
	400	40	13.49	232	7	5.02	12.60
PDCs (depth of 2895 m)	400	40	10.30	-	-	-	9.51
	400	40	10.40	153	4	5.02	5.80

were determined by the application of inductively coupled plasma optical emission spectrometry using ICP-OES JY 238 ULTRACE (Horiba Jobin Yvon). The analyzed samples were previously mineralized using HNO_3 (2:1 weight : volume) in Anton Paar-Multiwave 3000 microwave. The content of organic substances indicated as TOC was determined by TOC Analyzer SSM-5000A.

15.3 RESULTS AND DISCUSSION

15.3.1 *Properties of drill cuttings*

The examined drill cuttings were different both in terms of the physical and chemical parameters (Table 15.2). The bulk density of K-PDCs was lower compared to that of PDCs, which can be explained by different origin of the drilled rocks, which influences their mineralogical composition, specific gravity and grain size distribution, and by the depth at which the waste was taken out, which influences its compaction. The K-PDCs were derived from the drilling well located in Carpathian Foredeep Basin where the Miocene sediments were formed under the open sea conditions of the coastal facies. Clays with silt and sands inserts are deposited in this basin at the depths between 100 and 600 m (Laskowicz et al. 2007). The PDCs derived from the drilling well located in the western part of the Baltic Depression. The lower Paleozoic (Middle Cambrian) deposits such as sandstones, mudstones and claystones occur at the depth up to 3000 m (Modliński & Podhalańska 2010). The K-PDCs were taken out from 200 m, while PDCs – from 2895 m. It is known that the bulk density of the sediments, regardless of their grain size distribution, increases with depth, which is associated with an increase in the pressure of superimposed strata. Compaction increases the bulk density of the rocks and reduces porosity (Chapman 1983). The above-mentioned thesis was confirmed by the obtained results. The average value of specific gravity of PDCs was higher compared with the K-PDCs, which suggests the higher amount of quartz (2.6 - 2.65 g cm^{-3}, according to the Mineralogy Database) or secondary minerals just like kaolinite and illite (2.6 g cm^{-3}, 2.6 – 2.9 g cm^{-3}, respectively, according to Mineralogy Database), but the specific gravity of K-PDCs indicates the higher content of light minerals, such as vermiculite or montmorillonite (2.3 g cm^{-3} and 2.35 g cm^{-3}, respectively, according to Mineralogy Database).

The examined drill cuttings were characterized by strongly alkaline pH, and K-PDCs have significantly higher pH than PDCs. The alkalinity of drill cuttings resulted mainly from the composition of drilling fluids, which contain the pH control agents such as potassium hydroxide or caustic soda ash (Lyons et al. 2016; Skipin et al. 2017), and it was also observed in the case of other waste taken out from the drilled well (Kujawska et al. 2015; Skipin et al. 2017).

The waste was rich in organic substances, and the higher value (of 12%) of this parameter was observed in K-PDCs. Despite the high content of organic substances, the concentrations of nitrogen in the waste were low (<0.4% of dry organic mass) which proves the non-biological origin of these compounds. It indicates that high residues of drilling mud were the source of organics in the waste. Different groups of organic chemicals, such as glycol, xanthan gum, carboxymethylcellulose,

Table 15.2. Physical and chemical properties of the drill cuttings used in the experiment.

	Waste	
Parameter	K-PDCs	PDCs
Bulk density [g cm^{-3}]	0.91 ± 0.01	1.13 ± 0.03
Specific gravity [g cm^{-3}]	2.5 ± 0.14	2.60 ± 0.01
pH	13.44 ± 0.06	10.35 ± 0.07
Moisture [%]	49.94 ± 0.90	28.89 ± 2.37
Organic substances content [% d. m.]	12.12 ± 0.28	10.42 ± 0.06
Total Kjeldahl nitrogen content [% d. m.]	0.03 ± 0.02	0.04 ± 0.01
Carbonates content [%]	15.24 ± 2.49	24.23 ± 2.5

Explanations: d.m. – dry matter

polyanionic cellulose, starch or partially hydrolyzed polyacrylamide (PHPA) are used for mud production (Lyons et al. 2016). The differences in the content of organic compounds in the drill cuttings can result from the dissimilar amount of mud residuals and its composition. Taking into account that both the examined materials contained the water-based muds, the values of moisture should be positively correlated with the mud content in the waste. The moisture of K-PDCs was ca. 1.7-times higher compared to PDCs, this may explain the higher concentration of organic chemicals related to dry mass in K-PDCs. The high difference in moisture between the examined waste types can result from the use of various methods of its thickening. The type of drilled rock formations may also affect the considered parameters, determining their dewatering ability.

It is worth mentioning that the examined drill cuttings had different buffer capacity (Table 15.1). K-PDCs, which have lower carbonates content but higher pH (Table 15.2), exhibited greater buffer capacity than PDCs. The buffer capacity may result both from the rocks as well as drilling fluid properties. In the case of the drill cuttings where a polymer-potassium drilling fluid was used, strong opposition to the change in pH may be due to the more intense chemical reactions of acids with the bases. The K-PDCs have significantly higher pH than PDCs. Thus a larger amount of acids is needed to neutralize their water extract. The occurrence of aluminosilicate minerals in the drilled rocks could also be responsible for buffering capacity of the waste. These minerals that have large cation exchange capacity can adsorb the acidifying ions added into the water solution. These ions are replaced by a chemically equivalent amount of the other cations, e.g. Ca^{2+}, Na^{+}, Ba^{2+} which were previously adsorbed on the minerals. This mechanism led to increasing the consumption of the acids used for reaching the pre-set pH.

15.3.2 *Properties of the waste-derived eluates*

The results of the organic substances leachability are presented in Table 15.3. Generally, the water extract of K-PDCs contained less organics compared to PDCs, although the content of organic substances in the former was higher than in the latter (Table 15.2). The leachability of TOC from solid phase of K-PDCs were *ca.* 18-fold and 23-fold lower compared to PDCs, considering the eluates with unmodified and modified pH, respectively. The TOC values measured in the eluates with unmodified pH were higher than those in the modified ones. Acidification of the water solution to initial pH ~5 (Table 15.1) observed after 10 minutes of the batch test decreased the leachability of TOC from both the examined waste, but the decline in leachability was higher in the case of K-PDCs (*ca.* 24%) than PDCs (ca. 2.4%).

The results of the metal leaching test are presented in Tables 15.4 and 15.5. Barium and thallium exhibited the highest concentrations in the eluates among the analyzed elements. On the other hand, the concentrations of cobalt, copper, manganese and nickel were below the detection limits, both in the case of the samples with modified and unmodified pH. Low concentration of these four elements in the eluates obtained during batch leaching tests conducted on drilling cuttings extracted during unconventional shale gas production were also observed by Piszcz-Karaś et al. (2016).

Generally, the leachability of heavy metals related to dry mass of the waste was in the following order: Ba > Tl > Cr > Pb > Zn > Cd in K-PDCs, and Ba > Tl > Zn > Cr > Pb > Cd in PDCs. The highest leachability of Ba, and significant elution of Zn and Pb from drill cuttings with water-based mud was also observed by Kujawska et al. (2016). However, in contrast to our results, significant

Table 15.3. Leachability of organic substances from the examined drill cuttings.

Water eluates of drill cuttings	TOC Concentration in eluates [mg dm^{-3}]	Leachability [g kg^{-1} d.m.]
K-PDCs - unmodified pH	18.60	0.37
K-PDCs - modified pH	8.71	0.28
PDCs - unmodified pH	473.00	6.65
PDCs - modified pH	331.5	6.49

Table 15.4. Leachability of heavy metals from the examined drill cuttings.

Element	K-PDCs				PDCs				Limit value for leachability*** [mg kg^{-1}]		
	unmodified pH		modified pH		unmodified pH		modified pH				
	C* in eluates [μg dm^{-3}]	L** [mg kg^{-1}]	C* in eluates [μg dm^{-3}]	L** [mg kg^{-1}]	C* in eluates [μg dm^{-3}]	L** [mg kg^{-1}]	C* in eluates [μg dm^{-3}]	L** [mg kg^{-1}]	I	II	III
Ba	17790	355.44	19110	609.95	17450	245.43	17970	351.94	20	100	300
Cd	2.58	0.05	1.86	0.06	1.94	0.03	3.35	0.07	0.04	1	5
Co	<d.l.	-	<d.l.	-	<d.l.	-	<d.l.	-	n.l.	n.l.	n.l.
Cr	38.52	0.77	48.09	1.53	41.08	0.58	40.86	0.8	0.5	10	70
Cu	<d.l.	-	<d.l.	-	<d.l.	-	<d.l.	-	20	50	100
Mn	<d.l.	-	<d.l.	-	<d.l.	-	<d.l.	-	n.l.	n.l.	n.l.
Ni	<d.l.	-	<d.l.	-	<d.l.	-	<d.l.	-	0.4	10	40
Pb	5.04	0.10	6.67	0.21	5.55	0.08	6.11	0.12	0.5	10	50
Tl	259.24	5.18	203.05	6.48	203.38	2.86	155.87	3.05	n.l.	n.l.	n.l.
Zn	4.14	0.08	15.62	0.50	87.92	1.24	n.d.	n.d.	4	50	200

Explanations: *C - concentration,** L - leachability value determined as mg of metal per kilogram of dry matter of waste, *** limit values for metal leachability specified in the Regulation on the waste criteria and admission procedures for landfilling of a given type of waste (Journal of Laws 2015 item 1277) for: I – inert waste landfill, II – landfill of other than inert and hazardous waste, III – hazardous waste landfill, d.l. – detection limits: Co – 4 μg dm^{-3}, Cu – 18 μg dm^{-3}, Mn – 5 μg dm^{-3}, Ni – 5 μg dm^{-3}, n.l. – no limits for metal leachability, n.d. – no data

Table 15.5. Leachability of light metals from the examined drill cuttings.

Element	K-PDCs				PDCs			
	unmodified pH		modified pH		unmodified pH		modified pH	
	C* in eluates [$\mu g\ dm^{-3}$]	L** [$mg\ kg^{-1}$]	C* in eluates [$\mu g\ dm^{-3}$]	L** [$mg\ kg^{-1}$]	C* in eluates [$\mu g\ dm^{-3}$]	L** [$mg\ kg^{-1}$)]	C* in eluates [$\mu g\ dm^{-3}$]	L** [$mg\ kg^{-1}$]
Al	14.97	0.30	15.29	0.49	24.30	0.34	17.45	0.34
Ca	2893.67	57.80	4399.13	140.41	2362.81	33.23	2363.82	46.30
Mg	13.51	0.27	13.54	0.43	14.11	0.20	22.82	0.45
Sr	21.82	0.44	43.38	1.38	28.51	0.40	37.74	0.74
K	1477.43	29.51	1522.49	48.59	1522.49	21.41	1526.76	29.90
Na	2445.52	48.85	2474.50	78.98	7296.13	102.60	7693.65	150.86

Explanations: *C - concentration,** L - leachability determined as mg of metal per kilogram of dry matter of waste

mobility of Cu and Ni (greater than Cr), was observed by these authors. Another arrangement of the heavy metals leaching from drilling cuttings was noticed by Stuckman et al. (2015) who examined the waste derived from the Marcellus shale basin. Apart from Ba that ordinarily was the first on a list, the long-term release of Co, Ni, and Cu was observed. The considerable diversity of the results could be explained by a different chemical composition of the examined waste caused by the properties of rocks formation and the drilling fluid used.

The highest leachability of barium from drilling waste was also found in other studies (Kujawska et al. 2015; Piszcz-Karaś et al. 2016). Elevated barium concentrations in the eluates can be explained by a high content of this element in drilling waste, because of the widespread use of barite ($BaSO_4$) as a density increasing agent in many types of drilling fluids (Hartley 1996), especially in those used in deep wells, because the barite content in the drilling mud is increased with the depth of the drilled well (Neff et al. 1987).

Considering the effect of acid additions on the leachability of heavy metals, it can be said that the pH reduction (even by 1 unit observed at the end of the experiment in the case of eluate of K-PDCs) increased the mobility of all the examined heavy metals. This can be explained by the formation of water-soluble sulfates and nitrates of some metals due to the reaction with the acids or by cation exchangeability between the solid phase and water solutions. The change in pH more visibly influenced a mobility of Pb, Ba, Cr and Tl contained in K-PDCs than in PDCs (Table 15.4).

The thallium concentration in the eluates was high when compared to the examined heavy metals. This observation may cause concern because thallium is a strongly hazardous element. It is more toxic to humans than mercury, cadmium, lead, copper or zinc (Peter & Viraraghavan 2005). It is also toxic for other organisms, especially in the form of Tl^{3+} ions. Ralph & Twiss (2002) found that Tl^{3+} is 50,000 times more toxic to the *Chlorella* (algae) than Tl^{+}. However there is little information on the thallium applications in the drilling fluids production, Khan & Islam (2007) listed it among the pollutants contained in drilling muds. Additionally, this element is one of thirteen priority pollutant metals considered in the EPA regulations for drilling fluids used in oil and gas extraction (EPA 2000). Thallium occurs naturally in the Earth's crust (Hu & Gao 2008; Makowska et al. 2019). Thus, both drilling fluids and the drilled rocks are likely to be the sources of this element in the drill cuttings.

The data presented in Table 15.5 indicate that among the light metals Ca, Na and K were leached from the tested drill cuttings in the highest amounts, although their arrangement in a decreasing order was Ca > Na > K in the case of K-PDCs and Na > Ca > K in the case of PDCs. The leachability of Ca and K from the solid phase of K-PDCs (that contained polymer-potassium fluid) was significantly higher compared to PDCs (that contained polymer fluid), contrary to Na, the leachability of which was higher in the case of PDCs.

The other examined light elements were leached in the following order: Sr > Al > Mg, except for the PDCs sample extracted in the acid solution in which the leachability of Mg was higher than Al. High release of Na and Ca was also observed by Stuckman et al. (2015) in the leaching tests conducted on the drilling waste from unconventional gas exploration and by Kujawska et al. (2015) in the eluates obtained from leaching the drilling waste taken from the disposal site. High levels of the alkali metals in the eluates probably results from the usage of sodium, calcium and potassium hydroxides as well as salts in drilling muds production (Nyfors 2000), but it may also be caused by the release from the residues of drilled rocks.

The drop in the pH of the water extract in the case of K-PDCs from pH 13.5 to pH 12.6 due to the addition of acids significantly increased the leaching of all the light metals, but the highest increases in leachability 214% and 143% were observed for Sr and Ca, respectively. In the cases of the other elements, the increases were in range from 59% to 65%. In the case of PDCs, acidification of the water extract mostly increased the leaching of Mg and Sr (of 125 and 85%, respectively). The mobility of the other elements was increased in range of 0% (Al) to 47%, although the pH decrease of the water extract was more evident (from pH 10.4 in unmodified sample to pH 5.8 in acidified sample) when compared with K-PDCs. Thus, the change in the pH of eluates obtained for the examined drill cuttings did not influence the leaching of the examined metals to the same extent, which suggests that different mechanisms of metals bonding in solid phase of particular waste occurred.

Due to the frequently practiced storage of drilling waste at landfills it is worth to compare the obtained results of leaching test with the limits specified in the Regulation of the Minister of Economy of 16 July 2015 on the waste criteria and admission procedures for landfilling of a given type of waste (Journal of Laws 2015 item 1277). Meeting the criteria in the regulation mentioned is a condition for permitting the waste to be deposited on one of the three landfills types: hazardous, inert and other than inert and hazardous waste. The comparison of the results obtained in the leaching tests conducted in distilled water with the limit values given for selected metals in the mentioned Regulation (Table 15.4) revealed that K-PDCs cannot be deposited even in the hazardous waste landfill because of the exceeded limit for Ba (> 300 mg kg^{-1}), which indicate the need to look for alternative methods of its management. In contrast, PDCs can be stored in this type of landfill because the leachability of Ba (in distilled water) is below the limit value. Apart from the barium, Cr (in case of both the waste) and Cd (in the case of K-PDCs) were the reasons why the waste did not meet the standards for the inert waste, and why they should be stored in landfills designed for other than inert and hazardous waste. Piszcz-Karaś et al. (2016), who examined the drill cuttings generated from unconventional shale gas production, reached similar conclusions.

However, in addition to the results of leaching tests, the Regulation sets out additional conditions that must be met to allow the waste to be deposed in landfills. The limited content of organic substance expressed as LOI value is one of them. For hazardous waste landfills, it must not exceed 10% of the dry mass of the waste. PDCs slightly exceeded this value, which also excludes a possibility of their landfilling and indicates the necessity of searching for other methods of their disposal.

15.4 CONCLUSIONS

Considering the results pertaining to the analysis of the properties of the drilling waste generated during a shale gas exploration in two locations in Poland, and the leaching tests of this waste, it was concluded that:

- Both the examined waste types were alkaline and had similar organic matter content but differed in terms of specific gravity, bulk density, moisture, nitrogen and carbonates contents as well as buffer capacity.
- The tested drill cuttings did not pose a threat to the soil and water environment when the mobility of copper, cobalt, manganese, lead, nickel and zinc is considered. The leachability of these metals were below the limits given for the inert waste even in the acidified solutions.
- Barium, thallium, calcium, sodium, and potassium showed the highest leachability from both types of the waste.
- The addition of sulfuric and nitric acids (in the amount leading to obtain pH value of $\sim$5 in the initial phase of the test) reduced the leaching of organic substances but increased the leachability

of almost all the examined metals (except Al that was not influenced in case of one waste) from both waste types.

- High organic substances content (exceeding 10% of dry mass) in both types of drilling cuttings and high leachability of barium from the waste which contained polymer-potassium fluid, prevent their landfilling even in the hazardous waste landfill, which indicates the need of looking for alternative methods of managing this waste.
- High mobility of organic substances in the wastes containing polymer fluid suggests that flushing with water could be an effective pretreatment method that allows for its deposition in a hazardous waste landfill.

REFERENCES

Act of 10 July 2008 on mining waste. Journal of Laws of 2008, No.138, Item 865 with latter amendments.

Cel, W., Kujawska, J. & Wasąg, H. 2017. Impact of hydraulic fracturing on the quality of natural waters. *Journal of Ecological Engineering* 18(2): 63–68.

Chapman, R.E. (ed.) 1983. Compaction of Sediment and Sedimentary Rocks, and its Consequences. Developments in Petroleum Science. *Petroleum Geology* 16: 41–65.

Directive 2008/98/EC of the European Parliament and of the Council of 19 November 2008 on waste and repealing certain Directives (OJ L 312, 22.11.2008, p. 3).

Directive 2006/21/EC of the European Parliament and of The Council of 15 March 2006 on the management of waste from extractive industries and amending Directive 2004/35/EC (OJ L 102/15, 11.4.2006, p.15–34).

Duda, A., Żelazna, A. & Gołębiowska, J. 2016. Sustainable Development versus Prospecting and Extraction of Shale Gas. *Problemy Ekorozwoju – Problems of Sustainable Development 11* (1): 177–180.

EPA 2000. Development document for final effluent limitations guidelines and standards for synthetic-based drilling fluids and other non-aqueous drilling fluids in the oil and gas extraction point source category. EPA-821-B-00-013 December 2000, available online Apr 20, 2020.

Gaurina-Međimurec, N., Pašić, B., Mijić, P. & Medved, I. 2020. Deep underground injection of waste from drilling activities - An overview. *Minerals* 10(4): 303.

Hartley, J.P. 1996. Environmental monitoring of offshore oil and gas drilling discharges: a caution on the use of barium as a tracer. *Marine Polluion Bulletin* 32: 727–73.

Hu, Z. & Gao, S. 2008. Upper crustal abundances of trace elements: a revision and update. *Chemical Geology* 253: 205–221.

Jamrozik, A., Ziaja, J., Gonet, A. & Fijał, J. 2015. Selected aspects of drilling waste management in Poland. *AGH Drilling, Oil, Gas* 32(3): 565–573.

Khan, M.I & Islam, M.R. 2007. *True Sustainability in Technological Development and Natural Resource Management.* New York: Nova Science Publishers, Inc.

Kujawska, J., Cel, W. & Wasąg, H. 2016. Leachability of heavy metals from shale gas drilling waste. *Rocznik Ochrona środowiska - Annual Set The Environment Protection* 18(2): 909–919 (In Polish).

Kujawska, J., Pawłowska, M., Cel, W. & Pawłowski, A. 2015. Potential influence of grill cuttings landfill on groundwater quality-comparison of leaching tests results and groundwater composition. *Desalination and Water Treatment* 57(3): 1409–1419.

Laskowicz, I., Stec, B., Bliźniuk, A., Kwecko, P. & Tomassi-Morawiec, H. 2007. *Notes to the geo-environmental map of Poland 1: 50,000 Worksheet Rokietnica (1007)*. Warsaw: Polish Geological Institute (In Polish).

Lyons, W.C., Plisga, G.J. & Lorenz, M.D. (eds) 2016. *Standard Handbook of Petroleum and Natural Gas Engineering*. Elsevier.

Makowska, D., Strugała, A., Wierońska, F. & Bacior, M. 2019. Assessment of the content, occurrence, and leachability of arsenic, lead, and thallium in wastes from coal cleaning processes. *Environmental Science and Pollution Research* 26: 8418–8428.

Mikos-Szymańska, M., Rusek, P., Borowik, K., Rolewicz, M., Bogusz, P. & Gluzińska, J. 2018. Characterization of drilling waste from shale gas exploration in Central and Eastern Poland. *Environmental Science and Pollution Research* 25: 35990–36001.

Mineralogy Database, http://webmineral. Available on line April 20. 2020.

Modliński, Z. & Podhalańska, T. 2010. Outline of the lithology and depositional features of the lower Paleozoic strata in the Polish part of the Baltic region. *Geological Quarterly* 54 (2): 109–122.

Neff, J.M., Rabalais, N.N. & Boesch, D.F. 1987. Offshore oil and gas development activities potentially causing long-term environmental effects. In D.F. Boesch & N.N. Rabalais (eds), *Long-term environmental effects of offshore oil and gas development*: 149–173. London: Elsevier.

Nyfors, A. 2000. Skin diseases in Oil-Rig Workers. In L. Kanerva, P. Elsner, J.E. Wahlberg & H.I. Maibach (eds), *Handbook of Occupational Dermatology*: 1021–1027. Berlin, Heidelberg: Springer-Verlag.

Onwukwe, S.I. & Nwakaudu, M.S. 2012. Drilling Waste Generation and Management Approach. *International Journal of Environmental Science and Development* 3: 252–257.

Peter, A.L.J. & Viraraghavan, T. 2005. Thallium: a review of public health and environmental concerns. *Environment International* 31(4): 493–501.

Piszcz-Karaś, K., Łuczak, J. & Hupka, J. 2016. Release of selected chemical elements from shale drill cuttings to aqueous solutions of different pH. *Applied Geochemistry* 72: 136–145.

Ralph, L. & Twiss, M.R. 2002. Comparative toxicity of thallium(I), thallium(III), and cadmium(II) to the unicellular alga Chlorella isolated from Lake Erie. *Bulletin of Environmental Contamination and Toxicology* 68: 261–268.

Regulation of the Minister of Economy of 16 July 2015 on the waste criteria and admission procedures for landfilling of a given type of waste. Journal of Laws 2015, Item 1277.

Skipin, L.N., Petuhova, V.S. & Romanenko, E.A. 2017. Creation of Favourable Water – Physical Properties of Drill Cuttings with the use of Coagulants. *Procedia Engineering* 189: 593–597.

Stuckman, M., Lopano, C., Thomas, C. & Hakala, A. 2015. Leaching characteristics of drill cuttings from unconventional gas reservoirs; *Proceedings of Unconventional Resources Technology Conference*, San Antonio, 20–22 July, Texas, USA.

Śliwka, E., Kołwzan, B., Grabas, K., Klein, J. & Korzeń, R. 2012. Chemical composition and biological properties of weathered drilling wastes. *Environment Protection Engineering* 38(1): 129–138.

CHAPTER 16

Effect of copper ions on phenol biodegradation under anaerobic conditions

M. Zdeb
Lublin University of Technology, Lublin, Poland

ABSTRACT: The aim of the study was to determine the effect of copper addition in two forms: copper(II) sulfate and IDHA chelate, and in two concentrations, on the biodegradation of phenols under anaerobic conditions and on the efficiency of biogas production. The research was carried out in batch system in which the cereal straw was anaerobically digested. Phenol was added to obtain the initial concentration of 27 mg L^{-1}. The biogas production and methane concentration in biogas were analyzed. Additionally, the biochemical tests were performed. Studies have shown that the addition of copper in two different concentrations and chemical forms did not significantly affect the biogas potential of the digested feedstock. However, it was found that it influenced the degradation of phenol. It was concluded that the increase in copper concentration up to 60 mg L^{-1} stimulated the processes of anaerobic biodegradation of phenol without lowering the biogas production.

16.1 INTRODUCTION

Providing energy from renewable sources is very important nowadays, because it is environmentally friendly and requires minimal operation costs. Lignocellulosic biomass, being a renewable source consisting of cellulose, hemicellulose and lignin (Rowell 2012), coming from agriculture or forestry residues or energy crops, may be used for energy production *via* conversion into biogas in the anaerobic digestion. Hydrolysis, the first step of the process, is regarded as a rate-limiting step in biogas production. The process of organic substrate pretreatment is essential in speeding up the enzymatic hydrolysis and improving its efficiency. The methods of lignocellulosic biomass pretreatment result not only in the conversion of carbohydrate polymers to soluble monomeric sugars, but also in the release of various inhibitors. Thus, anaerobic digestion can be affected not only by the inhibitors contained in initial organic substrate, but also by these ones released during the process. Hydrolysis and pretreatment methods are regarded as the steps generating the highest amounts of inhibitors. The main factors determining the nature of the inhibitors are type of lignocellulosic substrate, composition of cell wall and thermochemical conditions of hydrolysis step. Inhibitors can be divided into a few groups: furan derivatives, weak organic acids, phenolic compounds and heavy metal ions (Chandel et al. 2011; Mussatto & Roberto 2004). They may cause a decrease in viability and fermentation productivity (Palmqvist & Hahn-Hägerdal 2000).

Phenolics are one of the main inhibitors produced during the alkaline pretreatment of biomass (Xie et al. 2018). They may inhibit microbial growth as well as product yield. Their negative influence may be connected with specific functional groups (Larsson et al. 2000). Phenolic compounds negatively influence the cell membranes by damaging their ability to act as selective barriers and enzyme matrices (Palmqvist & Hahn-Hägerdal 2000). However, in many cases their inhibitory mechanism has not been explained.

The phenolic compounds which remain in digestates obtained after anaerobic process can infiltrate into the soil when digestates are used as a fertilizer. They can inhibit enzymatic processes of soil microorganisms and can migrate with soil water; thus, they pose a long-term *threat* to *soil* and groundwater quality (Nyberg et al. 2006). Leven et al. (2006) noted the negative influence of phenols on the activity of the ammonia-oxidizing bacteria. Therefore, it is very important to decrease the

DOI 10.1201/9781003171669-16

Table 16.1. Parameters of cereal straw and inoculum used in the experiment.

Parameter	Unit	Cereal straw	Digestate (inoculum)
pH	–	–	7.77 ± 0.08
Total solids (TS)	$g\,TS\,kg^{-1}$	936.94 ± 0.67	41.61 ± 0.04
Volatile solids (VS)	$g\,VS\,kg^{-1}$	902.83 ± 4.53	27.31 ± 0.24
$COD_{dissolved}$	mgO_2L^{-1}	–	4896.0 ± 114.6
Phenolic compounds	$mg\,L^{-1}$	–	6.57 ± 0.15
Cu_{total}	$mg\,L^{-1}$	–	1.50
$Cu_{dissolved}$	$mg\,L^{-1}$	–	1.15

phenols concentrations during the anaerobic digestion in order to minimize their negative influence on the environment after using the digestate in agriculture.

Anaerobic biodegradation of phenols is a process in which numerous enzymes are involved. Their activity depends on the presence of some trace metals, acting as catalytic centers at active sites (Demirel & Scherer 2011; Glass & Orphan 2012). The lack of trace metals can result in low process efficiency which can be overcome by their addition.

The aim of the study was to determine the effect of copper (Cu) addition in two doses (to obtain the concentration in the feedstock of 20 $mg\,L^{-1}$ and 60 $mg\,L^{-1}$) of two forms (organic and inorganic) on the biodegradation of phenols under anaerobic conditions, and on the efficiency of biogas production during the cereal straw digestion.

16.2 MATERIALS AND METHODS

16.2.1 *Examined materials*

The biomass of cereal straw used in the test was collected in August 2018 from a private-owned farm localized in the Lubelskie Region (Poland). The straw was transported to the laboratory and stored in open bags, in the air-dried state. Before the digestion, the samples of straw biomass were mechanically fragmented (< 10 mm) and soaked in distilled water in the ratio 5 g : 100 mL to increase its digestibility. During soaking, the samples were shaken at speed of 100 rpm at 25°C (Stuart SI 500). After 22 h, the soaked straw samples were submitted to the anaerobic digestion process.

The digestate from an agricultural biogas plant in Siedliszczki (Lubelskie Region) was used as an inoculum. It was poured through a 2-mm sieve in order to avoid the errors resulting from the heterogeneous composition. The parameters of the materials used in the experiment are presented in Table 16.1.

Two forms of copper were used in the experiment: $CuSO_4$ as an inorganic form, and commercially available single ingredient foliar granulated fertilizer in Cu chelate form ADOB® Cu IDHA as an organic one. The commercially available starch and pure phenol (POCH S.A.) were also used in the experiment.

16.2.2 *Anaerobic digestion assays*

The samples of cereal straw soaked in distilled water were digested in the laboratory batch digesters. The experiment was conducted under mesophilic conditions (37 ± 1°C) in the BioReactor Simulator – BRS (Bioprocess Control, Sweden). The experimental setup device consisted of twelve glass bioreactors stirred in a semi-continuous mode (10 minutes of mixing at 80 rpm and 20 minutes break). Each bioreactor with 2 L of volume (1.8 L of working volume) was filled with 1000 mL of digestate (inoculum).

At the beginning of the experiment, the headspace of each bioreactor was flushed out for 2 minutes using nitrogen gas in order to eliminate the gases present in the headspace and ensure anaerobic

Table 16.2. Characteristics of the mixture of inoculum and WSCS.

Parameter	Unit	Value
pH	–	7.89 ± 0.0
Total solids (TS)	$g\ TS\ kg^{-1}$	44.58 ± 0.45
Volatile solids (VS)	$g\ VS\ kg^{-1}$	30.38 ± 0.23
$COD_{dissolved}$	$mgO_2\ L^{-1}$	4880 ± 36
Phenolic compounds	$mg\ L^{-1}$	7.59 ± 0.03
$Cu_{dissolved}$	$mg\ L^{-1}$	0.106

conditions. Afterwards, the digestate was submitted to 20 days of fermentation for ensuring its deep digestion until the residual daily biogas production of 0.01 NL d^{-1} was reached. Then, the water soaked cereal straw (WSCS) in ratio of TS in straw and digestate of 1:9 (w/w), and copper in two doses of two chemical forms: organic and inorganic, were added. Next, the bioreactors were tightly closed and purged with nitrogen gas once again, and phase I (also called an adaptation phase) of the targeted experiment was started.

Six various feedstocks of the reactors were prepared: 1 – digestate from agricultural biogas plant (inoculum), 2 (WSCS+$CuSO_4$ dose I) – mixture of inoculum and WSCS with an addition of $CuSO_4$(concentration of 20 mg $Cu^{2+}L^{-1}$), 3 (WSCS+$CuSO_4$ dose II) – mixture of inoculum and water soaked cereal straw with an addition of $CuSO_4$(concentration of 60 mg $Cu^{2+}L^{-1}$), 4 (WSCS+Cu IDHA dose I) – mixture of inoculum and water soaked cereal straw with an addition of ADOB Cu IDHA (concentration of 20 mg $Cu^{2+}L^{-1}$), 5 (WSCS+Cu IDHA dose II) – mixture of inoculum and water soaked cereal straw with an addition of ADOB Cu IDHA (concentration of 60 mg $Cu^{2+}L^{-1}$), 6 (WSCS without Cu addition) – mixture of inoculum and water soaked cereal straw (control). Considering the amount of copper in the inoculum, the concentration of this element in the feedstocks was 1.5 mg L^{-1} higher than it resulted from the Cu addition.

The production of the biogas was monitored automatically and registered on-line. The experiment was carried out in two replications. The parameters of the mixture of inoculum and water soaked cereal straw are presented in Table 16.2.

After 3 weeks of microorganisms adaptation to the new conditions, the phase II of the experiment was started, in which 10 g of starch, an easily biodegradable substrate for methanogens, was added to all bioreactors. At the same time, into the bioreactors enriched with Cu, pure phenol was added in an amount enabling to achieve the initial concentration in the feedstock of 27 mg L^{-1}.

16.2.3 *Analytical methods*

Total solids (TS) content was calculated as the weight of residue after water evaporation during drying the samples for 24 h at 105°C to constant weight in a drying chamber SUP-4 (Wamed, Poland). The content of volatile solids (VS) was determined on the basis of the results of burning the samples at 550°C for 24 h in a muffle furnace FCF 2.5 S (Czylok, Poland). The concentrations of COD and phenolic compounds were determined using HACH cuvette tests (COD 100–2000 mg L^{-1} LCK514, phenols 0.05–5 mg L^{-1} LCK345). The pH values were measured by using EasyPlus^{TM} (METTLER TOLEDO).

The copper concentrations were determined using 8900 ICP-MS Triple Quadrupole Agilent working in a helium mode with a plasma parameters: RF power 1500W, plasma gas flow rate 15.0 l min^{-1}, He flow rate 5.5 ml min^{-1}, nebulizer gas flow rates 1.05 l min^{-1} and MicroMist nebulizator.

The daily volume of the biogas produced from each bioreactor was measured on-line using the BRS-B unit and automatically standardized to normal conditions (0°C, 1 atm). The biogas composition was determined at the end of the experiment using Trace GC Ultra (Thermo Scientific) gas chromatograph with a RTQ-BOND column (30 × 0.25mm ID, d_f 10 μm), equipped with a TCD detector.

Microbial enzymatic activities of microorganisms in digestates were analyzed by using commercially available biochemical tests API ZYM (bioMérieux, France). They consist of a series of micro-cupules which contain dehydrated chromogenic substrates of 19 different enzymes and one control sample. The samples of bioreactor digestates obtained after the anaerobic process were centrifuged for 20 minutes by 4000 rpm (MPW-350 Med Instruments) and percolated through a filter paper (84 g m^{-2}). The API ZYM strips were inoculated with prepared feedstocks samples according to the manufacturer's instructions and incubated for 16 h at 37°C. After that time, the determination of the approximate concentration of 19 enzymes (in the range of 1–5) was performed, by comparing the intensity of color of micro-cupules with a color chart given by the manufacturer.

16.2.4 *Methods of the results elaboration*

The specific biogas potential (BP) and biochemical methane potential (BMP) were determined on the basis of Equations (1) and (2):

$$BP = \frac{V_B}{m_{VS}} \; [mL \, g_{VS}^{-1}] \tag{16.1}$$

$$BMP = \frac{V_B \cdot c_{CH4}}{m_{VS}} \; [mL_{CH4} \, g_{VS}^{-1}] \tag{16.2}$$

where V_B = volume of biogas produced during the experiment (mL); m_{VS} = mass of volatile solids added to the bioreactor with substrate (g); c_{CH4} = methane concentration in the biogas (–).

The removal efficiencies were calculated according to Equation (3):

$$\eta = \left(\frac{m_1 - m_2}{m_1} \right) \cdot 100\% \; [\%] \tag{16.3}$$

where m_1 = mass of parameter before the anaerobic digestion (mg); m_2 = mass of parameter after the anaerobic digestion (mg).

16.2.5 *Statistical analysis*

The means and standard deviations of the values of parameters determined in repetitions were calculated. The statistical analysis was performed by using Microsoft Excel 2003 for Windows

16.3 RESULTS AND DISCUSSION

16.3.1 *The biogas production and composition*

The curves illustrating the changes in a specific daily biogas production in the particular feedstocks are presented in Figure 16.1.

The daily biogas production has been changed through the adaptation phase. In the WSCS+$CuSO_4$ dose I, it ranged from 0.44 mL d^{-1} gVS^{-1} to 56.98 mL d^{-1} gVS^{-1}, while in the WSCS+$CuSO_4$ dose II – from 0 to 48.45 mL d^{-1} gVS^{-1}. The biogas production in the particular days of the experiment in the WSCS+Cu IDHA dose I ranged from 0.78 mL d^{-1} gVS^{-1} to 51.88 mL d^{-1} gVS^{-1} and in the WSCS+Cu IDHA dose II from 0.33 mL d^{-1} gVS^{-1} to 60.09 mL d^{-1} gVS^{-1}. This parameter in WSCS without the Cu addition (control) was within the limits of 0 and 45.68 mL d^{-1} gVS^{-1}. During that phase, the time-dependent changes of the daily biogas production of all the analyzed bioreactor feedstocks showed similar tendency with a first clearly visible peak observed on the 2nd–3rd day, whereby in the WSCS+$CuSO_4$ dose II and WSCS+Cu IDHA dose II the peaks appeared earlier than in the others. From that moment, the production of biogas started to decline gradually up the 22nd day.

In the second phase of the experiment (after starch and pure phenol addition), the biogas production in particular feedstocks was also diversified. WSCS+$CuSO_4$ dose I was characterized by the values from 0.14 mL d^{-1} gVS^{-1} to 123.12 mL d^{-1} gVS^{-1}, while in the WSCS+$CuSO_4$ dose II it ranged from 0 to 113.09 mL d^{-1} gVS^{-1}. The daily biogas production in the WSCS+Cu IDHA dose I was

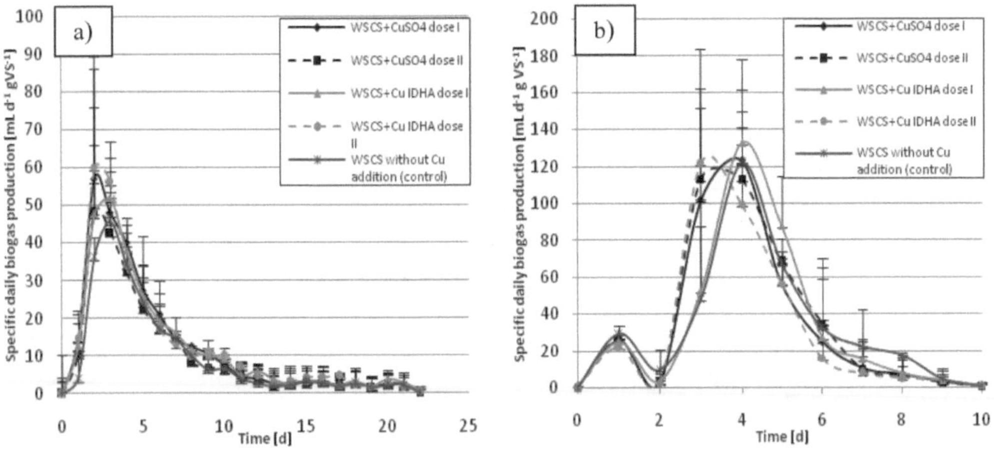

Figure 16.1. Specific daily biogas production in the experimental bioreactors in the first (a) and second (b) phase of the experiment.

within the limits of 0.24 and 132.70 mL d^{-1} gVS^{-1}, whereas in the WSCS+Cu IDHA dose II, it ranged from 0.10 mL d^{-1} gVS^{-1} to 121.30 mL d^{-1} gVS^{-1}. In the WSCS without the Cu addition, the biogas production was between 0 and 121.92 mL d^{-1} gVS^{-1}.

Two clearly visible peaks of this parameter values were observed: the first – one day after the conditions change and the second one – on the 3rd–4th day of the experiment. The first, lower peak, was observed in the same time in all the bioreactors. The second one, in which the highest values of daily biogas production were reached, was noted earlier in the feedstocks with higher doses of Cu, both organic and inorganic. In the feedstocks with lower doses of Cu and in the WSCS without Cu addition, the second peak was observed almost in the same time. From the 4th day onwards, the biogas production in all the analyzed feedstocks started to decrease till the last day of the experiment.

The highest maximum value of daily biogas production (132.70 mL d^{-1} gVS^{-1}) was observed in the WSCS+Cu IDHA dose I. This value was 17% higher than the maximum value of this parameter measured in the WSCS+$CuSO_4$ dose II and of about 9% higher than the maximum values measured in other bioreactor feedstocks. The curve of cumulative biogas production in the adaptation phase (up to the 22nd day) showed similar trends in all the analyzed bioreactor feedstocks. In the second stage of the experiment, this tendency was maintained with slightly lower values obtained in the control (Figure 16.2).

The value of 657.20 ± 99.33 mL gVS^{-1} observed in WSCS+$CuSO_4$ dose I turned out to be the highest of all. It was 12% higher than in the WSCS without the Cu addition, 4% higher than in the WSCS+$CuSO_4$ dose II and 3% higher than in the WSCS+Cu IDHA dose II.

Considering the values of cumulative biogas productions and average methane concentrations in the biogas produced in particular bioreactors feedstocks (which ranged from 47.98% to 59.60%), the biochemical methane potentials of the mixed substrate that consisted of straw and starch, were calculated (Table 16.3).

The biochemical methane potential calculated in the WSCS+Cu IDHA dose I was of 9% higher than in the control, of 7.5% higher than in the WSCS+$CuSO_4$ dose I and almost of 3% higher than in the feedstocks enriched with two forms of copper in the dose of 60 mg Cu^{2+} L^{-1}.

16.3.2 *Properties of the digestates*

The concentrations of TS and VS in the digestates obtained after anaerobic process are presented in Table 16.4. After comparing them with concentrations measured in initial feedstocks it was stated that the highest removal efficiencies in TS and VS, of 26% and 30%, respectively, were observed in the WSCS+Cu IDHA dose I. Only slightly lower (24% and 28%, respectively) were noted in the

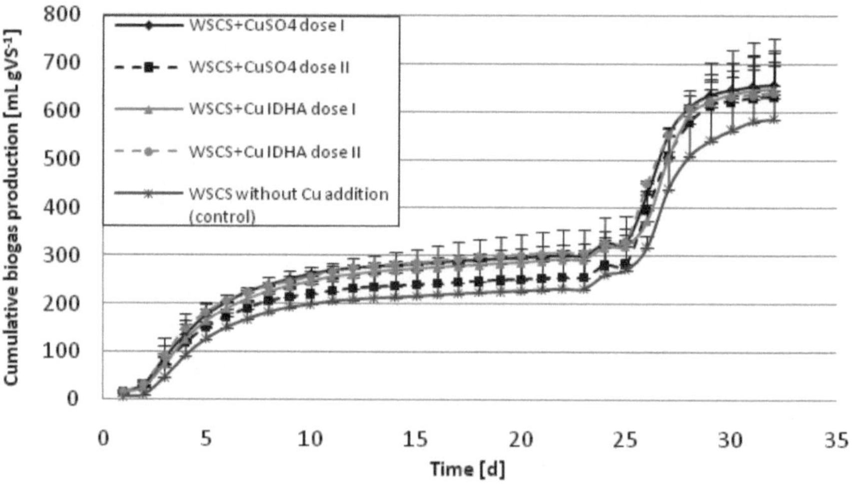

Figure 16.2. Cumulative biogas production of in the experimental bioreactors.

Table 16.3. Biogas potentials and biochemical methane potentials of the bioreactors feedstocks.

Parameter	Unit	WSCS+$CuSO_4$ dose I	WSCS+$CuSO_4$ dose II	WSCS+Cu IDHA dose I	WSCS+Cu IDHA dose II	WSCS without Cu addition
BP	mL gVS^{-1}	657.20 ±99.33	632.23 ±93.43	648.65 ±53.72	641.34 ±68.78	586.37 ±147.00
BMP	mL_{CH4} gVS^{-1}	350.62 ±52.99	367.45 ±54.30	376.93 ±31.22	368.77 ±39.55	337.75 ±84.67

WSCS+$CuSO_4$ dose I and dose II. The lowest ones were observed in WSCS without Cu addition (19% and 24%, for TS and VS, respectively).

The phenol concentrations in the final products of digestion varied between the particular bioreactors (Table 16.4). In the WSCS without Cu addition (control) their concentration increased from 7.59 mg L^{-1} to 19.1 mg L^{-1} during the anaerobic digestion. In the case of this sample, phenols were probably produced as a result of the lignocellulose contained in the straw biomass biodegradation. Contrary to these observations, the phenol concentration decreased due to digestion in the feedstocks enriched initially with copper, to which phenol was added after the adaptation phase (WSCS+$CuSO_4$ dose I and dose II, WSCS+Cu IDHA dose I and dose II) in comparison to the initial concentrations of 27 mg L^{-1} obtained on the 23rd day. The highest removal efficiency (42%) of these inhibitors was observed in the WSCS+Cu IDHA dose I. The values of this parameter in other bioreactors feedstocks were as follows: 38%, 31% and 26%, in WSCS+Cu IDHA dose II, WSCS+$CuSO_4$ dose II and in WSCS+$CuSO_4$ dose I, respectively. The final phenol concentrations in the WSCS+$CuSO_4$ dose II, WSCS+Cu IDHA dose I and WSCS+Cu IDHA dose II were lower than in the WSCS without Cu addition, although its initial concentrations were higher than in control. It could be the evidence that the copper addition had a positive influence on phenol decomposition, probably by affecting the enzymatic activity of enzymes responsible for phenols degradation. Moreover, no significant dependency of these inhibitors removal efficiency on copper form and dose was observed.

The role of Cu in methanogenesis is not clearly determined. Generally, it had been stated that many alkali metals and metalloids play an important role in methanogenesis; however, the Cu-dependent methanogenesis enzymes have not been identified so far (Glass & Orphan 2012). Copper was found in many methanogenic bacteria strains (Myszograj et al. 2018).

Table 16.4. Characteristics of the digestates obtained after the anaerobic process.

Parameter	Unit	WSCS+ $CuSO_4$ dose I	WSCS+ $CuSO_4$ dose II	WSCS+ Cu IDHA dose I	WSCS+ Cu IDHA dose II	WSCS without Cu addition
Total solids (TS)	g TS kg^{-1}	33.96 ±0.18	33.68 ±0.74	32.96 ±0.55	34.13 ±0.65	35.89 ±0.77
Volatile solids (VS)	g VS kg^{-1}	21.99 ±0.09	21.85 ±0.55	21.21 ±0.55	22.07 ±0.77	23.08 ±0.59
Phenols	mg L^{-1}	20.1 ±4.95	18.7 ±5.16	15.7 ±7.57	16.7 ±5.3	19.1 ±4.38

Dokulilová et al. (2018) who tested the influence of copper addition in 5 concentrations (from 100 to 1000 mg L^{-1}) on the anaerobic stabilization of sewage sludge observed that until the highest introduced Cu concentration, no significant inhibition of biogas production was observed. However, the significant methane production inhibition was observed at the copper concentration of 600 mg L^{-1}. In the described experiment, the addition of 20 mg Cu^{2+} L^{-1} in a $CuSO_4$ form caused 12% increase of the total biogas production in comparison to the biogas production from the WSCS without the Cu addition. It can indicate its positive effect on the biogas production efficiency. Moreover, increasing the Cu dose, both in the organic and inorganic form (from 20 mg Cu^{2+} L^{-1} to 60 mg Cu^{2+} L^{-1}), resulted in earlier achievement of maximal daily production of the biogas in comparison to other feedstocks, which can suggest the copper stimulating activity. However, neither chemical form, nor two different doses of Cu had a significant influence on the biogas yield of digested feedstock in the bioreactors enriched with copper.

Regarding gas production, there was no regularity between the biogas potential and biochemical methane potential; the highest values were not observed in the same type of feedstock: the highest BP was observed in the WSCS+$CuSO_4$ dose I and the highest BMP was noted in WSCS+Cu IDHA dose I. Therefore, the Cu influence on the methane production cannot be clearly determined.

The inhibiting effect of Cu on the anaerobic digestion process was observed in many more experiments (Nguyen et al. 2019; Zayed & Winter 2000). Copper, just like other heavy metals, is non-biodegradable and can accumulate to potentially toxic concentrations (Nayono 2009). It was stated that Cu may have a negative influence on the abundance of the microorganisms acting in the anaerobic process. Huang (2008) observed its toxicity towards predominant archaeal species (e.g. *Methanothrix soehngenii, Methanosaeta concilii* and uncultured *euryarchaeota*) during 2-chlorophenol anaerobic degradation. Yu & Fang (2001) noted the inhibiting influence of copper on dairy wastewater acidogenesis at all studied concentrations: from 5 to 400 mg L^{-1}. The inhibition effect of Cu during the anaerobic digestion of sewage sludge was observed also in the range of concentrations from 70 to 400 mg L^{-1} (Ahring & Westermann 1985).

Nguyen et al. (2019), who studied the heavy metals effects on the anaerobic co-digestion of waste activated sludge and septic tank sludge observed that the TS and VS removal efficiencies were decreasing along with the increasing Cu concentration added to the feedstock. At the highest concentration of the copper addition (80 mg L^{-1}), the lowest TS and VS removal efficiencies (4.12% and 9.01%, respectively), in comparison to the efficiencies calculated in the samples with 19, 40 and 60 mg L^{-1} of Cu addition, were observed. Additionally, at that dose of Cu, the inhibition of microorganisms growth was observed. However, a significant decrease of the TS and VS removal efficiency was observed already at 40 mg L^{-1}, similarly to the results of the study of Zayed & Winter (2000). In the experiment described in this article, the TS and VS decreases observed during the anaerobic process were mainly the result of the cereal straw digestion (at the first phase) and starch biodegradation (in a second one). The lowest TS and VS removal efficiencies were noted in the feedstock without the Cu and phenols addition. It may suggest that one of the feedstocks additions introduced to other bioreactors: copper or phenol, was the factor positively influencing the total and volatile solids removal efficiency indicating organic compounds degradation.

The lowest biogas potential observed in the WSCS without the Cu addition (control) in comparison to other feedstocks may indicate pure phenol positive influence on the biogas production. However, because of the same concentration of this inhibitor in all feedstocks enriched with copper, the phenol effect on the biogas potential cannot be clearly determined, and further studies are needed.

There are some literature reports which confirmed that phenols may be finally used by methanogens and converted into methane (Fang et al. 2004; Fedorak & Hrudey 1984). This can explain the significant biogas production rise in the feedstocks with the addition of phenols and Cu compared to the control in my experiment. Fedorak & Hrudey (1984) studied the influence of phenol and seven alkylphenols(*o*, *m*- and *p*-cresol, 2.5-, 2.6-, 3.4- and 3,5-dimethylphenol) on the methane production from domestic anaerobic sludge. The transformations of phenolic compounds added at various concentrations were analyzed in order to check if they were digested to CH_4 or if they showed an inhibitory effect on the process. It was stated that phenol and *p*-cresol were converted into methane. The reduction of the rate and the volume of produced methane was noted at 500 mg L^{-1} of 2,5-, 3,4- and 3,5-dimethylphenol. In the case of *o*-, *m*- and *p*-cresol, the inhibitory effect was observed at the concentration of 1000 mg L^{-1}.

Chapleur et al. (2016) observed that after the increase in phenol concentration (0–4.0 g L^{-1}) during the anaerobic digestion of cellulose, the different steps of the process were affected, whereby methanogenesis was the most sensitive. The rate and volume of biogas production were decreasing with phenol concentration increase. However, these authors also suggested that the cellulose-degrading microbial consortium can adapt to the presence of phenol and conduct cellulose degradation till the threshold phenol concentration.

It is known that phenolics affect microbial cells by modifying the permeability of its membranes, resulting in the leakage of intracellular contents and disturbance of the cell enzymatic systems (Palmqvist & Hahn-Hägerdal 2000). However, the mechanism of their inhibition has not been thoroughly investigated, mainly because of the lack of adequate analyses (qualitative and quantitative). Many studies of phenolic inhibition on anaerobic digestion were conducted by using far higher concentrations than are practically present in the substrate (Delgenes et al. 1996; Mikulásová et al. 1990).

16.3.3 *Enzymatic activity of microorganisms in the digestates*

The API ZYM system is a semi-quantitative micro-method allowing for rapid determination of 19 enzymatic reactions, applied especially in medicine. So far, little is known about using it in other fields e.g. in analyzing the enzymes responsible for the reactions occurring during anaerobic digestion.

The extracellular enzyme activity of the bioreactors feedstocks was defined. The results obtained after using the 5-point scale of the API ZYM system are presented in Table 16.5.

For the most of studied enzymes (alkaline phosphatase, esterase (C4), esterase lipase (C8), lipase (C14), leucine arylamidase, valine arylamidase, cystine arylamidase, trypsin, α-chymotrypsin, acid phosphatase, β-galactosidase, α-glucosidase and β-glucosidase) no significant diversity in the intensity of reactions in particular micro-cupules were observed. Acid and alkaline phosphatases catalyze the hydrolysation of phosphoric esters, according to their pH optima (Bergmeyer 2012). However, both forms were presented in the examined digestates in high concentration despite their alkaline pH. Esterase participates in ester bonds hydrolysation, while lipases catalyze triacylglycerol hydrolysis and enable amides and peptides synthesis (Kołodziejska et al. 2013). Arylamidases catalyze hydrolysis of some peptides (BRENDA). Trypsins catalyze hydrolysis of bonds in proteins and peptides (Simpson 2006). *β*-galactosidases catalyze hydrolysis of galactose residues (BRENDA). Glucosidases participate in hydrolyzing glucosides. β-glucosidase, taking part in cellulose hydrolysis (Malý & Siebielec 2016), was the one of the three enzymes for which the highest activity was stated in all the analyzed samples.

Only in the case of four enzymes high variation in activity was noted: for naphthol-AS-BI-phosphohydrolase (from 3 to 5), for α-galactosidase (from 0 to 3), for β-glucuronidase (from 1 to 4) and for n-acetyl-β-glucosaminidase (from 2 to 4). Considering naphthol-AS-BI-phosphohydrolase activity, slight decrease after Cu IDHA chelate addition was observed, simultaneously with its activity growth after $CuSO_4$ dose I addition, in comparison to the feedstock without copper. Phosphohydrolase catalyzes phosphates hydrolysis (BRENDA). In turn, *α*-galactosidase catalyzes galactose residues

Table 16.5. Read-outs of intensity of colorations in particular micro-cupules of plastic gallery of API ZYM strip according to comparative table from the manufacturer for bioreactors feedstocks (the means of three separate read-outs)

	Enzymes																			
Reactor	1	2	3	4	5	6	7	8	9	10	11	12	13	14	15	16	17	18	19	20
Inoculum		5	3	4	5	5	4	3	4	4	5	5	1	5	4	4	5	4	0	0
WSCS +$CuSO_4$ dose I		5	4	4	5	5	4	3	4	4	5	5	3	5	3	4	5	4	0	0
WSCS +$CuSO_4$ dose II		5	3	4	5	5	3	2	4	4	4	4	0	5	2	3	5	3	0	0
WSCS +Cu IDHA dose I	control	4	3	4	5	4	3	2	4	4	4	3	2	5	1	3	5	2	0	0
WSCS +Cu IDHA dose II		4	3	4	5	4	3	2	4	4	4	3	2	5	2	4	5	2	0	0
WSCS without Cu addition		5	3	4	5	5	3	2	4	3	4	4	2	5	2	4	5	3	0	0

Explanations: color intensity scale: $5 \geq 40$ nM of hydrolyzed substrate; $4 \geq 30$ nM of hydrolyzed substrate; $3 \geq 10$ nM of hydrolyzed substrate; $2 \geq 5$ nM of hydrolyzed substrate; $1 < 5$ nM of hydrolyzed substrate; 0 no activity. Enzymes: 2 – alkaline phosphatase, 3 – esterase (C4), 4 – esterase lipase (C8), 5 – lipase (C14), 6 – leucine arylamidase, 7 – valine arylamidase, 8 – cystine arylamidase, 9 – trypsin, 10 – α-chymotrypsin, 11– acid phosphatase, 12 – naphthol-AS-BI-phosphohydrolase, 13 – α-galactosidase, 14 – β-galactosidase, 15 – β-glucuronidase, 16 – α-glucosidase, 17 – β-glucosidase, 18 – n-acetyl-β-glucosaminidase, 19 – α-mannosidase, 20 – α-fucosidase.

hydrolysis e.g. galactooligosaccharides and polysaccharides (Shabalin et al. 2002). In the case of α-galactosidase, its activity increases after $CuSO_4$ dose I and it decreases after $CuSO_4$ dose II addition was noted (in comparison to the feedstock without Cu). β-glucuronidase participates in the hydrolysis of β-glucuronoside, while n-acetyl-β-glucosaminidase catalyzes the hydrolyzation of hexosamine residues (BRENDA). Cu IDHA chelate addition decreased the β-glucuronidase activity, whereas the $CuSO_4$ addition caused its activity growth after comparison to WSCS without Cu addition. In the case of n-acetyl-β-glucosaminidase, slight activity decrease after Cu IDHA chelate addition was observed and its growth in WSCS+$CuSO_4$ dose I, in comparison to the feedstock without copper.

The feedstocks activity of α-mannosidase and α-fucosidase was not exhibited in any of the bioreactors. These enzymes take part in hydrolysis of the saccharides which contain mannose and fucose, respectively (BRENDA).

16.4 CONCLUSIONS

The studies showed that the addition of copper in two different concentrations and chemical forms did not significantly affect the biogas potential of the digested feedstock. The concentrations of phenols in the digestates obtained after fermentation of the feedstocks enriched with Cu and pure phenol, measured at the end of the experiment, did not differ significantly from those obtained in digestates to which no phenol was added. Thus, it was concluded that the increase in the copper concentration up to 60 mg L^{-1} stimulated the processes of anaerobic biodegradation of phenol without lowering the biogas production. However, the addition of copper did not influence the activity of most of the examined enzymes. Only four out of nineteen enzymes used in the standard biochemical tests API ZYM showed the visible differentiation in terms of the concentration. There were the enzymes involved in the decomposition of monosaccharide – galactose, complex carbohydrates – β-glucuronoside, hexosamine residues and phosphates.

It was stated that further studies on the influence of Cu concentration in a wider range on biodegradation of different types of phenolic compounds during the digestion of other types of feedstock are needed to confirm these observations.

ACKNOWLEDGEMENTS

This work has been prepared as a part of the implementation of statutory tasks No. FNM/130-12/2019 of the Faculty of Environmental Engineering of Lublin University of Technology.

REFERENCES

Ahring, B.K. & Westermann, P. 1985. Sensitivity of thermophilic methanogenic bacteria to heavy metals. *Current Microbiology* 12:273–276.

Bergmeyer, H.-U. 2012. *Methods of Enzymatic Analysis V2, 2.* Elsevier.

BRENDA, The Comprehensive Enzyme Information System, https://www.brenda-enzymes.info

Chandel, A.K., Silva, S.S., Singh, O.V. 2011. Detoxification of lignocellulosic hydrolysates for improved bioconversion of bioethanol. In Bernardes MAS (ed.), *Biofuel production-recent developments and prospects. InTech*: 225–246. Rijeka: In Tech.

Chapleur, O., Madigou, C., Civade, R., Rodolphe, Y., Mazéas, L., Bouchez, T. 2016. Increasing concentrations of phenol progressively affect anaerobic digestion of cellulose and associated microbial communities. *Biodegradation* 27: 15–27.

Delgenes, J.P., Moletta, R., Navarro, J.M. 1996. Effects of lignocellulose degradation products on ethanol fermentations of glucoseand xylose by Saccharomyces cerevisiae, Pichia stipitis, and Candida shehatae. *Enzyme and Microbial Technology* 19: 220–225.

Demirel, B. & Scherer, P. 2011. Trace element requirements of agricultural biogas digesters during biological conversion of renewable biomass to methane. *Biomass & Bioenergy* 35(3): 992–998.

Dokulilová, T., Koutný, T., Vítěz, T. 2018. Effect of zinc and copper on anaerobic stabilization of sewage sludge. *Acta Universitatis Agriculturae et Silviculturae Mendelianae Brunensis* 66(2): 357–363.

Fang, H.H.P., Liu, Y., Ke, S.Z., Zhang, T. 2004. Anaerobic degradation of phenol in wastewater at ambient temperature. *Water Science and Technology* 49(1): 95–102.

Fedorak, P.M. & Hrudey, S.E. 1984. The effects of phenol and some alkyl phenolics on batch anaerobic methanogenesis. *Water Research* 18(3): 361–367.

Glass, J.B. & Orphan, V.J. 2012. Trace metal requirements for microbial enzymes involved in the production and consumption of methane and nitrous oxide. *Frontiers in Microbiology* 3: 61.

Huang, A. 2008. Effects of Cd(II) and Cu(II) on microbial characteristics in 2-chlorophenol-degradation anaerobic bioreactors. *Journal of Environmental Sciences* 20: 745–752.

Kołodziejska, R., Karczmarska-Wódzka, A., Tafelska-Kaczmarek, A., Studzińska, R., Dramiński, M. 2013. Enantioselective enzymatic desymmetrization catalyzed in the presence of lipase. Part I. Prochiral compounds. *Wiadomości chemiczne* 67: 7–8.

Larsson, S., Quintana-Sáinz, A., Reimann, A., Nilvebrant, N.-O., Jönsson, L.J. 2000. Influence of lignocellulose-derived aromatic compounds on oxygen-limited growth and ethanolic fermentation by Saccharomyces cerevisiae. *Applied Biochemistry and Biotechnology* 84: 617–632.

Leven, L., Nyberg, K., Korkea-aho, L., Schnürer, A. 2006. Phenols in anaerobic digestion processes and inhibition of ammonia oxidising bacteria (AOB) in soil. *Science of The Total Environment* 364(1–3): 229–238.

Malý, S. & Siebielec, G. 2015. Badania egzogennej materii organicznej w celu bezpiecznego stosowania do gleby, Publisher: Ústřední kontrolní a zkušební ústav zemědělský, Brno.

Mikulásová, M., Vodny, S., Pekarovicová, A. 1990. Influence of phenolics on biomass production by Candida utilis and Candida albicans. *Biomass* 23: 149–154.

Mussatto, S.I. & Roberto, I.C. 2004. Alternatives for detoxification of diluted-acid lignocellulosic hydrolyzates for use in fermentative processes: a review. *Bioresource Technology* 93: 1–10.

Myszograj, S., Stadnik, A., Płuciennik-Koropczuk, E. 2018. The influence of trace elements on anaerobic digestion process. *Civil and Environmental Engineering Reports* 28(4): 105–115.

Nayono, S.E. 2009. *Anaerobic digestion of organic solid waste for energy production.* Karlsruhe: KIT Scientific Publishing.

Nguyen, Q.-M., Bui, D.-C., Phuong, T., Doan, V.-H., Nguyen, T.-N., Nguyen, M.-V., Tran, T.-H., Do, Q.-T. 2019. Investigation of Heavy Metal Effects on the Anaerobic Co-Digestion Process of Waste Activated Sludge and Septic Tank Sludge. *International Journal of Chemical Engineering* 1: 1–9.

Nyberg, K., Schnurer, A., Sundh, I., Jarvis, A., Hallin, S. 2006. Ammonia-oxidizing communities in agricultural soil incubated with organic waste residues. *Biology and Fertility of Soils* 42(4): 315–323.

Palmqvist, E. & Hahn-Hägerdal, B. 2000. Fermentation of lignocellulosic hydrolysates II: inhibitors and mechanism of inhibition review. *Bioresource Technology* 74: 25–33.

Rowell, R.M. 2012. *Handbook of wood chemistry and wood composites, 2nd edition.* Boca Raton FL: CRC Press.

Shabalin, K.A., Kulminskaya, A.A., Savel'ev, A.N., Shishlyannikov, S.M., Neustroev, K. N. 2002. Enzymatic properties of α-galactosidase from *Trichoderma reesei* in the hydrolysis of galactooligosaccharides. *Enzyme and Microbial Technology* 30(2): 231–239.

Simpson, R.J. 2006. Fragmentation of protein using trypsyn. *Cold Spring Harbor Protocols* 5

Xie, Y., Hu, Q., Feng, G., Jiang, X., Hu, J., He, M., Hu, G., Zhao, S., Liang, Y., Ruan, Z., Peng, N. 2018. Biodetoxification of Phenolic Inhibitors from Lignocellulose Pretreatment using *Kurthia huakuii* LAM0618^T and Subsequent Lactic Acid Fermentation. *Molecules* 23(10): 2626.

Yu, H.Q. & Fang, H.H.P. 2001. Inhibition on acidogenesis of dairy wastewater by zinc and copper. *Environmental Technology* 22: 1459–1465.

Zayed, G. & Winter, J. 2000. Inhibition of methane production from whey by heavy metals – protective effect of sulfide. *Applied Microbiology and Biotechnology* 53(6): 726–731.

CHAPTER 17

Current research trends in the application of hydrodynamic cavitation in environmental engineering

M. Bis
Lublin University of Technology, Lublin, Poland

ABSTRACT: Hydrodynamic cavitation (HC) is a phenomenon of a sudden decrease in the pressure inside a liquid medium causing the formation of vapour and gas bubbles having a tremendous force of implosion. The basic concept of practical application of the cavitation is to employ these specific conditions and huge energy released by cavities in a positive way. The paper provides an overview of the latest applications of hydrodynamic cavitation in environmental engineering. The major applications of HC in wastewater treatment and degradation of different organic compounds (i.e. pesticides, pharmaceuticals, dyes) were presented. The challenges and prospects of HC applications in decomposition of lignocelluloses biomass, a pretreatment of biomass for biofuel production, and also water and wastewater disinfection were also discussed. Further, the optimization of geometrical and operating parameters was summarize and various designs of cavitation devices that affect the cavitation effectiveness were shown.

17.1 INTRODUCTION

Hydrodynamic cavitation (HC) is a very good example of how the human ingenuity can employ the undesirable phenomena in a positive way. Initially, the need for the cavitation research was forced by the failures in the introduction of propellers for the propulsion of ships. In 1754, Leonhard Euler reported the heterogeneity in liquids resulting from air cavities when accelerated liquids under negative pressure. In this way, he is considered to be the first person to postulate the possibility of cavitation (Chen et al. 2011). The first basic research and description of this phenomenon were presented by Osborne Reynolds in 1894, whereas Parsons conducted the first significant applied research. He identified the cavitation in a special tunnel with a special stroboscope of his own design and proposed a solution by developing the concept of using a turbine to propel ships (Tropea et al. 2007). Due to the destructive effects of cavitation on machine components and flow systems, the early research has focused on limiting the negative consequences of HC. The practical use of this phenomenon has been started only in the last 30 years. One of the first practical applications of cavitation concerned the destruction of yeast (Save et al. 1994) and microbial cells (Save et al. 1997). Pandit & Joshi (1993) studied hydrolysis of fatty acids, while Suslick et al. (1997) showed the chemical effect of cavitation by reaction of the decomposition of KI. Nowadays, HC is used in water and wastewater treatment, disintegration of activated sludge, microalgal cell disruption, and extraction of bio-components, freezing and gene transfer into cells or tissues. The latest applications include preparation of liquid emulsion membrane (LEM), internal surface finishing process or HC-assisted beer brewing. Cavitation is used in the sustainable production of environmentally-friendly fuels like biogas, biodiesel and bioethanol from different types of wastes in order to conduct their pretreatment. Currently HC is considered an innovative and energy efficient technique that has great prospects for development (Panda et al. 2020).

The term "cavitation" comes from the Latin word "cavitas" (emptiness, cavity) and is defined as the formation of a gas phase in a liquid. In other words, cavitation is a process whereby the liquid structure is ruptured to form a "hole" (Zhou et al. 1994). It is a phenomenon of phase change similar to boiling, differing in that boiling occurs as a result of temperature rise, while cavitation as a result of pressure drop. Cavitation occurs when the local fluid pressure takes a critical value, i.e. the saturated vapor

DOI 10.1201/9781003171669-17

pressure of the fluid. The required pressure drop could be provided by acoustic wave propagation through the working fluid, hydrodynamic pressure drop due flow to constrictions known as orifice and venturi as well as rotations in turbomachinery such as pumps or propellers (Gevari et al. 2020). As a result of collapsing bubbles, the local hot and high pressure spots of 5000 K and 500 atm are generated and a huge amount of energy is released. In addition, the generated local turbulence and aggressive shock waves upon collapse produce active molecules, which are capable of accelerating chemical reactions (Gevari et al. 2020).

In general, two mechanisms of cavitation can be distinguished, the physical one, in which the physical processes and thermolysis prevail, as well as the chemical effect, where the activity of free radicals is the most important. This mechanical effect of cavitation causes intensification of the physical processing application and enhances the rates in the transport processes, whereas the chemical effect results in the intensification of the chemical processing applications (Gogate & Patil 2015). On the basis of the generation method, cavitation can be divided into two main categories: acoustic and hydrodynamic. Acoustic cavitation is generated by means of ultrasound waves with a frequency in the range of 16 kHz – 2 MHz, while the hydrodynamic cavitation is the result of pressure variations of the liquid flowing the constriction. Compared to acoustic cavitation, HC is characterized by higher energy efficiency, ability to scale up and lower operating costs (Thanekar & Gogate 2018).

The paper presents the basic mechanism of hydrodynamic cavitation and the geometrical and operating parameters influencing the cavitation performance. Further, various designs of HC devices were shown and compared. Finally, the practical applications of hydrodynamic cavitation and the recent development of HC in environmental engineering are discussed.

17.2 FACTORS INFLUENCING CAVITATION PROCESS

Hydrodynamic cavitation is the occurrence of vapor cavities inside an initially homogeneous flowing liquid. Due to local high velocity, low pressures are induced, which "break" the liquid medium, initiating the formation of single cavitation bubbles and then bubble clouds. Then, the generated bubbles collapse rapidly, during which an intensive shock wave is emitted (Dular et al. 2016). Consequently, the hot spots, local turbulence and highly reactive free radicals are created and a very high temperature and pressure conditions are generated locally (Gevari et al. 2020). Under such conditions, it is difficult to determine what from a physical point of view is actually responsible for the obtained cleaning effect. Comparing the cavitation results obtained by different researchers with each other is difficult due to the large number of parameters affecting HC, including the parameters of liquid like: temperature, pH, kind and initial concentration of contamination, and process parameters like inlet pressure and time. In addition, various reactor designs, system volumes, parameters and construction of cavitation inductors generate a lot of possibilities for system modification. Under such conditions, studying and describing the phenomenon of cavitation is a challenge. In order to compare the results of cavitation, some researchers use the concept of cavitation efficiency, relating the obtained effect to the cost of energy.

17.2.1 *Temperature*

The influence of the operating temperature on the cavitation performance has been investigated in numerous studies. According to literature reports, with the increase of temperature, the generation of cavities becomes more intense and the solvent vapor pressure inside the bubble increased. This leads to the lower energy of cavities collapse, which results in an attenuating effect and, consequently, lower cavitation efficiency (Gogate & Patil 2015; Joshi & Gogate 2012). A reduction of the cavitation performance caused by the higher volatility of the substance was observed by Wu et al. (2007) during chlorocarbons degradation as well as by Joshi & Gogate (2012) during pesticide (dichlorvos) removal. Dular et al. (2016) reported that the optimal operating temperature to pharmaceuticals removal was 50°C. An increase of temperature to 68°C resulted in a decrease of removal efficiency, because the water vapor fills the cavitation bubble, which amortizes cavitation bubbles collapse. For this reason, many cavitation reactors were equipped with a cooling system to counteract the temperature rise.

However, when the boiling point of the contamination is higher than that of water, the effect of vapor pressure inside the bubble is insignificant. Therefore, only a slight influence of temperature changes was observed during the degradation the imidacloprid, examined by Patil et al. (2014). However, on the other hand, an increase in temperature increases the kinetic rates of reaction. For these cavitation applications, when kinetics is more important than the number of cavitation events, a higher temperature improves the rate of the degradation process (Gągol et al. 2018). While some of the tests were prepared in such a way that the process was carried out at a constant temperature to exclude the effect of the thermal effect, Sun et al. (2020) indicated, that the thermal effect was a dominant factor affecting on the disinfection efficiency. The increase in temperature of the medium during cavitation provides additional energy that can be used to improve the disinfection effect. Additionally, the temperature change affects the cavitation intensity due to changes in the physicochemical properties of the liquid (Panda et al. 2020).

17.2.2 *Inlet pressure*

The operating inlet pressure in hydrodynamic cavitation reactors indicate energy supplied for the process (Dhanke & Wagh 2020). Many researchers agree that the optimal value of inlet pressure should be used, which depends on the type and design of the hydrodynamic reactor and the type of contaminants being removed (Bokhari et al. 2016; Gogate & Patil 2015). When the inlet pressure increases beyond the optimal value, the super cavitating conditions will occur, which disturb the growth of the bubbles and result in reduced cavitational intensity (Dhanke & Wagh 2020).

Several studies reported that an increase of the inlet pressure to a certain point contributes to an increase in the rate of decomposition reaction. Patil & Gogate (2012) reported that the extent of diclofenac sodium degradation increases along with the inlet pressure from 2 to 3 bar and decreases with the further increase in the pressure to 4 bar. The authors attribute the initial increase to a decrease in the cavitation number, which in turn increases the number density of cavities. However, the further decrease in the extent of degradation was attributed to the condition of choked cavitation i.e. cavities form a larger bubble, thus reducing the cavitational intensity due to incomplete or cushioned collapse of the cavities. Bokhari et al. (2016) found the optimal inlet pressure of 3 bar for the pretreatment of rubber seed oil in the HC reactor, while Bagal & Gogate (2014) reported a decrease of 2,4-dinitrophenol degradation beyond the optimal value of inlet pressure of 4 bar.

The efficiency of the hydrodynamic cavitation reactor depends on the energy dissipated in the reaction volume, which is associated with the cavitation intensity. The relationship of the pressure and flow parameters with the cavitation intensity is described by a dimensionless parameter called cavitation number, determined by Equation (17.1) (Rajoriya et al. 2017):

$$C_v = \frac{P_2 - P_v}{\frac{1}{2}\rho \upsilon^2} \tag{17.1}$$

where P_2 is the static pressure in undisturbed flow (recovered outlet pressure), [Pa] P_υ is the saturated vapor pressure of water at a given temperature, [Pa], ρ is the liquid density, [kg m^{-3}], υ is the fluid velocity in a constriction, [m s^{-1}].

Typically, as the operating pressure increases, the fluid flow velocity in the system and the throat velocity also increase; in turn, this reduces the number of cavitation. It is assumed that bubbles can be generated under the conditions where the $C_v \leq 1$, so lower cavitation numbers give better results (Arrojo & Benito 2008). Šarc et al. (2017) point to the fact, that the cavitation number cannot be used as a single parameter that gives the cavitation condition. During a series of tests with various parameter values (cavitating geometry, flow rate, pressure, temperature, and water-gas content) they show differences in cavitation size, appearance, and aggressiveness. They prove that the cavitation condition can be altered only by the slight change of one parameter, even if the cavitation appears visually the same.

17.2.3 *Initial concentration of contamination*

It is generally believed that using HC, the degradation rate decreases with the increase of the initial concentration of contamination. Khajeh et al. (2020) demonstrated that with the increase in the initial concentration of Direct Red (DR89) dye from 30 to 90 mg/L, the removal efficiency decreased from 36.3% to 17.5%. The effect was attributed to the fact that the quantity of pollutant increases along with the initial concentration, whereas the total concentration of OH radicals remains constant. Similar trends were demonstrated by Patil et al. (2014) for the effect of imidacloprid degradation. The authors associate it with the fact that the presence of a higher concentration of pollutants may also reduce the efficiency of the OH radicals formation. Patil & Gogate (2012) also reported similar results of the initial concentration of methyl parathion – the maximum extent of degradation was obtained at the lowest initial concentration. The authors explain that the pollutant and byproducts formed compete for the OH radicals in the solution.

An analogous trend was reported in the HC disinfection studies (Loraine et al. 2012). The higher initial concentrations found to delay the onset of disinfection, thereby reducing its effectiveness. This effect was attributed to higher viscosity or clumping of the biomass at concentrations above ~106 CFU/ml. In disinfection, the mechanical effects of the shock waves are more important than the chemical mechanisms with OH radicals (Arrojo et al. 2008).

17.2.4 *Solution pH*

Solution pH is a very important parameter related to the extent of degradation in the cavitation process. According to the literature, acidic pH of the solution favor the generation of hydroxyl radicals, and also the oxidation capacity of hydroxyl radicals is higher. Therefore, for the pollutants present in the molecular form, better degradation results are obtained at lower pH values of the solution. The higher removal efficiency of the COD and TOC removal from the kitchen waste effluent was observed at the solution pH of 3 (Mukherjee et al. 2020). Similarly, an optimal pH of 3 was found on removal of the DR89 dye (Khajeh et al. 2020). Bagal & Gogate (2014) investigated the effect the solution pH on the extent of diclofenac sodium degradation and obtained the best results at pH of 4. At a lower pH, diclofenac sodium occurs in a molecular form, and due to the hydrophobicity it can penetrate the gas-water interface of the cavities and react with the hydroxyl radicals at an increasing rate.

In turn, Rajoriya et al. (2017) studied the degradation of cationic dye Rhodamine 6G (Rh6G) and reported that extend of decolorization increased with increasing solution pH from 2.0 to 10.0. They explained that an Rh6G molecule becomes hydrophobic under basic conditions and locate itself at the cavity-water interface, where the maximum concentration of OH radicals and therefore higher decolorization rate was obtained. On the other hand, under acidic pH, an Rh6G molecule becomes hydrophilic and remains in the bulk solution, where the concentration of OH radicals is minimal. Therefore, lower decolorization rate was achieved. Similarly, Wang & Zhang, (2009) studied the degradation of alachlor and reported an increase of the rate constant along with the increasing pH. Thus, the effect of solution pH on the degradation also depends on the investigated compounds and their individual chemical properties.

17.2.5 *Process time*

Time of cavitation process is associated with the number of cavitation cycles (passes thought the cavitation zone), which related with the volume of the entire system and flow rate. The greatest performance is obtained in the first minutes of cavitation. Sun et al. (2020) studied disinfection using HC and achieved the effect after 8–14 min. depending on the flow. The organic pollutants resistant to oxidation require longer residence time in the cavitation zone (Gągol et al. 2018). Wang et al. (2020) during the treatment of explosive wastewater, rich in organic pollutants, such as benzene derivatives, found moderate efficiency of COD removal during the first 10 min. of the process. During the degradation of large molecules of organic matter, the initial stages of oxidation involve decomposition into smaller molecules and a ring-opening reaction. In many studies, the process time did not exceed 60 min. (Wang et al. 2020), but in a few studies, it reached even 130 min. (Khajeh et al. 2020).

Extending the cavitation time is associated with higher energy expenditure; therefore, the benefits of cavitation should be considered in terms of energy.

17.3 CAVITATION CHAMBER DESIGN

The most commonly used method for generating hydrodynamic cavitation is the passage of the liquid through a constriction (single and multiple orifice or venturi). In that method, the cross-sectional area is changing, compressing and accelerating passing fluids, cause the static pressure are converted to kinetic energy (Panda et al. 2020). The intensity of the cavity collapse and the cavitational yield depend on the geometry of the device: geometrical configuration, throat perimeter and area, size and shape of the throat. Because of their simple construction and ease of use, they have been widely employed for the effectiveness and mechanism research of the cavitation technology (Sun et al. 2020). The main types of cavitation devices are shown in Table 17.1.

Cavitation can also be generated by the different type of rotational reactors, in an area swept by high-speed rotors. In the study by Mezule et al. (2010) cavitation took place in a liquid layer within the space between rotor and the plate. In turn, Petkovšek et al. (2013) and Zupanc et al. (2014) investigated the shear-induced cavitation generator, consist of two facing rotors. By turning the rotors in opposite directions, many zones of low pressure are formed, which leads to more aggressive cavitation extents over a larger volume. So-called "shear cavitation" forms between the aligned counter-teeth of the rotors when the space between them resembles the Venturi geometry (Dular et al. 2016).

In contrast, in a device known as the rotor-stator, a rotor moves inside a concentric stator and both are equipped with a variable number of vanes on their outer and inner surfaces. The liquid flow is similar to that of many Venturi constrictions in series, and changing the rotation speed and the number of rotor and stator vanes allows modifying the frequency of cavitation events (Cerecedo et al. 2018).

Wang et al. (2008; 2014) generated the cavitation in a swirling jet reactor. In this device, liquid is injected into the swirling cavitation chamber through the injection ports and passes through the chamber forming a swirling stream in which cavitation bubbles are generated. Then, the swirling jet hits the bottom surface of the connected chamber, where the cavitation bubbles collapse. The swirling jet-induced cavitation reactors can operate at lower pressure compared to conventional one. A device called Ecowirl, patented by Econovation GmbH, Germany is based on a similar operating concept (Mancuso et al. 2016). The Ecowirl consist of a frustum-conical pre-swirling chamber with 6 injection slots, which divide the stream into six single vortices generating a helical stream, where cavitation bubbles are generated. Then, the braided stream is ejected to the double cone chamber and impact to an insert, where the pressure is risen rapidly causing the cavitation bubbles collapse.

In the devices described above, the entire stream was cavitated directly, passed through the cavitation device. Zezulka et al. (2020) applied indirect cavitation, in which the cavitating stream is directed against a contaminated stream, thus producing a cavitation effect at the point of contact. In this way, a smaller cavitation stream can effectively clean a larger volume of liquid or run continuously.

The treatment effect and rate, as well as the economic efficiency of a reactor largely depend on the cavitator design. Some researchers point out that the HC-based processes need to be improved for industrial applications (Mancuso et al. 2016). Therefore, the design of high-performance reactors has recently become one of the most extensively studied issues (Sun et al. 2020).

Cavitation can also be generated by the different type of rotational reactors, in an area swept by high-speed rotors. In the study by Mezule et al. (2010) cavitation took place in a liquid layer within the space between rotor and the plate. In turn, Petkovšek et al. (2013) and Zupanc et al. (2014) investigated the shear-induced cavitation generator, consist of two facing rotors. By turning the rotors in opposite directions, many zones of low pressure are formed, which leads to more aggressive cavitation extents over a larger volume. So-called "shear cavitation" forms between the aligned counter-teeth of the rotors when the space between them resembles the Venturi geometry (Dular et al. 2016).

In contrast, in a device known as the rotor-stator, a rotor moves inside a concentric stator and both are equipped with a variable number of vanes on their outer and inner surfaces.

Table 17.1. The main types of cavitation devices.

				Reactor device	References
Direct cavitation-liquid passing through HC device	Non-rotational (convergence)	Orifice plate	single hole	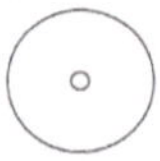	(Patil & Gogate, 2012) (Dhanke & Wang, 2020)
			mutli hole	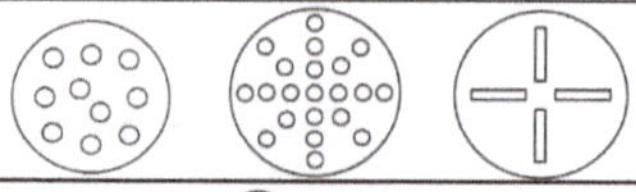	(Arrojo et al. 2008) (Dhanke & Wang, 2020)
		Venturi	Circular	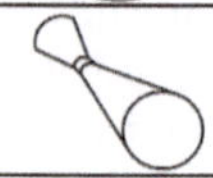	(Badve et al. 2015) (Rajonya et al. 2017)
			Rectangular	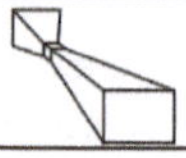	(Badve et al. 2015) (Rajonya et al. 2017)
			Venturi Array	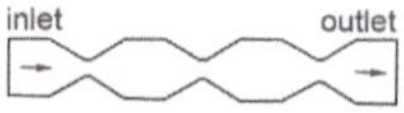	(Ladino et al. 2016)
	Rotational	Rotors	Shear-induced HC generator (two rotors)	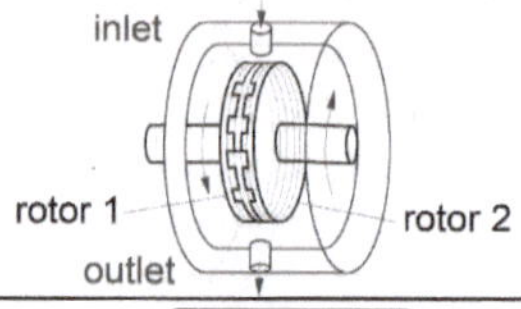	(Petkovšek et al. 2013) (Zupanc et al. 2014) (Dular et al. 2016)
			Rotor-stator	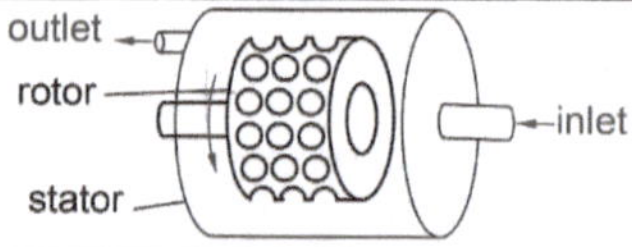	(Badve et al. 2013) (Cerecedo et al. 2018)
		Swirling device	Swirling chamber	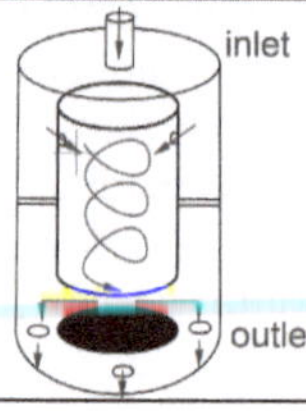	(Wang et al. 2008) (Wang et al. 2014)
			Ecowirl	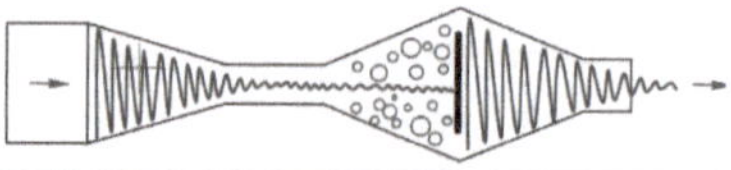	(Mancuso et al. 2016)
Indirect cavitation - high pressure jet stream is directed against its flow			High-pressure jet nozzle	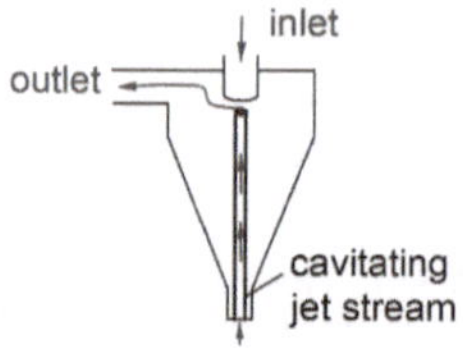	(Zezulka et al. 2020)

The liquid flow is similar to that of many Venturi constrictions in series, and changing the rotation speed and the number of rotor and stator vanes allows modifying the frequency of cavitation events (Cerecedo et al. 2018).

Wang et al. (2008; 2014) generated the cavitation in a swirling jet reactor. In this device, liquid is injected into the swirling cavitation chamber through the injection ports and passes through the chamber forming a swirling stream in which cavitation bubbles are generated. Then, the swirling jet hits the bottom surface of the connected chamber, where the cavitation bubbles collapse. The swirling jet-induced cavitation reactors can operate at lower pressure compared to conventional one. A device called Ecowirl, patented by Econovation GmbH, Germany is based on a similar operating concept (Mancuso et al. 2016). The Ecowirl consist of a frustum-conical pre-swirling chamber with 6 injection slots, which divide the stream into six single vortices generating a helical stream, where cavitation bubbles are generated. Then, the braided stream is ejected to the double cone chamber and impact to an insert, where the pressure is risen rapidly causing the cavitation bubbles collapse.

In the devices described above, the entire stream was cavitated directly, passed through the cavitation device. Zezulka et al. (2020) applied indirect cavitation, in which the cavitating stream is directed against a contaminated stream, thus producing a cavitation effect at the point of contact. In this way, a smaller cavitation stream can effectively clean a larger volume of liquid or run continuously.

The treatment effect and rate, as well as the economic efficiency of a reactor largely depend on the cavitator design. Some researchers point out that the HC-based processes need to be improved for industrial applications (Mancuso et al. 2016). Therefore, the design of high-performance reactors has recently become one of the most extensively studied issues (Sun et al. 2020).

17.4 HYDRODYNAMIC CAVITATION APPLICATIONS IN ENVIRONMENTAL ENGINEERING

17.4.1 *Disinfection*

Disinfection is the inactivation or destruction pathogenic microorganisms, which is a necessary stage of water treatment process. The chemical disinfection methods such as chlorination and ozonation are commonly used and effective techniques; however, they can produce dangerous or inconvenient organic disinfection byproducts (DBPs). Moreover, they are less effective for bacteria that hide in loose deposits or biofilms (Sun et al. 2020). Hydrodynamic cavitation, as a physical process, does not introduce any new chemicals to water which might change the properties, as smell or taste and affect the environment after water is released into the environment (Dular et al. 2016). Furthermore, it is also a cheaper method compared to most of the conventional ones (Gevari et al. 2020).

Many researchers have confirmed the high effectiveness of using HC to destroy various microorganisms. Dalfré Filho et al. (2015) reached an inactivation of more than 90% of bacteria *Escherichia coli* after 15 min. at a pressure of 10 MPa. Kosel et al. (2017) have shown that cavitation is sufficient for the inactivation of enteric virus MS2 and reached more than 4 logs reductions of viral infectivity. Sawant et al. (2008) tested the utility of cavitation for zooplankton deactivation in seawater and achieved over 80% disinfection efficiency. In turn, Mezule et al. (2010) achieved the reduction of Escherichia coli ability to divide by over 75% after just 3 min. of cavitation.

However, due to the fact that many processes are harmful to microorganisms occur during HC, the exact mechanism of bacterial inactivation is not understood (Pandur et al. 2020). It is generally assumed that the extreme conditions during bubble collapse and mechanical mechanisms of shock waves damage the cell membrane of bacteria. Moreover, some bacteria tend to form clusters to act as protection against biocides, and cavitation shock waves destroy the clusters and isolate the bacteria, which increases the effectiveness of additionally added disinfectants (Joyce et al. 2003). Loraine et al. (2012) examined the disinfection rates of gram-positive and gram-negative bacteria and showed that the gram-negative species with thinner cell walls were more amenable to hydrodynamic cavitation.

In this way, they confirm that the destructive mechanism of microorganisms is the damage of the cell wall. They also proved that higher initial concentrations of bacteria found to delay the onset of disinfection, thereby reducing its effectiveness, which is due to higher viscosity or clumping of the biomass at the concentrations above ~106 CFU/ml.

In turn, Cerecedo et al. (2018) applied transmission electron microscopy (TEM) for identification and interpretations of the cavitation mechanisms that contribute do the *E.coli* bacterial elimination. The researchers first observed morphological changes of the bacteria in a form of coagulation of cytoplasmic matter or absence of matter in the periplasmic space, followed by membrane ruptures and release of intracellular components. They explained the obtained results in the field of cytoplasmic matter coagulation with shock waves propagating to the liquid and high temperature at bubble implosion (hot spots). The membrane ruptures were mainly attributed to the extremely high shear stresses of micro-jets.

Loraine et al. (2012) investigated liposomes as model systems of study leakage and stability of lipid bilayers due to their similarity to the biological cell membrane. They reported that the lipid bilayer integrity is crucial for cell viability, and bacterial inactivation by HC is at least in part due to membrane disruption. They also note that extremely adverse local environment can result in interference with osmotic responsiveness, loss of intercellular material, and disruption of protein synthesis. In turn, Pandur et al. (2020) also applied hydrodynamic cavitation to destroy lipid bilayers (liposomes) and thus recognize several mechanisms of microorganism's destruction. They identified the role of highly reactive free radicals in the process, which oxidize phospholipids, leading to pore formation, destabilization, and disintegration of bilayers.

Many authors agree that it cannot be said with certainty what, from a purely mechanical point of view, is harmful to microorganisms. This is because, in HC reactors, besides imploding bubbles many other processes are triggered as well (Burzio et al. 2020). Šarc et al. (2014) proved that, in the disinfection process, sudden pressure changes are even more effective than imploding bubbles. Arrojo et al. (2008) have shown that more favorable conditions for disinfection occur in Venturi than in orifice plates, when larger bubbles are formed and more cavitation events and also extended pressure oscillations take place. In turn, Mezule et al. (2010) indicated that for reducing the metabolic activity of *E. coli*, the energy input of cavitation is more important than the exposure time.

Therefore, it would be extremely important to quantify the sensitivity of microorganisms to fluid shear and normal stresses when designing and optimizing the HC reactors (and to the non-dimensional parameters that control the magnitude of such stresses) (Burzio et al. 2020).

17.4.2 *Wastewater treatment*

The wastewater from agricultural and domestic activities, as well as from industrial processes, contains large amounts of recalcitrant organic compounds, such as textile dyes, pharmaceuticals, aromatic and phenolic compounds and chlorinated hydrocarbons. Conventional biological methods are not able to completely degrade them due to their biorefractory character and toxicity to microorganisms (Bagal & Gogate 2014). Effective, simple and cheap technologies are needed to decompose these biorefractory compounds. One of such methods may be hydrodynamic cavitation. It has been successfully applied for the degradation of various organic pollutants in wastewater such as dyes (Dhanke & Wagh 2020; Saharan et al. 2011; Rajoriya et al. 2017; Wang et al. 2014), pesticides (Gogate & Patil 2015), insecticides like imidaclopid (Patil et al. 2014), methyl parathion (Patil & Gogate 2012,) and alachlor (Wang & Zang 2009), pharmaceuticals (Bagal & Gogate 2013; Braeutigam et al. 2012; Musmarra et al. 2016; Rajoriya & Saharan, 2014) and phenol compounds (Bagal & Gogate 2013).

The decomposition of organic compounds can occur in three places: in the gas phase inside the bubble, at the gas-liquid interface and in the liquid volume phase (Arrojo & Benito 2008). Very high temperatures and pressures occur in the gas phase inside the cavitation bubble, which create ideal conditions for the pyrolysis and direct thermal reactions of volatile organic compounds (Figure 17.1). At the gas-liquid interface high temperature and pressure gradients exist, creating the conditions for the decomposition of semi-volatile, non-volatile and hydrophobic compounds. However, in the liquid volume phase at ambient temperature non-volatile and hydrophilic substances are decomposed (Dular et al. 2016; Wang et al. 2008).

Inside the cavitation bubbles, free radicals are formed which participle in the oxidation reaction mainly at the gas-liquid interface and only a small part of them penetrate into the bulk liquid (Tao et al. 2016).

At the same time, it has been reported that the mechanism of the cavitation-induced wastewater treatment should be attributed to the mechanical, chemical and thermal effects induced by bubble collapse. The mechanical effect like strong shear stress can break down the carbon-carbon bond allowing the breakdown of macromolecules into low-molecular organic compounds. During the so-called liquid-phase combustion process, under high-temperature conditions of the gas inside the bubble

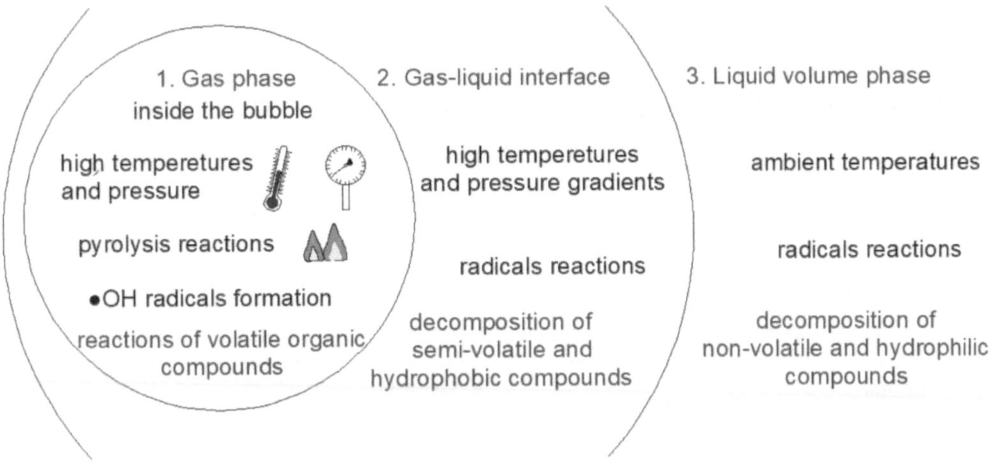

Figure 17.1. The scheme of the places of decomposition of organic compounds.

or at the gas-liquid interface, the organic molecules can be directly decomposed during bubble collapse (Tao et al. 2016).

The degradation process of organic compounds is determined by their structure and chemical properties. Dular et al. (2016) studied the resistance of various pharmaceutical compounds to hydrodynamic cavitation and observed different removal efficiencies under identical operating parameters. These studies proved that the individual chemical properties of the compounds played an important role. Due to low vapor pressure, the Rhodamine B dye is destructed at the gas-liquid interface or inside the liquid phase, by separating the polyaromatic rings from the chromophore by the hydroxyl radicals (Wang et al. 2008). Similarly, the degradation of ibuprofen occurs mainly by reaction with hydroxyl radicals at the bubble-liquid interface by the hydroxylation process followed by de-methylation or de-carboxylation (Musmarra et al. 2016). In the case of chitosan degradation, β-(1,4) glycosidic linkages were broken, but there was no modification of the chemical structure of chitosan (Huang et al. 2013). Some other studies show that hydroxyl radicals are largely responsible for the decomposition of organic compounds during HC (Patil & Gogate 2012; Rajoriya et al. 2017).

17.4.3 *Biomass pretreatment method for biofuel production*

The production of a biodegradable and renewable biofuels like biogas and bioethanol could limit the harmful environmental impact of conventional fossil fuels (Panda et al. 2020). Agricultural waste and herbaceous energy crops containing lignocellulose biomass are considered to be a favorable substrate used in the biofuel production processes due to the renewability and high energy value (Zieliński et al. 2019). However, effective decomposition of lignocellulose biomass into fermentable monomers is inhibited due to a complex structure composed of cellulose, hemicellulose, and lignin (Kim et al. 2015). Therefore, before further processing, pretreatment that releases carbohydrates and breaks them down into fermentable monomers is necessary. The commonly used methods of biological, physical or chemical pretreatment require expensive enzymes, high energy or chemical input. Additionally, the inhibitory by-products could also be created (Zieliński et al. 2019).

Hydrodynamic cavitation was successfully applied as a pretreatment for bioethanol production from reed (Kim et al. 2015). Similarly, HC was successfully used for *Sida hermaphrodita.* Silage, applied for 5 min. allowed to obtain an energy profit of 0.17 Wh/g TS (Zieliński et al. 2019). A similar result was reported for other potential applications of HC in improving the efficiency of alkaline pretreatment of sugarcane bagasse (SCB) (Terán Hilares et al. 2016) or pretreatment of rubber seed oil via esterification reaction (Bokhari et al. 2016).

In the process of HC-assisted pretreatment, the compositional changes of biomass take place, like lignin cellulose and hemicellulose degradation, resulted in a higher content of carbohydrates in the liquid (glucose release increase from 16 to 85%). In addition, the structural changes (increase in porosity) take place, which increases the efficiency of enzymatic hydrolysis. These changes are attributed to the physical effects of cavitation such as shockwaves generation during the violent collapse of cavities, which increase the specific surface area and porosity of biomass (Terán Hilares et al. 2020).

17.5 CONCLUSIONS

Hydrodynamic cavitation is of increasing interest in a variety of new applications in environmental engineering. Modern reactor designs are developed and optimal operating parameters are selected in order to achieve the maximum effects of disinfection, wastewater treatment, and biofuel production. The collapsing bubbles release a huge amount of energy and highly reactive hydroxyl radicals, which allows for the achievement of degradation results that cannot be achieved with other traditional methods.

REFERENCES

Arrojo, S. & Benito, Y. 2008. A theoretical study of hydrodynamic cavitation. *Ultrasonics Sonochemistry* 15: 203–211.

Arrojo, S. Benito, Y. & Martinez Tarifa, A. 2008. A parametrical study of disinfection with hydrodynamic cavitation. *Ultrasonics Sonochemistry* 15(5): 903–908.

Bagal, M.V. & Gogate, P.R. 2013. Degradation of 2,4-dinitrophenol using a combination of hydrodynamic cavitation, chemical and advanced oxidation processes. *Ultrasonics Sonochemistry* 20: 1226–1235.

Bokhari, A. Chuah, L.F. Yusup, S. et al. 2016. Optimisation on pretreatment of rubber seed (Hevea brasiliensis) oil via esterification reaction in a hydrodynamic cavitation reactor. *Bioresource Technology* 199: 414–422.

Braeutigam, P. Franke, M. Schneider, R.J. et al. 2012. Degradation of carbamazepine in environmentally relevant concentrations in water by hydrodynamic-acoustic cavitation (HAC). *Water Research* 46: 2469–2477.

Burzio, E. Bersani, F. Caridi, G.C.A. Vesipa, R. Ridolfi, L. & Manesa C. 2020. Water disinfection by orifice-induced hydrodynamic cavitation. *Ultrasonics Sonochemistry* 60: 104740.

Cerecedo, L.M. Dopazo, C. & Gomez-lus, R. 2018. Water disinfection by hydrodynamic cavitation in a rotor-stator device. *Ultrasonics Sonochemistry* 48: 71–78.

Chen, D. Sharma, S.K. & Mudhoo, A. Handbook on Applications of Ultrasound. Sonochemistry for Sustainability, July 26, 2011 by CRC Press.

Dalfré Filho, J.G. Assis, M.P. & Genovez, A.I.B. 2015. Bacterial inactivation in artificially and naturally contaminated water using a cavitating jet apparatus. *Journal of Hydro-environment Research* 9(2): 259–267.

Dhanke, P.B. & Wagh, S.M. 2020. Intensification of the degradation of Acid RED-18 using hydrodynamic cavitation. *Emerging Contaminants* 6: 20–32.

Dular, M. Griessler-Bulc, T. Gutierrez-Aguirre, I. Heath, E. et al. 2016. Use of hydrodynamic cavitation in (waste) water treatment. *Ultrasonics Sonochemistry* 29: 577–588.

Gągol, M. Przyjazny, A. & Boczkaj, G. 2018. Wastewater treatment by means of advanced oxidation processes based on cavitation – A review. *Chemical Engineering Journal* 338: 599–627.

Gevari, M.T. Abbasiasl, T. Niazi, S. Ghorbani, M. & Kosar, A. 2020. Direct and indirect thermal applications of hydrodynamic and acoustic cavitation: A review. *Applied Thermal Engineering* 171: 115065.

Gogate P.R. & Patil P.N. 2015. Combined treatment technology based on synergism between hydrodynamic cavitation and advanced oxidation processes. *Ultrasonics Sonochemistry* 25: 60–69.

Huang, Y. Wu, Y. Huang, W. Yang, F. & Ren, X.E. 2013. Degradation of chitosan by hydrodynamic cavitation. *Polymer Degradation and Stability* 98: 37–43.

Joshi, R.K. & Gogate, P.R. 2012. Degradation of dichlorvos using hydrodynamic cavitation based treatment strategies. *Ultrasonics Sonochemistry* 19(3): 532–539.

Joyce, E. Phull, S.S. Lorimer, J.P. & Mason, T.J. 2003. The development and evaluation of ultrasound for the treatment of bacterial suspensions. A study of frequency, power and sonication time on cultured Bacillus species. *Ultrasonics Sonochemistry* 10(6): 315–318.

Khajeh, M. Amin, M.M. Taheri, E. Fatehizadeh, A. & McKay, G. 2020. Influence of co-existing cations and anions on removal of direct red 89 dye from synthetic wastewater by hydrodynamic cavitation process: An empirical modeling. *Ultrasonics Sonochemistry* 67: 105133.

Kim, I. Lee, I. Jeon, S.H. Hwang, T. & Han, J.-I. 2015. Hydrodynamic cavitation as a novel pretreatment approach for bioethanol production from reed. *Bioresource Technology* 192: 335–339.

Kosel, J. Gutiérrez-Aguirre, I. Rački, N. Dreo, T. Ravnikar, M. & Dular, M. 2017. Efficient inactivation of MS-2 virus in water by hydrodynamic cavitation. *Water Research* 124: 465–471.

Ladino, J.A. Herrera, J. Malagón, D. et al. 2016. Biodiesel production via hydrodynamic cavitation: numerical study of new geometrical arrangements. *Chemical engineering transactions* 50: 319–324.

Loraine, G. Chahine, G. Hsiao, C.T. et al. 2012. Disinfection of gram-negative and gram-positive bacteria using DynaJets® hydrodynamic cavitating jets. *Ultrasonics Sonochemistry* 19: 710–717.

Mancuso, G. Langone, M. Laezza, M. et al. 2016. Decolourization of Rhodamine B: a swirling jet-induced cavitation combined with NaOCl. *Ultrasonics Sonochemistry* 32: 18–30.

Mezule, L. Tsyfansky, S. Yakushevich, V. & Juhna, T. 2010. A simple technique for water disinfection with hydrodynamic cavitation: effect on survival of Escherichia coli. *Desalination* 251: 152–159.

Mukherjee, A. Mullick, A. Teja, R. Vadthya, P. Roy, A. & Moulik, S. 2020. Performance and energetic analysis of hydrodynamic cavitation and potential integration with existing advanced oxidation processes: A case study for real life greywater treatment. *Ultrasonics Sonochemistry* 66: 105116.

Musmarra, D. Prisciandaro, M. Capocelli, M. Karatza, D. et al. 2016. Degradation of ibuprofen by hydrodynamic cavitation: reaction pathways and effect of operational parameters. *Ultrasonics Sonochemistry* 29: 76–83.

Panda, D. Saharan, V.K. & Manickam, S. 2020. Controlled Hydrodynamic Cavitation: A Review of Recent Advances and Perspectives for Greener Processing. *Processes* 8: 220.

Pandit, A.B. & Joshi, J.B. 1993. Hydrolysis of Fatty Oils: Effect of Cavitation. *Chemical Engineering Science* 48(19): 3440–3442.

Pandur, Ž. Dogsa, I. Dular, M. & Stopar, D. 2020. Liposome destruction by hydrodynamic cavitation in comparison to chemical, physical and mechanical treatments. *Ultrasonics Sonochemistry* 61: 104826

Patil, P.N. Bote, S.D. & Gogate, P.R. 2014. Degradation of imidacloprid using combined advanced oxidation processes based on hydrodynamic cavitation. *Ultrasonics Sonochemistry* 21: 1770–1777.

Patil, P.N. & Gogate, P.R. 2012. Degradation of methyl parathion using hydrodynamic cavitation: effect of operating parameters and intensification using additives. *Separation and Purification Technology* 95: 172–179.

Petkovšek, M. Zupanc, M. & Dular, M. et al. 2013. Rotation generator of hydrodynamic cavitation for water treatment. *Separation and Purification Technology* 118(30): 415–423.

Rajoriya, S. Bargole, S. & Saharan, V.K. 2017. Degradation of a cationic dye (Rhodamine 6G) using hydrodynamic cavitation coupled with other oxidative agents: reaction mechanism and pathway. *Ultrasonics Sonochemistry* 34: 183–194.

Rajoriya, S. & Saharan, V.K. 2014. Degradation of diclofenac sodium salt using hydrodynamic cavitation. G.C. Mishra (Ed.), Energy Technology and Ecological Concerns: A Contemporary Approach 82–86.

Saharan, V.K. Badve, M.P. & Pandit, A.B. 2011. Degradation of Reactive Red 120 dye using hydrodynamic cavitation. *Chemical Engineering Journal* 178: 100–107.

Šarc, A. Oder, M. & Dular, M. 2014. Can rapid pressure decrease induced by supercavitation efficiently eradicate Legionella pneumophila bacteria? *Desalination and Water Treatment* 57: 2184–2194.

Šarc, A. Stepišnik-Perdih, T. Petkovšek, M. Dular, M. 2017. The issue of cavitation number value in studies of water treatment by hydrodynamic cavitation *Ultrasonics Sonochemistry* 34: 51–59.

Save, S.S. Pandit, A.B. & Joshi, A.B. 1994. Microbial cell disruption: role of cavitation, The Chemical Engineering *Journal and the Biochemical Engineering Journal* 55(3): B67–B72.

Save, S.S. Pandit, A.B. Joshi, A.B. 1997. Use of Hydrodynamic Cavitation for Large Scale Microbial Cell Disruption, *Food and Bioproducts Processing* 75(1): 41–49.

Sawant, S.S. Anil, A.C. Krishnamurthy, V. Gaonkar, C. et al. 2008. Effect of hydrodynamic cavitation on zooplankton: a tool for disinfection. *Biochemical Engineering Journal* 42: 320–328.

Sun, X. Liu, J. Ji, L. Wang, G. Zhao, S. Yoon, J.-Y. & Chen S. 2020. A review on hydrodynamic cavitation disinfection: The current state of knowledge. *Science of The Total Environment* 737: 139606.

Suslick, K.S. Mdleleni, M.M. Ries, J.T. 1997. Chemistry Induced by Hydrodynamic Cavitation *J. Am. Chem. Soc.* 119(39): 9303–9304.

Tao, Y. Cai, J. Huai, X. et al. 2016. Application of Hydrodynamic Cavitation to Wastewater Treatment [J]. *Chemical Engineering & Technology* 39(8): 1363–1376.

Terán Hilares, R. Dionízio, R.M. Muñoz, S.S. et al. 2020. Hydrodynamic cavitation-assisted continuous pre-treatment of sugarcane bagasse for ethanol production: Effects of geometric parameters of the cavitation device. *Ultrasonics Sonochemistry* 63: 104931.

Terán Hilares, R. Dos Santos, J.C. Ajaz, & Ahmed, M. 2016. Hydrodynamic cavitation-assisted alkaline pretreatment as a new approach for sugarcane bagasse biorefineries. *Bioresource Technology* 214: 609–614.

Thanekar, P. & Gogate, P. 2018. Application of Hydrodynamic Cavitation Reactors for Treatment of Wastewater Containing Organic Pollutants: Intensification Using Hybrid Approaches. *Fluids* 3(4): 98.

Tropea, C. Yarin, A.L. & Foss J.F. (Eds.) 2007. *Springer Handbook of Experimental Fluid Mechanics*, Springer-Verlag Berlin Heidelberg.

Wang, J. Guo, Y. Guo, P. Yu, J. et al. 2014. Degradation of reactive brilliant red K-2BP in water using a combination of swirling jet-induced cavitation and Fenton process. *Separation and Purification Technology* 130: 1–6.
Wang, K. Jin, R. Qiao, Y. Wang, X. Wang, Ch. & Lu, Y. 2020. 2,4,6-Triamino1,3,5-Trinitrobenzene Explosive Wastewater Treatment by Hydrodynamic Cavitation Combined with Chlorine Dioxide. *Propellants Explosives Pyrotechnics* 45(8): 1243–1249.
Wang, X. Wang, J. Guo, P. Guo, W. & Li, G. 2008. Chemical effect of swirling jet-induced cavitation: Degradation of rhodamine B in aqueous solution. *Ultrasonics Sonochemistry* 15(4): 357–363.
Wang, X. & Zhang, Y. 2009. Degradation of alachlor in aqueous solution by using hydrodynamic cavitation. *Journal of Hazardous Materials* 161: 202–207.
Wu, Z-L. Ondruschka, B. & Bräutigam, P. 2007. Degradation of Chlorocarbons Driven by Hydrodynamic Cavitation. *Chemical Engineering & Technology* 30(5): 642–648.
Zezulka, Š. Maršálková, E. Pochylý, F. Rudolf, P. Hudec, M. & Maršálek, B. 2020. High-pressure jet-induced hydrodynamic cavitation as a pre-treatment step for avoiding cyanobacterial contamination during water purification. *Journal of Environmental Management* 255: 109862.
Zieliński, M. Rusanowska, P. Krzywik, A. Dudek, M. et al. 2019. Application of Hydrodynamic Cavitation for Improving Methane Fermentation of Sida hermaphrodita Silage. *Energies* 12: 526.
Zupanc, M. Kosjek, T. Petkovšek, M. Dular, M. et al. 2014. Shear-induced hydrodynamic cavitation as a tool for pharmaceutical micropollutants removal from urban wastewater *Ultrasonics Sonochemistry* 21: 1213–1221.
Zhou, Z.A. Xu, Z. & Finch, J.A. 1994. On the role of cavitation in particle collection during flotation-a critical review. *Minerals Engineering* 7(9): 1073–1084.

CHAPTER 18

Indoor air quality in fitness facilities

A. Staszowska & M.R. Dudzińska
Department of Indoor and Outdoor Air Quality, Lublin, Poland

ABSTRACT: The aim of this study was to assess the indoor air quality in selected rooms of three popular fitness facilities located in Lublin, Poland. Determination of the bacterial and fungal bioaerosol concentrations as well as carbon dioxide, temperature and relative humidity were carried out as part of the study. Parallel to the indoor air testing, the same parameters were monitored in the outdoor air. The concentration of carbon dioxide in each of the objects were correlated with the number of exercisers. The total number of bacteria (14-16255 CFU /m^3) increased during the measurement day along with the number of people in gym rooms and fitness rooms, and was higher than in the outdoor air. An inverse relationship was observed for the fungal aerosol (0–199 CFU /m^3), the concentrations of which depended on the season of the year in which the measurements were carried out.

18.1 INTRODUCTION

Millions of inhabitants of developed countries regularly use the services of professional gyms and fitness clubs. According to sociologists and economists, the popularity of this form of spending free time is associated with the growing awareness related to the need of leading the healthy lifestyle (WHO 2010). Among the most frequently chosen trainings, short but intense cardio or aerobic exercises are dominant. It is well-known that the indoor air quality at leisure and sports facilities can affect the health and well-being of occupants (Andrade & Dominski 2018; Mukherjee et al. 2014). The main source of biological pollution are the exercisers themselves. They are also responsible for the increase of air temperature, humidity and the CO_2 concentration indoors (Małecka-Adamowicz et al. 2019; Onchang and Panyakapo 2014). Good air quality in these facilities should be ensured by the properly designed and constructed ventilation and air conditioning installations. Unfortunately, these systems are often inefficient because ensuring the appropriate quality of indoor air in recreation and sports rooms is still a great challenge for sanitary engineers (Meadow et al. 2014). To date, there is no thermal comfort and air quality standard addressing fitness facilities. The difficulties in the design and implementation of ventilation or air conditioning installations in sports facilities are also caused by the high dynamics of changes in the indoor air parameters. This is connected to the variable number of exercising people and the intensity of performed workout (Slezakova et al. 2018a, 2018b). Hence, many of the facilities that operate on the recreational services market struggle with the problem of poor air quality, which can affect the effectiveness of the performed exercises and even the occupants' health (Ramos ae al. 2014 and 2015).

The most popular criterion for assessing the operation of a ventilation and air conditioning system is the measurement of the indoor CO_2 concentration. The concentrations higher than in the outside air testify to the air quality class (IDA1-4). Additionally, the study on the bacterial and fungal bioaerosol is also used in the indoor air quality assessment. Under normal conditions, most biological particles in the air do not pose a health risk.

However, some of them may have allergenic or toxic properties or cause infections of the upper and lower respiratory tract. Bearing in mind the popularity of the fitness facilities among the society, conducting the indoor air quality assessment research in these places is justified and even necessary because the available literature data are still very scarce.

The aim of this study was to assess the indoor air quality in the selected rooms (gym, fitness room, men's and women's locker rooms) of three popular fitness facilities located in the Lublin voivodship,

DOI 10.1201/9781003171669-18

Poland. The air quality in the rooms of the selected sports buildings was assessed on the basis of: (i) CO_2 concentration, (ii) bacterial and fungal aerosol concentration, and (iii) results of surveys conducted among users of sports facilities.

18.2 MATERIALS AND METHODS

18.2.1 *Sampling sites*

The present study was carried out in three fitness facilities (Facility 1, 2 and 3) located in the city of Lublin, Lublin voivodship, Poland. They are very popular among the residents of Lublin, which also affects their high occupancy rate. During the conducted research, the turnout among the exercisers was from 85 to 100% of the available places. The following room types were selected for testing in each of the facilities: gym, fitness room for group and individual exercises, women's and men's locker. All studied rooms (n=12) were located in the buildings constructed between 2010–2015. All buildings are insulated with glass wool, foamed polystyrene and plastered. In addition, the buildings are equipped with new window frames, and most rooms have gravitational ventilation with exception of the gyms boasting air conditioning systems. In the windows of aerobic exercise rooms, fresh air is supplied through window ventilators. During the initial inspection, there were no signs of moisture in the buildings, i.e. mold, efflorescence on the walls, ceilings or floor. The hygienic condition of buildings were good. All facilities are non-smoking and using electronic cigarettes is forbidden. The only complaint pertained to the blocked ventilation grilles in the lockers in Facility 1 and dusty wall air conditioners in all studied gyms.

Parallel to the indoor air quality testing, the same outdoor air parameters were monitored. The outdoor air quality was assessed on the basis of tests carried out at three measuring points. Their location was variable during subsequent research series, depending on meteorological conditions (mainly wind direction). The outdoor meteorological data came from local weather station. The research series took place in March, July and September 2017.

18.2.2 *CO_2 measurements*

The measurement of carbon dioxide concentration in both outdoor and indoor air was performed with the 160 IAQ (Testo) equipment. The measurements were continuous, carried out 24 hours a day. In the case of rooms, the number of loggers per one measuring point was 3 pieces. The loggers were placed at the height of the breathing zone of the exercisers, i.e. about 1.2 m. Additionally, the numbers of exercisers were monitored.

18.2.3 *Bacterial and fungal aerosol air sampling*

Individual exposure to the indoor bioaerosols in non-occupational environments is a subject of concern due to the related adverse health effects such as allergic diseases, asthma and respiratory tract infections. The microbiological air sampling was performed by MAS 100 Eco (Merck) air sampler using Petri dishes with the diameter of 90 mm. The air aspiration volume (depending on the expected contamination level) was 50–100 L for 1 minute for each sampling. The microbiological tests for the presence of bacteria and fungi were carried out in accordance with the recommendations contained in the Polish standards PN-89-Z-04111/02, PN-89-Z-04111/03, using appropriate media for the cultivation and identification of microorganisms. The conditions for the cultivation of microorganisms, reading the results and interpretation regarding the assessment of the indoor microbial pollution was carried out in accordance with the provisions contained in these standards. Each measurement was performed in three replications. In each air sample, the total number of bacteria, the number of mannitol positive staphylococci, the number of *Actinomycetes*, the total number of fungi, including mold and yeast were determined. The following cultural media types (BTL, Poland) were used in this study: TSA agar medium for bacteria (Trypticase Soya Agar), Sabouraud agar medium with chloramphenicol for fungi/ molds, Chapman medium for staphylococci, and Pochon medium for *Actinomycetes*.

For each plate, the calculation of CFU/m^3 (colony forming units per cubic meter) was obtained according to the device manufacturer guidelines.

As part of each of the three research series, five measurements were conducted during the day in three replications each. They were marked as: 1 – before the facility was opened (7 a.m.), 2 – at 12 p.m., 3 – at 4 p.m., 4 – at 6 p.m., 5 – after the facility was closed (at 9 p.m.), respectively. Each measurement series lasted 2 weeks. The bacterial and fungal bioaerosol measurements were performed on Mondays, Wednesdays and Fridays.

18.2.4 *Perceived indoor air quality*

For the purposes of this study, a questionnaire form was developed, which was voluntarily completed by the users of sports facilities before and after the exercises. Due to the fact that most of the respondents used several exercise rooms during their stay in the facility – the survey results were averaged. The survey form consisted of a series of questions that concerned: perceived air quality – odors, thermal comfort – temperature, humidity, and ventilation system performance. The total number of people who completed the surveys during the study was 214, including 142 women.

18.3 RESULTS AND DISCUSSION

18.3.1 *CO_2 concentration*

The concentration of carbon dioxide in the indoor air in each of the facilities changed throughout the day. The lowest values were recorded at night and in the morning and were numerically similar to the values measured in the outside air (450–520 ppm). However, the highest carbon dioxide concentration values (2864 ppm) occurred in the afternoon and were maintained until the facilities were closing – only applies to gyms and fitness rooms. This trend can be explained by the increasing number of people exercising indoors and, apparently, problems with the ventilation system, which was not able to provide IDA 1 class air in any of the examined facilities (Satish et al. 2012). On the basis of the

Table 18.1. Indoor air quality in studied Facilities 1–3 based on the fluctuations of the CO_2 concentration.

Category of indoor air quality	Sampling points			
	fitness room	gym	locker	
			D*	M*
FACILITY 1				
IDA 1 (high)	10%	7%	51%	56%
IDA 2 (medium)	28%	12%	45%	34%
IDA 3 (moderate)	3%	8%	4%	10%
IDA 4 (low)	59%	73%	–	–
FACILITY 2				
IDA 1 (high)	4%	2%	33%	28%
IDA 2 (medium)	25%	7%	55%	52%
IDA 3 (moderate)	45%	11%	12%	20%
IDA 4 (low)	26%	80%	–	–
FACILITY 3				
IDA 1 (high)	7%	9%	20%	22%
IDA 2 (medium)	26%	17%	68%	59%
IDA 3 (moderate)	53%	26%	12%	19%
IDA 4 (low)	24%	48%	–	–

*D-women's locker, M-men's locker

carbon dioxide concentrations in the indoor air of studied rooms, the percentage shares that allowed determining the air quality characteristics were calculated (Table 18.1) (EN 13779:2007).

In the premises of Facility 1 the air met the criteria of IDA class 1 only in the lockers. For fitness rooms and gyms, the air quality was classified to IDA class 4, which means that the measured concentration of carbon dioxide exceeds the value of the hygiene requirement recommended by the World Health Organization (WHO) and ASHRAE, i.e. 1000 ppm. As in the case of Facility 1, high and medium air quality was found in the rooms of women's and men's lockers of Facility 2. The lowest air quality was recorded in the gym. In the case of Facility 3, the air quality assessed on the basis of an increase in the concentration of carbon dioxide in the indoor air, belongs to IDA 3 and IDA 4 classes for most samples. While the situation in the case of a fitness room and gym can be explained by an inefficient ventilation system, in the case of both lockers with gravitational ventilation, it was caused by the possibly obstructed ducts and obstructed air flow resulting from the sealing of the ventilation grilles.

18.3.2 *Microbiological air quality*

The presence of bacterial microflora diversified in terms of quantity and quality was observed in the examined rooms. This composition was less varied in the morning and afternoon, which can be explained by the greater number of people exercising at that time. The air purity standards were most often exceeded by bacteria and *Actinomycetes*.

Table 18.2. Total number of bacteria (range in CFU/m^3).

Sampling time	FACILITY 1			
	fitness room	gym	D*	M*
7 a.m.	321–455	118–225	16–113	14–246
12 p.m.	148–606	139–369	124–397	333–450
4 p.m.	1112–3560	144–980	245–358	452–561
6 p.m.	3154–9880	230–1599	249–2697	888–1504
9 p.m.	10190–12430	1381–14057	2001–3400	3893–4300
outdoor air		48–52		
Sampling time	**FACILITY 2**			
	fitness room	gym	D	M
7 a.m.	56–44	236–667	26–88	33–148
12 p.m.	22–53	260–1088	120–365	226–561
4 p.m.	1147–1422	2600–4608	270–399	405–1669
6 p.m.	1204–11257	8016–12069	455–1106	562–1788
9 p.m.	3254–8924	11066–11806	908–2300	499–2978
outdoor air		39–57		
Sampling time	**FACILITY 3**			
	fitness room	gym	D	M
7 a.m.	169–351	366–507	33–154	49–191
12 p.m.	229–547	499–699	79–299	159–208
4 p.m.	304–1579	1564–2990	360–588	460–880
6 p.m.	608–2273	4560–11800	1440–1708	1900–2640
9 p.m.	2402–9877	6880–16255	2518–3084	2480–4990
outdoor air		42–114		

*D-women's locker, M-men's locker

Table 18.3. Concentrations of viable fungi aerosols in fitness facilities (range in CFU/m^3).

Sampling time	FACILITY 1			
	fitness room	gym	D*	M*
7 a.m.	15–22	22–43	14–18	16–22
12 p.m.	11–36	25–39	10–19	0–28
4 p.m.	40–68	41–88	16–31	0–44
6 p.m.	39–71	40–199	21–30	16–46
9 p.m.	34–73	28–61	22–40	29–59
outdoor air	25–45			
Sampling time	**FACILITY 2**			
	fitness room	gym	D	M
7 a.m.	22–38	12–34	8–21	16–26
12 p.m.	16–27	10–28	12–18	12–30
4 p.m.	16–24	25–44	20–38	20–41
6 p.m.	35–49	36–87	24–40	32–51
9 p.m.	44–82	67–142	36–128	88–102
outdoor air	44–389			
Sampling time	**FACILITY 3**			
	fitness room	gym	D	M
7 a.m.	10–22	12–20	12–19	22–28
12 p.m.	33–42	12–28	20–25	20–32
4 p.m.	16–38	14–57	20–34	15–30
6 p.m.	44–68	40–61	29–59	14–24
9 p.m.	31–80	31–110	43–128	33–92
outdoor air	37–351			

*D-women's locker, M-men's locker

Tables 18.2 to 18.5 present the concentrations of the monitored microbe groups in the indoor air. The range values from all study series are shown together with the results of the outdoor air measurements.

The total number of bacteria (Table 18.2) determined in the samples of the indoor air was greater than in the ambient air in all rooms of the tested facilities in each of the measurement series. This relationship is consistent with the current knowledge about the concentration of bacterial bioaerosol in this type of room, because the basic source of bacteria indoors are their users, i.e. in this case the persons exercising (Jo & Seo 2005). The concentration of the bacterial bioaerosol increased during the measurement day along with the number of occupants. This concentration can also be explained by the inefficiency of the ventilation system in individual rooms, which could not keep up with the supply of fresh air stream to refresh it. Another explanation may be the hygienic condition of the ventilation ducts. A cursory inspection carried out before the start of the tests showed strong dust accumulation and difficult flow through the clogged ventilation grilles, e.g. in lockers,. The condition of the air conditioners also raised objections, showing biological growth and heavy dust load. The measured bacterial aerosol concentration indicates that by 4 p.m. the air in the examined rooms was clean from the sanitary point of view, but close to the recommended standard limits which is 1000 CFU/m^3. After 4 p.m., the air quality deteriorated rapidly, reaching the criterion of heavily polluted air.

An inverse relationship was observed for the fungal bioaerosol. Its concentrations were higher in rooms than in ambient air in series No. 1 in March. In the research series No. 2 and 3 (July and September), lower concentrations of fungi in the room than in atmospheric air were noted periodically.

Table 18.4. Concentrations of *Staphylococci* mannitolo(+) (range in CFU/m^3).

Sampling time	FACILITY 1			
	fitness room	gym	D*	M*
7 a.m.	0–5	0–19	0	0
12 p.m.	0–20	5–33	0	0
4 p.m.	0–19	17–40	0–15	0–22
6 p.m.	22–52	30–62	0–22	4–18
9 p.m.	19–55	12–58	2–36	22–43
outdoor air		0		
Sampling time	**FACILITY 2**			
	fitness room	gym	D	M
7 a.m.	0	0	0	0
12 p.m.	0–4	0–9	2–5	5–11
4 p.m.	0–12	5–14	12–17	10–25
6 p.m.	10–33	22–29	16–26	19–27
9 p.m.	15–46	12–40	24–33	22–49
outdoor air		0		
Sampling time	**FACILITY 3**			
	fitness room	gym	D	M
7 a.m.	0	0	0	0
12 p.m.	2–8	0–15	0–6	4–10
4 p.m.	2–6	4–9	3–15	6–22
6 p.m.	11–24	2–28	16–18	9–27
9 p.m.	10–36	13–44	10–28	16–46
outdoor air		0		

*D-women's locker, M-men's locker

This is a known relationship, because in summer and autumn the concentration of fungi in the outdoor air naturally increases in tepid climate (Jones & Harrison 2004).

The concentrations of staphylococci indicate the category of air pollution from clean to heavily polluted, with a predominance of air with a high pollution (Table 18.4).

Exercisers should be seen as a source of these bacteria in the indoor air. The highest concentrations of staphylococci were noted during the research series No. 1, i.e. at the turn of winter and spring. This period is particularly conducive to the bacterial infections of the upper and lower respiratory tract. Many of the practitioners reported problems like a runny nose, cough or hoarseness during the survey. However, in their opinion, this was not a reason to give up training. Staphylococci were not found in the outside air.

18.3.3 *Perceived indoor air quality*

For the needs of the research, a questionnaire form was developed, which was completed by the users of sports facilities before and after the exercises. Due to the fact that most of the respondents used several exercise rooms during their stay in the facility, the survey results were averaged. The survey form consisted of a series of questions that concerned: (i) perceptible air quality – odors, (ii) thermal comfort – temperature, humidity, (iii) ventilation system performance. The total number of people who completed the surveys during the study was 214, including 142 women.

Table 18.5. Concentrations of *Actinomycetes* (range in CFU/m^3).

Sampling time	FACILITY 1			
	fitness room	gym	D*	M*
7 a.m.	0	0	0	0–14
12 p.m.	0	0	7–16	2–35
4 p.m.	0–5	0	12–44	38–40
6 p.m.	3–7	2–15	19–34	22–51
9 p.m.	5–20	10–32	38–45	43–59
outdoor air	32-105			
Sampling time	**FACILITY 2**			
	fitness room	gym	D	M
7 a.m.	0	0	0	0
12 p.m.	0–7	0–12	0	15–26
4 p.m.	0–12	0–21	6–24	18–39
6 p.m.	4–14	12–19	10–37	46–52
9 p.m.	9–18	11–23	22–51	37–58
outdoor air	5–44			
Sampling time	**FACILITY 3**			
	fitness room	gym	D	M
7 a.m.	0	0	0	0
12 p.m.	0–7	0–12	14–22	21–28
4 p.m.	0–22	0–16	23–37	19–46
6 p.m.	6–18	0–24	31–74	24–99
9 p.m.	4–36	7–25	19–66	16–84
outdoor air	6–61			

*D-women's locker, M-men's locker

In the morning, with incomplete occupancy of the exercise rooms, the users had no comments regarding the air quality and did not raise any objections. This concerned both the feelings before and after training.

In the afternoon, with the increase in the number of people exercising, the number of people complaining about the air quality in exercise rooms has increased rapidly. The main objections concerned unpleasant odors as well as the "stale" and heavy air in the rooms. The respondents also pointed out that they were getting tired faster by doing the same series of exercises (Revel & Arnesano 2014).

The temperature in the exercise rooms also raised reservations among the respondents. According to 82% of the people who finished their training, the temperature and humidity were too high. The most critical comments about the poor air quality were made by the exercising women (91%). Among men, this percentage was lower and amounted to 64%.

18.4 CONCLUSIONS

- the carbon dioxide concentration in the indoor air indicates moderate and low indoor air quality
- the ventilation system does not fulfill its role in any of the studied buildings

- most of the measured carbon dioxide concentrations exceed the WHO recommended hygiene standard of 1000 ppm
- the air in gyms was heavily contaminated with the bacterial aerosol, the main source of which were the occupants
- mycologically, the air quality can be considered as a clean
- a sanitation and maintenance program for the monitored facilities is needed to ensure healthier space for the indoor physical activity.

ACKNOWLEDGEMENTS

This work was supported by the Ministry of Science and Higher Education in Poland within funding FN-85/IŚ/2020.

REFERENCES

Andrade, A., Dominski, F.H. 2018. Indoor air quality of environments used for physical exercise and sports practice: systematic review. *Journal of Environmental Management* 206: 577–586.

EN 13779:2007. Ventilation for non-residential buildings. Performance requirements for ventilation and room-conditioning systems.

Frankel, M., Bekö, G., Timm, M., Gustavsen, S., Hansen, E.W., Madsen, A.M. 2012. Seasonal variations of microbial exposure and their relation to temperature, relative humidity, and air exchange rate. *Applied and Environmental Microbiology* 78: 8289–8297.

Jo, W.K., Seo, Y.Y. 2005. Indoor and outdoor air bioaerosol levels at recreation facilities, elementary schools, and homes. *Chemosphere* 61: 1570–1579.

Jones, A.M., Harrison, R.M. 2004. The effects of meteorological factors on atmospheric bioaerosol concentrations. *Science of the Total Environment*: 326: 151–180.

Małecka-Adamowicz, M., Kubera, Ł., Jankowiak, E., Dembowska, E. 2019. Microbial diversity of bioaerosol inside sport facilities and antibiotic resistance of isolated *Staphylococcus spp*. *Aerobiologia* 35: 731–742.

Meadow, J.F., Altrichter, A.E., Kembel, S.W., Kline, J. et al. 2014. Indoor airborne bacterial communities are influenced by ventilation, occupancy, and outdoor air sources. *Indoor Air*: 24, 41–48.

Mukherjee, N., Dowd, S.E., Wise, A., Kedia, S., Vohra, V., Banerjee, P. 2014. Diversity of bacterial communities of fitness center surfaces in a U.S. Metropolitan Area. *International Journal of Environmental Research and Public Health* 11: 12544–12561.

Onchang, R., Panyakapo, M. 2014. The physical environments and microbiological contamination in three different fitness centers and participants' expectations: measurements and analysis. *Indoor and Built Environment* 25: 213–228.

Polish Standard PN-89/Z-04111/02. Air purity protection. Microbiological testing. Determination of the bacteria number in the outdoor air (emission) with sampling by aspiration and sedimentation method. National Standards Body (PKN), Warszawa. Poland.

Polish Standard PN-89/Z-04111/03. Air purity protection. Microbiological testing. Determination of the fungi number in the outdoor air (emission) with sampling by aspiration and sedimentation method. National Standards Body (PKN), Warszawa. Poland.

Ramos, C.A., Wolterbeek, H.T., Almeida, S.M. 2014. Exposure to the indoor air pollutants during physical activity in fitness centers. *Building and Environment* 82: 349–360.

Ramos, C.A., Viegas, C,, Verde S.C., Wolterbeek H.T., Almeida S.M. 2015. Characterizing the fungal and bacterial microflora and concentrations in fitness centers. *Indoor and Built Environment* 25: 872–882.

Revel, G.M., Arnesano, M. 2014. Perception of the thermal environment in sports facilities through subjective approach. *Building and Environment* 77: 12–19.

Satish, U., Mendell, M.J., Shekhar, K., Hotchi, T., Sullivan, D., Streufert, S., Fisk, J.W. 2012. Is CO_2 an indoor air pollutant? Direct effects of low-to-moderate CO_2 concentrations to human decision-making performance. *Environmental Health Perspectives* 120: 1671–1677.

Slezakova, K., Peixoto, C., Oliveira, M., Delereu-Matos, C., Pereira, M.C., Morais, S. 2018a. Indoor particulate pollution in fitness centers with emphasis on ultrafine particles. *Environmental Pollution* 233: 180–193.

Slezakova, K., Peixoto, C., Pereira, M.C., Morais, S. 2018b. Indoor air quality in health clubs: impact of occupancy and type of performed activities on exposure levels. *Journal of Hazardous Materials* 359: 56–66.

World Health Organization. 2010. Global recommendations on physical activity for health, WHO, Copenhagen.

CHAPTER 19

Lower brominated PBDE congeners in the dust from automobile compartments in Eastern Poland

M.R. Dudzińska & A. Staszowska
Department of Indoor and Outdoor Air Quality, Lublin University of Technology, Lublin, Poland

ABSTRACT: The automobile cabin is an important part of the living environment. Particulate matter collected in automobile compartments may contain elevated levels of semi volatile compounds (SVOCs), such as PBDE (polybrominated diphenyl ethers). Therefore, the exposure in transport vehicles is of growing importance. Dust samples were collected from private cars in the city of Lublin, Poland. Lower brominated congeners were found in all the samples, with BDE-99 being the most abundant congener. Higher concentrations of particular congeners in cars equipped with AC were found, which might be related mainly to indoor sources of PBDE. Higher odometer readings might show the impact of outdoor sources, although some transformation of congeners during usage and sun light influence might be a reason. The highest concentration of BDE-138 were found in the automobiles produced outside the EU, although the highest concentrations of BDE-99 were observed in cars manufactured in the EU.

19.1 INTRODUCTION

Due to the fact that people spend more time in living premises, indoor air has higher impact on people's health and well-being than the atmospheric air. By indoor air we understand all closed environments which are not under any regulation related to occupational safety. With increasing time spent indoors, the indoor air may exert a substantial influence on people's health, their living conditions and prosperity. In studies on indoor air; the places of residence, offices, schools and education premises are usually the subjects of the main focus. However, in the modern societies, the automobile cabin is an important part of the living environment. In many countries, time spent on commuting to a workplace is growing and easily exceeds one hour. Statistics on the U.S. citizens' way of living show that they spend 87% of their time indoors, 8% outdoors and 5% in transportation (by car, bus, train or plane). When it comes to Europe, similar tendencies are observed: 90% indoors, 6% outdoors and 4% transportation (Abadie et al., 2004; Choi et al. 2017). There is no such evaluation for Poland, but due to the increased traffic, time spent in either private car or public transport vehicle increased significantly in the recent years. For a person whose occupation requires spending longer periods of time inside a vehicle (policemen, taxi, bus and truck drivers, servicemen, sales representatives), the relative contribution of in-vehicle exposures to total is greater than 30%. Toxic substances originate from interior materials, gasoline loss, and infiltration of outdoor air pollutants (Olukunle et al. 2015).

Yoshida and Matsunaga (2005) identified over 160 substances in a cabin of a new car. Some of the published in-vehicle studies have identified the elevated levels of many unleaded petrol and diesel fuel related pollutants, such as volatile organic compounds (VOCs), carbon monoxide, and particulate matters compared to other indoor environments (Knibbs et al., 2009). Particulate matter collected in automobile compartments may contain elevated levels of semivolatile compounds (SVOCs), such as polyaromatic hydrocarbons (PAHs) or brominated flame retardants (BFRs), including PBDE (polybrominated diphenyl ethers). Polybrominated diphenyl ethers are a group of synthetic organobromine compounds, which have been used as additive flame retardants in combustible materials to increase their fire resistance. PBDEs are present in high-impact plastics, electronics and textiles, as well as polyurethane foams, polyesters and phenolic and epoxy resins, so they might be found in domestic

DOI 10.1201/9781003171669-19

and industrial appliances and equipment, e.g. computers, printers, TV sets, mobile phones, mattresses and insulation boards. PBDE concentrations in plastic that are flameproof by PBDE range typically from 0.5 to 30% w/w (Alaee et al., 2003). Therefore, the highest PBDE levels were detected in indoor samples, namely in dust, at concentrations which were often 50-fold higher than the one reported for the outdoor samples (Rudel et al., 2003; Kemmelein et al., 2003, Staszowska et al., 2008). There are 209 possible PBDE congeners but commercial PBDE products used widely in the past have been marketed as three technical mixtures of several congeners, labelled as Penta-BDE, Octa-BDE and Deca-BDE. The Penta mixture contains mainly BDE 47, 99, 100 with smaller contributions from BDE 28, 153 and 154; whereas the Octa mixture contains BDE 153, 154, 183 and the Deca mixture consists mainly of BDE 209 (La Guardia et al., 2006). Pentabromodiphenyls are banned from being used in new EEE (electronic and in the EU Directive 2002/95/EC). They have been added to the Stockholm list in 2009 and forbidden in many countries. At present Deca-BDE (BDE 209) is the only permitted compound in the EU and the USA. Concentration of BDE-209 (Deca-BDE) cannot exceed 0.1% in EEE in the EU. However, Deca-BDE might decay to lower brominated congeners, including Penta due to its photolability (Stapleton and Dodder, 2008). Lower brominated diphenyl ethers are of higher toxicological concern than higher brominated diphenyl ethers (Birnbaum and Staskal, 2004; Cristale et al. 2018). Dust analysis from vehicles is difficult to conduct due to the fact that dust may originate from outdoor aerosols, soil from users' shoes or cigarette smoke; rather than from the automobile interior components. The first attempt to measure PBDE in the dust samples collected from cars were made by Harrad et al. (2008) and Lagalante et al. (2009). Both research groups continued measurements, studying degradation of BDE-209 in the dust collected from personal automobiles (Lagalante, 2011; Harrad and Abdallah, 2011; Besis et al. 2017). As far as Poland is concerned, the first attempt to measure PBDE in Polish automobiles was performed in 2019. The evaluation of the exposure to different pollutants in public and private cars was performed in Lublin, a city located in the eastern Poland. Therefore, in accordance with Stockholm convention, the particular attention was devoted to lower brominated PBDEs that were banned.

19.2 MATERIALS AND METHODS

Dust samples were collected by vacuum cleaner with HEPA filter from a number of private cars in Lublin, south-eastern Poland. Lublin, the city of 400 thousand inhabitants suffers from traffic problems and a lot of cars and bus users complain about the amount of time spent in vehicles that is constantly extending. Lublin is situated in the south-eastern, less developed, part of Poland and one of the poorest regions of the EU. A lot of cars used by private owners are second hand cars, which are older than 10 years. Private automobiles varied in terms of model, place of production, kilometrage or the presence of air conditioning system. Owners have been asked not to vacuum them for one week before sampling. Dust was collected from all interior surfaces of the cars (excluding trunks). Thorough cleaning procedure of all not disposable parts of vacuum cleaner followed in order to eliminate the contamination of samples. Dust samples were sieved and only fractions up to 125 μ m were analyzed. All chemicals used for PBDE analysis were of high purity. Toluene and dichloromethane (Ultra-Resi Analysed) were purchased from T.J. Baker (Germany). PBDE analytical mixture standards series BDE-CVS-F containing BDE 28, 47, 85, 99, 100, 153, 154, 183 congeners; as well as mass labeled (13C) PBDE recovery solution BDE-MFX containing BDE 77L and 138L were obtained from Wellington Laboratories (Ontario, Canada). Further standard solutions of native PBDE (BDE-MXF) containing BDE 28, 47, 85, 99, 100, 153 congeners at a concentration 2000 ng/cm3 were obtained from Wellington Laboratories and from Accustandards (mixture BDE-527). Entire glass equipment applied for the PBDE analysis was made of amber glass. Dust samples were being extracted with toluene in a Soxhlet apparatus for 16 hours in a dark room. The extract was evaporated under a gentle nitrogen stream and then the solvent was changed to dichloromethane. Further purification was performed using size exclusion chromatography (SEC) Brezee 1525, (Waters) system equipped with a double pump system, 2000 μ L loop, combined 19x150 mm and 19x300 mm Envirogel-GPC®cleanup columns (Waters, USA), tunable UV-VIS detector Waters M2487 working @ 254 nm, FC III fraction collector (Waters, USA); system control and data acquisition Breeze 3.30SPA software. Dichloromethane was

Table 19.1. The concentrations of BDE-28, 47, 66, 99 100, 101, 138, 153 and 154 in automobile dust (ng/g).

	Min	Max	Mean	Median
BDE-28	<dl	269	35.8	<dl
BDE-47	<dl	5476	1233.3	210
BDE-66	<dl	5593	497.3	<dl
BDE-99	<dl	14447	2600	230
BDE-100	<dl	4042	680.1	54.5
BDE-101	<dl	5673	620.5	<dl
BDE-138	<dl	4169	844	550.5
BDE-153	<dl	492	153.1	<dl
BDE-154	<dl	1033	177.3	<dl

used as a mobile phase with the flow rate 5 cm^3/min. Purified extracts were being collected at the window from 11 to 17 minutes. This method is usually used to clean-up samples for PAHs analysis (Gevao et al., 2006) but it has also been applied for other compounds e.g. pharmaceuticals (Debska et al., 2004), chlorinated and brominated compounds (Jansson et al., 1991), and also for PBDE (Saito et al., 2004; Dudzinska and Czerwinski, 2011). Final chromatographic analysis was done on the Trace Ultra – Polaris Q GC-MS system. The GC-MS technique which is cheaper and is more suitable for analyses of low substituted PBDE congeners than LC-MS has been used for PBDE analyses by many researchers (de Boer and Cofino, 2002; Hyötylainen and Hartonen, 2002). The parameters of the GC-MS system were published by Dudzinska and Czerwinski (2011). Our research was focused on the Penta mixture, which has been banned in the EU since 2009. Therefore, 2,4,4′-TriBDE (BDE 28), 2,2′,4,4′-TeBDE (BDE 47), 2,2′,4,4′,6-PeBDE (BDE 99), 2,2′,4,4′,5-PeBDE (BDE 100), 2,2′, 3,4,4′,5′-HxBDE (BDE 138), and 2,2′,4,4′,5,5′-HxBDE (BDE 153) were measured. The detection limit for this method was evaluated on 0.5 ng/L ($S/N \geq 3$) in extracts for PBDE 28 and PBDE 100, and 1.2 ng/L for PBDE 153. The quantification limit was evaluated on the level of 10 pg/injection ($S/N \geq 5$). Linearity was better than 10–250 for all congeners except for BDE-153 (better then 10–200). 74 to 91% recovery was obtained in triplicate analysis while the relative standard deviation for triplicate analysis of real samples ranged from 12% for PBDE 28 up to 27% for PBDE 153.

19.3 RESULTS AND DISCUSSION

Lower brominated congeners have been detected in all the examined vehicles, although in a few samples, some congeners were found below the detection limits. In Table 19.1 the results for BDE-28, 47, 66, 99, 100, 101, 138, 153, 154 183 congeners (minimum, maximum, mean and median) are presented.

The highest concentrations among targeted congeners were found for BDE-99 (with the highest measured value of 14447 ng/g) and for Tetra-BDE. Other Penta-BDEs have been present at considerably high levels, but revealed larger discrepancies. Odometer readings, the air conditioning system (AC) and years in service as well as the place of production (in the EU and outside the EU) have been considered as possible parameters influencing PBDE concentrations in the indoor dust. The mean concentrations of BDE-28, 47, 66, 99 100, 101, 138, 153 and 154 in cars older than 5 years compared with cars younger than 5 years are presented in Figure 19.1.

For most of the examined congeners, higher mean concentrations were found in older cars, excluding BDE-99 and BDE-153. BDE-99 was present on the highest levels in the most of the examined vehicles, rising up to 14 447 ng/g. The second congener, according to concentration, was BDE 47 with the maximum value of 5476 ng/g of dust. The most significant differences between 'old' and 'new' cars were noticed for BDE-66 and BDE-100. The presence of BDE-99 and BDE-47 in 'old' cars might have originated from the mixture itself. However, the statement that these congeners are at high levels

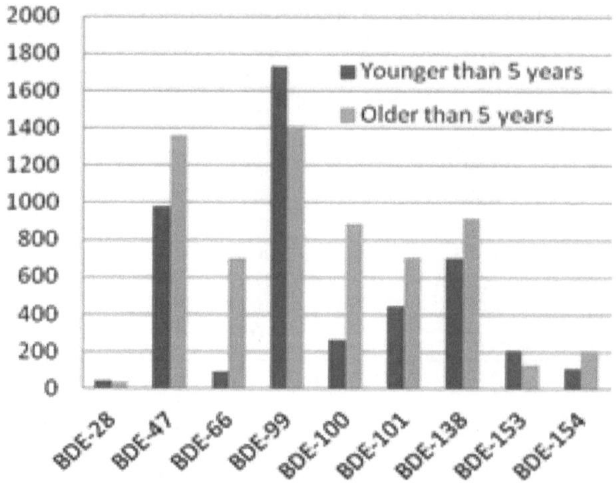

Figure 19.1. Mean concentrations of BDE congeners for 'old' and 'new' vehicles (ng/g).

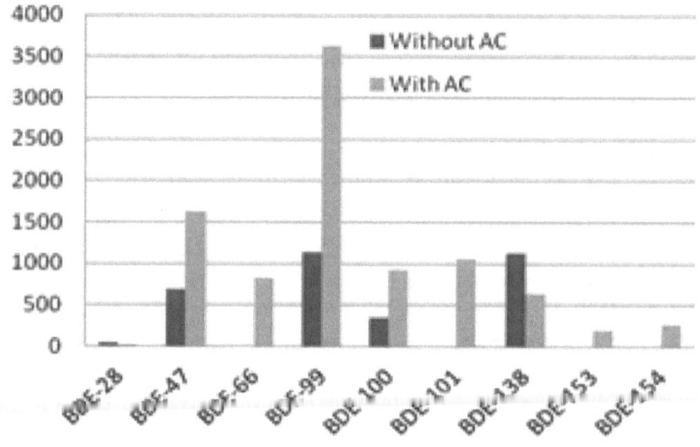

Figure 19.2. Mean concentrations of BDE congeners (ng/g of dust) in the cars equipped with AC system and without AC system.

in 'younger' vehicles cannot be proved too. Hence, decay via debromination might be a similarly possible explanation.

In Figure 19.2 the mean concentration of BDE in cars with and without air conditioning (AC) was also compared. In the majority of cases, cars with AC had higher mean concentration, excluding BDE-28 and BDE-138. The highest level of mean concentration was reported for BDE-99 and reached 3638 ng/g. For cars without AC; BDE-66, BDE-153, BDE-154 and BDE-183 were under the detection level. The highest difference between mean concentration for cars with and without AC was noticed for BDE-99 and BDE-47.

The air conditioning system circulates air inside the automobile compartment and causes the increase in the concentration of pollutants. Therefore, PBDE level might also be elevated.

Kilometrage was also a criterion (Figure 19.3). In this case, the highest mean concentration was for BDE-99 that reached 3999 ng/g. All types of congeners were detected in both cases. As in the case of a car with/without AC, the biggest differences were noticed for BDE-99 and BDE-47. Higher mean concentration of almost every congener, except for BDE-128, were found in cars with kilometrage greater than 100 000 km. Although measurements were limited, odometer reading and the presence of the air conditioning system were the most noticeable parameters influencing PBDE concentrations in the indoor dust. Car age also had an impact on the level of congeners, although not so significant.

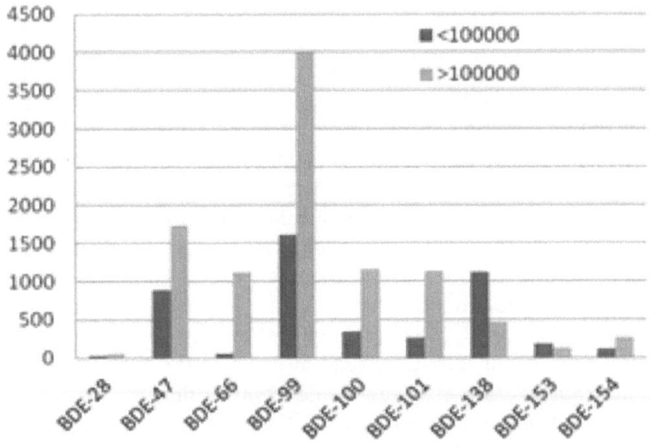

Figure 19.3. Mean concentrations of BDE congeners (ng/g of dust) in relation to odometer reading (below and above 100 000 km).

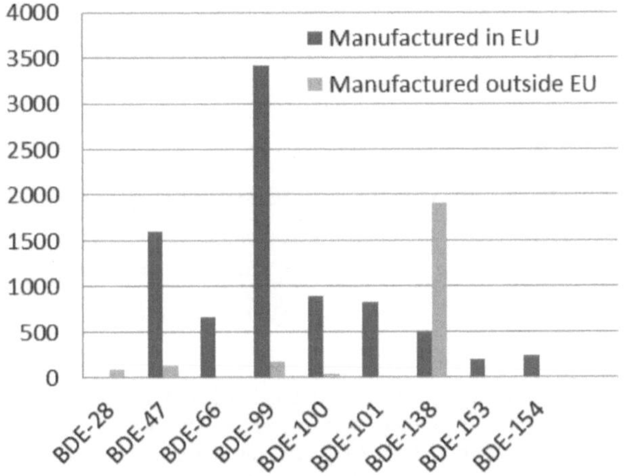

Figure 19.4. Mean concentrations of BDE congeners (ng/g of dust) in the automobiles manufactured in the EU versus manufactured outside the EU.

Figure 19.4 presents the mean concentrations of the measured congeners in the car manufactured in the EU or outside it. The highest mean concentration in the dust sampled in the cars manufactured in the EU for BDE-99 was 3408 ng/g, while the second result was for BDE-47 (1600 ng/g). These congeners were also found in the samples from the cars manufactured outside the EU, although at noticeably lower levels. In the cars produced outside the EU, the elevated levels were measured for BDE-138 (mean concentration equals to 1903 ng/g).

The results might be surprising as higher levels were found in the cars produced in the EU (at least such information was given in the car documents). However, it does not mean that all components were produced in Europe, as electronic equipment is manufactured mainly in the Asian countries.

The highest concentration of BDE-138, the main component of Octa-mixture, which was found in the dust collected from the cars produced outside the EU might have mainly originated from the mixture itself.

19.4 CONCLUSIONS

Lower brominated congeners were found in all of the examined samples, with highest concentrations of congeners which were used in Penta-BDE mixture in the past. For older cars, they may have originated directly from the indoor materials and electronic devices. However, the higher concentration was found mainly in newer cars, equipped with the air conditioning system and with high odometer reading. Although measurements were limited, odometer reading and the presence of the air conditioning system were the most noticeable parameters influencing PBDE concentrations in the indoor dust. Car age also had an impact on the level of congeners, albeit not so significant.

The higher concentrations of particular congeners in cars equipped with AC relate mainly to indoor sources of PBDE. The higher odometer readings might show the impact of outdoor sources, although some transformation of congeners during usage and sun light influence might also be a reason. PBDEs in older cars, but with lower odometer readings due to the time spent in a garage (without direct sunlight), might be not so easily decayed via debromination.

The high concentrations of BDE-99 and BDE-49 have been found in the dust collected in cars produced in the EU, but these cars were, generally 'older', compared to the examined cars produced outside the EU. The highest concentration of BDE-138, the main component of Octa-mixture was found in the dust from the cars produced outside the EU and these congeners originated probably from the mixture itself.

Further studies to evaluate exposure, especially of people spending more than 8 hours per day (policemen, sales representatives) and children travelling longer distances with their parents, are necessary.

ACKNOWLEDGEMENTS

This work was supported by the Ministry of Science and Higher Education in Poland within funding FN-85/IŚ/2020.

REFERENCES

Abadie, M., Liman, K., Bouilly, J., Genin, D. 2004. Particle pollution in the French high-speed train (TGV) smoker cars: measurements and prediction of passenger exposure. *Atmospheric Environment* 38:2017–2027.

Alaee, M., Arias, P., Sjodin, A., Bergman, A. 2003. An overview of commercially used brominated flame retardants, their applications, their use patterns in different countries/regions and possible modes of release. *Environment International* 29:683–689.

Besis, A., Christia C., Poma, G., Covaci, A. 2017. Legacy and novel brominated flame retardants in interior car dust – implications for human exposure. *Environmental Pollution* 230: 871–881.

Birnbaum, L.S., Staskal, D.F. 2004. Brominated Flame Retardants: Cause for Concern? *Environmental Health Perspectives* 112:9–17.

de Boer, J., Cofino, W.P. 2002. First world-wide interlaboratory study on polybrominated diphenylethers (PBDEs). *Chemosphere* 46: 625–633.

Choi, J., Jang, Y.C., Kim J.G. 2017. Substance flow analysis and Environmental releases of PBDEs in life cycle of automobiles. *Science of the Total Environment* 574: 1085–1094.

Cristal, J., Bele, T.G.A., Lacorte, S., de Marcji, M.R.R. 2018. Occurrence and human exposure to bromianted and organophosphorus flame retardants via indoor dust in a Brazilian city. *Environmental Pollution* 237: 695–703.

Debska, J., Kot-Wasik. A., Namieśnik, J. 2004. Fate and analysis of pharmaceutical residues in the aquatic environment. Critical Reviews in *Analytical Chemistry* 34:51–67.

Dudzińska, M.R., Czerwiński, J. 2011. Persistent organic pollutants (POPs) in leachates from municipal landfills. *International Journal of Environmental Engineering* 2013:253–263.

Gevao, B., Al-Bahloul, M., Al-Ghadban, A.N., Al-Omair. A., Ali, L., Zafar, J., Helaleh, M. 2006. House dust as a source of human exposure to polybrominated diphenyl ethers in Kuwait. *Chemosphere* 64:603–608.

Harrad, S., Abdallah, M.A.-E. 2011. Brominated flame retardants in dust from UK cars – Within-vehicle spatial variability, evidence for degradation and exposure implications. *Chemosphere* 82(9):1240–1245.

Harrad, S., Ibarra, C., Abdallah, M.A.E., Boon, R., Neels, H., Covaci, A. 2008. Concentrations of brominated flame retardants in dust from United Kingdom cars, homes, and offices: Causes of variability and implications for human exposure. *Environment International* 34(8): 1170–1175.

Hyötylainen, T., Hartonen, K. 2002. Determination of brominated flame retardants in environmental samples. *Trends in Analytical Chemistry* 21(1): 13–29.

Jansson, B., Andersson, R., Asplund, L., Bergman, A., Litzen, K., Nylund, K., Reutergardh, L., Seelström, U., Uvemo, U.-B., Wahlberg, C., Wideqvistt, U. 1991. Multiresidue method for the gas- chromatographic analysis of some polychlorinated and polybrominated pollutants in biological samples. *Fresenius' Journal of Analytical Chemistry* 340: 439–445.

Kemmelein , S., Hahn, O., Jann, O. 2003. Emissions of organophosphate and brominated flame retardants from selected consumer products and building materials. *Atmospheric Environment* 37: 5485–5493.

Knibbs, L.D., de Dear, R.J., Atkinson, S.E. 2009. Field study of air change and flow rate in six automobiles. *Indoor Air* 19: 303–313.

Lagalante, A.F, Oswald, T.D, Calvosa, F.C. 2009.Polybrominated diphenyl ether (PBDE) levels in dust from previously owned automobiles at United States dealerships. *Environment International* 35(3):539–544.

Lagalante, A.F, Shedden, C.S, Greenbacker, P.W. 2001. Levels of polybrominated diphenyl ethers (PBDEs) in dust from personal automobiles in conjunction with studies on the photochemical degradation of decabromodiphenyl ether (BDE-209). *Environment International* 37(5):899–906.

La Guardia, M.J, Hale, R.C, Harvey, E. 2006. Detailed Polybrominated Diphenyl Ether (PBDE) congeners composition of the widely used Penta-, Octa-, and Deca-PBDE technical flame-retardant mixtures. *Environment Science & Technology* 40:6247–6254.

Olukunle, O.I., Okonkwo, O.J., Wase A.G. 2015. Polybrominated diphenyl ethers in car dust in Nigeria: concentrations and implications for non-dietary human exposure. *Microchemical Journal* 123: 99–104.

Rudel, R.A., Camann, D.E., Spengler, J.D., Korn, L.R., Brody, J.G. 2003. Phthalates, alkylophenols, pesticides, polybrominated diphenyl ethers and other endocrine disrupting compounds in indoor air and dust. *Environment Science & Technology* 37:4543–4553.

Saito, K., Sjödin, A., Sandau, C.D., Davis, M.D., Nakazawa, H., Matsuki, Y., Patterson D.G. Jr. 2004. Development of an accelerated solvent extraction and gel permeation chromatography analytical method for measuring persistent organohalogen compounds in adipose and organ tissue analysis. *Chemosphere* 57(5):373–381.

Stapleton, H.M, Dodder, N.G. 2008. Photodegradation of decabromodiphenyl ether in house dust by natural sunlight. *Environmental Toxicology and Chemistry* 27:306–312.

Staszowska, A., Połednik, B., Dudzińska, M.R., Czerwiński, J. 2008. Commercial Penta-BDE mixtures in dust sample from indoor environments in Lublin, Poland – a case study. *Archives of Environmental Protection* 34:239–247.

Yoshida, T., Matsunaga, I. 2005. A case study on identification of airborne organic compounds and time courses of their concentrations in the cabin of a new car for private use. *Environment International* 32:58–79.

CHAPTER 20

Microbiological contamination in school rooms before and after thermo-modernization – a case study

S. Dumała
Lublin University of Technology, Lublin, Poland

ABSTRACT: The research was carried out in a school located in the province of Lublin, built in the 1980s, before and after thermal modernization. During the measurement period, the temperature ranged between 20.8–28.5°C, and humidity – 18–55%. The measurement of the CO_2 concentration allowed determining the class of the room. In all cases, the CO_2 concentration exceeded 1000 ppm after only 20 minutes of the first class; therefore the rooms were qualified to the 3rd and 4th class rooms (IDA). During the measurement period, the CO_2 concentration in the outside air was 500 ppm for winter and 350 ppm for summer. The dominant bacteria and molds isolated from the tested air samples are: *Bacillus lentus, Bacillus licheniformis, Bacillus pumilus, Bacillus cereus, Pseudomonas stut-zeri, Micrococcus ssp., Staphylococcus xylosus, Staphylococcus saprophyticus, Staphylococillus haemillium, Aspergilus acridonium, Aspergilus haemillium Alternaria, Cladosporium, Epicocum, Mucor, Penicillinium.* The data showed that the bacteria and fungi levels varied between seasons, with the non-heating season concentrations being higher than those in winter. The bacteria concentrations varied in the range from 1727 to 3670 CFU/m^3, while the fungi levels varied from 64 to 489 CFU/m^3. The indoor bacteria levels were between two and three times higher than the outdoor levels, while the I/O ratio for fungi was <1.

20.1 INTRODUCTION

According to the World Health Organization (WHO), indoor air has a greater impact on our health and well-being than outdoor air. Despite the recent reports on air pollution in Poland, mainly with PM particles and bioaerosols, this claim remains true, because the threat to the human health depends not only on the concentration, but most of all on the time of exposure to the harmful factor, and people stay indoors for almost 90% of life (Raczkowski et al. 2010).

The exposure to harmful substances is particularly important for young organisms, and adolescents often spend more than 8 hours a day (a third of the time) in school classes. Many years of research have proven that the indoor air quality has a significant impact not only on the health of users, but also on the ease of acquiring knowledge, remembering and general well-being (Raczkowski et al. 2010; Wargocki 2005).

Most schools in our country are two-story buildings with gravity ventilation and a traditional heating system based on panel radiators. The height of the gravity ventilation ducts, especially for the upper storey, is very small (1–2 m), which means that in periods with no wind, the value of active pressure is low and prevents the inflow of air into the room.

One of the elements having a significant impact on the air quality is the presence of aerosol and bioaerosol pollutants of plant, animal or mineral origin.

Bioaerosols are a mixture of solid or liquid particles, including from biological material (viruses, bacteria, fungal spores, mycelial fragments, etc.) (An et al. 2004), products of their metabolism (endotoxins, entertotoxins, enzymes, mycotoxins (Agranovski et al. 2002; Dudzińska 2013; Law et al. 2001) or pollen, plant debris, animal dandruff, animal and human epidermis (Maus et al. 2001). Bioaerosols can also be single bacterial cells with the size of 0.5–2.0 μ m aerosol particles e.g. *Bacillus*, *Pseudo-monas*, *Xanthomonas*, *Arthrobacter*, similarly to the spores of mold fungi

DOI 10.1201/9781003171669-20

Aspergillus fumigatus (2.5–5.0 μ m), *Trichoderma harizanum* (2.8–3.2 μm) *Aspergillus nigeri* (3,0–4.5 μ m), *Cladosporium macrocarpum* (5–8 μ m) *Penicillium brevicompactum* (7–17 μ m) (Bonetta et al. 2009; Dudzińska 2013; Menetrez et al. 2007).

The abundance and composition of the microflora present in the air is influenced by the type of emission source, emission level, distance from the emission source, microbial survival, temperature, humidity, exposure to light and UV rays (Barabasz et al. 2007; Breza-Boruta 2010; Hameed et al. 2012; Kiliszczyk et al. 2013).

The microbial air contamination in closed spaces comes from both internal sources and the atmospheric air (endogenous and exogenous). The primary source of microorganisms in closed spaces is the user and his microflora from the upper respiratory tract, microorganisms from exfoliated epidermis, bacterial dust (from floors, clothes, etc.) and house fungus spores. The habitat of microorganisms may be building partitions, items in the room, floor finishes, dust, ventilation systems. Bioaerosols are transmitted with the movement of people and animals and under the influence of air currents in the room. The biological pollutants of indoor air are less exposed to the action of meteorological factors or UV radiation; therefore, their survival period and number are not subject to seasonal fluctuations, as in the case of the microorganisms present in the outdoor air. The species composition is also more stable. Among the fungi, representatives of the *Penicillium*, *Aspergillus*, and *Cladosporium* genera are dominant (Cabral 2010; Flannigan et al. 2011; Krzyśko-Łupicka 2010).

The saprophic microflora predominates among the indoor air pollutants. The most common bacteria belong to the *Micrococcus* (more than 42%), *Staphylococcus* (38.8%), and *Bacillus* (10%) genera, which constitute 68 ÷ 80% of all microorganisms in this environment. Other types of bacteria are also encountered, such as: *Ochrobactrum*, *Pseudomonans*, *Aeromora*, *Xartonans*, *Pasteurella*, *Sphingomous*. *Actinomycetes* in the amount of 7 to $5.3*10^4$ CFU/m^3 of air may be present in the air with high relative humidity (above 70%). The most common species belong to the *Streptomces*, and *Nocordia* genera. *Actinomycetes* can constitute even 25 ÷ 40% of the total number of bacteria in the air (Libudzisz 2009). Own research conducted in the schools in the Lublin region confirmed the presence in the air of the following genera: *Bacillus lentus, Bacillus licheniformis, Bacillus pumilus, Bacillus cereus, Pseudomonas stutzeri, Micrococcus ssp., Staphylococcus xylosus, Staphylococcus saprophyticus, Staphylococcus saprophyticus, Staphylococus, Staphylococus, Asphyticium Aspergillus. Aspergillus niger, Alternaria, Cladosporium, Epicocum, Mucor,* and *Penicilinium* (Dumała & Dudzińska 2013).

The presence of fungi in the indoor air is possible due to their ability to adapt to the conditions in a given room. The development of fungi on building partitions favors the domination of these microorganisms in the indoor air. In residential premises without mold symptoms, their number in the air fluctuates around $5.0*10^2$ CFU/m^3. However, in the rooms with mycosis of building partitions, the number increases and is in the range of $4.0*10^3$–$1.4*10^4$ CFU/m^3.

In the air of rooms with fungus, the most common are the molds causing allergies from the types of *Penicillinium* (45–90% depending on the type of room), *Cladosporium* (22–73%), Aspergillus (17–52%), *Alternaria* (3–5%) and others (Libudzisz 2009).

The type of room is the main factor influencing the qualitative and quantitative composition of the air present in it. In public utility rooms, such as lecture halls, classrooms, libraries, the amount of bacteria is quite significant. The main source of their origin is the surface of the human body. High concentrations of bacteria are found on the skin, hair, or in the respiratory tract. Air-conditioning devices, contaminated ventilation ducts, finishing materials and furniture are often sources of bacterial emissions into the air of school or office rooms.

Like the amount of microorganisms in the air, their type depends mainly on the type of room. Mainly gram-negative bacteria occur in residential and livestock rooms, while streptococci and staphylococci occur in school and office rooms.

Fungi are found in all types of rooms. In the rooms with relative air humidity, the xerophilic *Aspergillus* and *Penicillinium* fungi are found most often

20.1.1 *Effect of bioaerosols on the human health*

With each inhalation, we are exposed to the inhalation of bioaerosols, because they can account for 5 to 34% of the indoor air pollution. The particles with sizes lower than 7 μm or 10 μm (depending on the author of the research) are a particular threat. While the particles with a diameter of 4.7–7 μm are deposited in the throat 3.3–4.7 μm reach the trachea and primary bronchi, 1.1–3.3 μm can reach the secondary and terminal bronchi, and these less than 1.1 μm to pulmonary bronchioles. The most dangerous particles are considered to be smaller than 2.5 μm (Górny 2004b; Stetzenbach et al. 2004).

The risk is posed not only by the presence of microorganisms, but also by metabolites of microorganisms such as endotoxins and mycotoxins. They play a significant role in inflammatory responses and contribute to the deterioration of lung function and organism infections (Stentzenbach 1998). Excessive amount of saprophytic microorganisms is also harmful, especially when their composition is not diversified and organisms of one species dominate (Cabral 2010; Flannigan et al. 2011; Gładysz et al. 2010).

Airborne pathogens are the pathogens that are produced in the respiratory system and released in the exhaled air. In a review article, Morawska (2005) describes the mechanism of generation and the place of formation of pathogen droplets.

It also takes into account the factors influencing this process and the fate of droplets released from the respiratory tract. She claims that there is still a need for additional research related to understanding the mechanism of pathogen transmission in the rooms intended for permanent or temporary residence of people. The respiratory tract has 4 parts where microbes can multiply and be dispersed in the exhaled air: the nose, mouth, throat and lungs. Each provides a different habitat to which various pathogens have adapted: tuberculosis in the lungs, Streptococcus agalactiae in the throat, etc. The generation of pathogens occurs primarily through the nose and mouth, most often with saliva during conversation, coughing or sneezing (Lidwell 1974). Fiegel et al. (2006) identified the deep lung region as the main source of bioaerosol generation and the associated transport of airborne pathogens. She concluded that the use of "salt therapy" (immobilizing bioaerosols in the lungs) would reduce the generation of airborne pathogens, enabling the natural clearance mechanisms. It is clear that the contamination depends on the preferred habitat of the pathogen: coughing will produce droplets with deep lung pathogens, while talking, sneezing, etc. will disperse pathogens inhabiting mainly the mouth, nose or throat.

According to Tellier (2006), medical findings suggest that the influenza A virus enters the air by spraying from the lower lung region. Airborne transmission has been shown to be dominant in three respiratory diseases: measles, varicella and tuberculosis (Qian et al. 2006). When people cough, sneeze, speak or breathe, they generate particles of different sizes, with various air flows, and with diversified properties. Nicas et al. (2005) summarized the data on the breathing aerosol size distribution taking into account water loss. When evaporation is complete, the molecule is about half its original diameter. However, the authors are skeptical about the experimental data on droplet size distribution. They propose a new logarithmic model to describe the particle size distribution during coughing based on the results of Loudon and Roberts (1967). According to this model, the distribution of "small" particles with a geometric diameter (GM) has a standard deviation (GSD) of 9.8 μm and 9.0 μm, "large" GM particles 160 μm and a GSD of 1.7 μm, respectively.

Small particles account for 71% of all particles emitted by a cough. The particles with the diameter of 10 μm and smaller are able to penetrate the lungs (Hinds 1999). Therefore, airborne droplets with a diameter of up to 20 μm are considered a hazard, as they reach a diameter of 10 μm or less when the water contained in them is fully evaporated. An experiment by Yang et al. (2007) shows that the distribution of droplet size when coughing is in the range of 0.62–15.9 μm. In the experiment, they used two methods of identifying droplet sizes. In the first method, the drops were mixed in a test column with clean air of low relative humidity (35%). In order to avoid interference from the surrounding environment, subjects wore a mask with a P100 filter that was attached to a test column. In the second method, the subjects coughed directly into the bag. However, it was found that compared to the first method, more particles were retained in the bag. The influence of age and gender on droplet generation was also investigated. No significance was found in both cases ($p > 0.1$). Their findings are consistent with the conclusions of Nicas et al. (2005) but only for the small particle range.

This was probably due to the fact that larger particles were trapped in the filter medium or were stuck on the walls of the bags, which means that there is a need to conduct additional research related to the size of the generated droplets and their distribution.

Airborne pathogens must survive in the surrounding environment, making such factors as air temperature and relative humidity very important in assessing their life cycle. It has been shown that moderately humid conditions (40–60%) are more lethal for non-pathogenic bacteria (Hatch & Wolochow 1969). The viruses with more lipids tend to be more stable at lower relative humidity, while the viruses with lower or zero lipid content are more stable at higher relative humidity (Assai & Block 2000). Loosli et al. (1943) showed that a humidity level in the range of 80–90% maintained for 30 minutes ensures that the influenza virus does not infect mice, while at lower humidity levels (17–24%) it has the highest infectivity. Lowen et al. (2007), when conducting experiments on guinea pigs in a test chamber, confirmed that the efficiency of transmission of the influenza A virus depends on relative humidity. Four infected animals were separated from four healthy animals: each animal was kept in a cage to avoid any contact. The only possible route of transmission of the pathogen was by air. At low relative humidity (20% or 35%), the pathogen was most likely to infect. Only one animal was infected at 50% and no virus transmission was observed at 80% relative humidity. The temperature was kept constant at 20°C in all cases. The authors suggest that dry air may dry out the nasal mucosa, damage the epithelium, and/or reduce the ability of mucociliary clearance, making the host vulnerable to respiratory infections. This, of course, also depends on the stability of the virus as well as the droplet nucleation mechanism: at low humidities, the droplets evaporate faster, shrink and change their size, if their diameter is less than 10 μm, the lung deposit capacity increases.

In other studies, survival of some viruses has been shown to be independent of relative humidity (Elazhary & Derbyshire 1979). Miller and Artenstein (1967) showed that picornaviruses and adenoviruses survive better in high relative humidity. Measles and enveloped viruses exist best in low RH aerosols (Hemmes & Winkler 1960; Jong & Winkler 1964).

Virus survival is influenced not only by relative humidity, but also by temperature. At 20°C and moderate humidity, the coronavirus was most stable, but also proved to be relatively stable at low humidity (Ijaz et al. 1985). The same study confirmed a similar virus survival at 6°C and 80% humidity. Lower temperatures have also been shown to increase the survival of rhinoviruses at high relative humidities (Karim et al. 1985).

20.1.2 *Legal regulations related to microbiological contamination*

The issues related to the control of microbiological air purity in global and Polish legislation are still unregulated today, despite the fact that the first attempts were made over a hundred years ago by Odon Bujawid, who proposed the criterion of permissible microbial contamination for living quarters at the level of 50 bacteria in 1 liter of air. Further suggestions appeared only in the 1970s (Górny 2004a; Krzysztofik 1992; Chmiel et al. 2015).

Currently, in many countries there are already quantitative standards or acceptable numerical values that are helpful in interpreting the obtained measurement data. Standards for residential and non-industrial premises for bacteria can be found, for example, in a 1993 report by the Commission of the European Communities – CEC (Table 20.1).

As with other types of air pollution, there are many more regulations for workplaces. Currently in Poland, the Regulation of the Minister of Health of April 22, 2005 on harmful biological agents for health in the work environment and health protection of workers professionally exposed to these factors is legally valid.

In Poland, there are no recommendations related to the criterion for assessing the exposure to biological agents, reference values or methodological recommendations for the indoor air in residential buildings.

Due to unregulated national legal acts, in order to assess the internal environment, one should refer to arbitrary normative values, which are usually determined based on the analysis of research conducted by researchers over several years available in the literature (Table 20.2, 20.3). They define the acceptable levels of pollution in given environments. In the case of schools, there are few articles specifying the concentration limits for the microorganisms that may be present in these rooms. Table 20.2 summarizes the values proposed by some global government agencies for the concentration of bacteria in selected rooms.

Table 20.1. Acceptable number of bacteria according to the Report of the Commission of the European Communities – CEC (1993).

CFU/m^3	Home	Non nonindustrial area
Very small pollution	$<1.0*10^2$	$<5.0*10^1$
Small pollution	$<5.0*10^2$	$<1.0*10^2$
Medium/increased pollution	$<2.5*10^3$	$<5.0*10^2$
Large Pollution	$<1.0*10^4$	$<2.0*10^3$
Very large pollution	$>1.0*10^4$	$>2.0*10^3$

Table 20.2. Bacterial concentration limits for selected rooms (Górny 2004).

Normative values (reference CFU/m^3)	Document	Organization/institution or state
0 CFU/m^3 – there is no safe level for pathogenic microorganisms	Biohazards reference manual 1986	AIHA (American Industrial Hygiene Association)
$<1,0\cdot10^2$ (houses) i $<5,0\cdot10^1$ (non-industrial premises) – very little pollution $<5,0\cdot10^2$ (houses) i $<1,0\cdot10^2$ (non-industrial premises) – low pollution $<2,5\cdot10^3$ (houses) i $<5,0\cdot10^2$ (non-industrial premises) – moderate/increased pollution $<1,0\cdot10^4$ (houses) i $<2,0\cdot10^3$ (non-industrial premises) – high pollution $>1,0\cdot10^4$ (houses) i $>2,0\cdot10^3$ (non-industrial premises) – very high pollution	Biological particles in indoor environment Raport No. 12 1993	CEC (Commission of the European Communities)
$4,0\cdot10^3$ CFU/m^3 – limit value for classrooms (GB9668-1996)	Chines National Standards 1996	Chiny

Table 20.3. Acceptable level of microbiological air pollution in utility rooms (Krzysztofik 1992).

	Permissible number of microorganisms in 1 m^3 of air		
Type of utility room	Whole number Microorganisms on the MPA substrate	number microorganisms hemolytic on blood agar	total number of fungi on the substrate Sabouraud
Outdoor air	$3,0\cdot10^3$	$1,0\cdot10^2$	$1,0\cdot10^3$
Healthcare rooms:			
– operating room	$1,0\cdot10^2$		
– treatment room	$1,5\cdot10^2$		$5,0\cdot10^1$
– hospital room	$1,0\cdot10^3$	$5,0\cdot10^1$	$2,0\cdot10^2$
Rooms of residential houses:			
– kitchen and dining room	$2,0\cdot10^3$	$1,0\cdot10^2$	$3,0\cdot10^2$
– salon	$1,5\cdot10^3$	$5,0\cdot10^1$	$2,0\cdot10^2$
– bedroom	$1,0\cdot10^3$	$5,0\cdot10^1$	$1,0\cdot10^2$
School rooms:			
– lecture halls	$\mathbf{1,5\cdot10^3}$	$\mathbf{5,0\cdot10^1}$	$\mathbf{2,0\cdot10^2}$
– auditoriums	$\mathbf{2,0\cdot10^3}$	$\mathbf{1,0\cdot10^2}$	$\mathbf{2,0\cdot10^2}$
– gyms	$\mathbf{3,0\cdot10^3}$	$\mathbf{1,5\cdot10^2}$	$\mathbf{3,0\cdot10^2}$

Table 20.4. Description of the school covered by the study.

School location	Age	Class location	Windows	Equipment	Ventilation system central heating	Plants	Others
In the middle of the estate, on a busy street	Built in the years 80's	On the ground floor, in the middle of the school	Double glazed, surface 11 m^2, facing east	15 double tables, desk, wardrobes, blackboard	Gravity ventilation/ radiators	Yes	A classroom for grades I-III

In Poland, in the 1970s, the acceptable level of indoor air pollution with microorganisms was determined, which for school rooms was set at 1.5–3.0*10^3 CFU/m^3 for bacteria and 2.0–3.0*10^2 CFU/m^3 for fungi – see Table 20.3 (Krzysztofik 1992).

20.2 METHOD

The research was carried out in a school located in the province of Lublin, built in the 1980s, before and after thermal modernization. Detailed information about the facility is presented in Table 20.4. The classrooms in the building were subject to the same cleaning schedule, some daily cleaning in corridors and common spaces were made during children's occupancy in the classrooms and the thorough cleaning was conducted daily after classes. The building is equipped with a central heating system and have kitchen with a gas stove. The kitchen is located on the ground floor.

20.2.1 *Measurement of thermo-hygrometric parameters and carbon dioxide*

The thermo-hygrometric parameters (air temperature, relative humidity) were electronically recorded at one-minute intervals using a four-channel Hobo 12 recorder with 12-bit resolution, capable of storing up to 43 000 measurements. The recorder uses the USB interface to start and read measurements. The temperature measurement range is −20°C–70°C (with an accuracy of +/−0.35°C in the range from 0°C to 50°C), while for a relative humidity of 5%–95% (with an accuracy of +/−2.5% in the range from 10% to 90%).

Carbon dioxide was measured with the Vaisala GMD20 series carbon dioxide transmitter. The measuring range of the transmitter is 0–2000 ppm with an accuracy of +/−2%. The recorder software and data reading was performed with the Hoboware Pro program.

The measurements of the thermo-hygrometric parameters and the carbon dioxide concentration were carried out 24/7 for two weeks, in winter and summer.

20.2.2 *Measurement of microbiological contamination*

The microbiological assessment of the air was carried out on the basis of the qualitative and quantitative analysis of bacteria and fungi.

The air samples were collected with an impactor, which was located in the central part of the room at a height of 1–1.5 m.

The sampling time was 10 minutes. Microorganisms were collected in petri dishes with an appropriate microbial medium. For fungi, Sabourand agar with chloramphenicol was used as a medium, and for bacteria, typical soy agar. The petri plates were incubated 48h at 36 ± 1°C for bacteria and 14 days at 27 ± 1°C for fungi. After the end of the incubation period, the quantitative analysis of bacteria and fungi was performed by counting and correcting based on the Andersen's correlation tables.

Then, the concentration of microorganisms was calculated in CFU – colony forming units on a solid medium in 1 m^3 of air taken [CFU/m^3]. This relationship is presented by the formula:

$$CFU/m^3 = number\ of\ colonies/V \qquad (20.1)$$

Where V is anair volume [m^3].

In order to obtain detailed information on the existing air condition inside the classrooms, selected colonies of bacteria and fungi were subjected to a thorough qualitative analysis. The qualitative analysis consisted in performing tests enabling the identification of the species of microorganisms present in a given air. The identification process consisted in determining the morphological features, cell wall structure (in the case of bacteria) and biochemical features of the tested bacteria and fungi. These features make it possible to assign microorganisms to a given species, as well as to determine their properties, including pathogenic potential.

The study used API 20 NE tests to identify gram-negative rods, API STAPH to identify the *Staphylococcus, Micrococcus* and *Kocuria* genera, and API 50 CH and API 50 CHB/E Medium to identify *Bacillus*.

In addition, macroscopic and microscopic analysis of the molds grown was performed.

20.3 RESULTS AND DISCUSSION

During the measurement period, the temperature ranged between 20.8–28.5°C, and humidity 18–55%. The measurement of the CO_2 concentration allowed determining the class of the room. In all cases, the CO_2 concentration exceeded 1000 ppm after only 20 minutes of the first class; therefore, the rooms were qualified to the 3rd and 4th class rooms (IDA). During the measurement period, the CO_2 concentration in the outside air was 500 ppm for winter and 350 ppm for summer.

The dominant bacteria and molds isolated from the tested air samples are: *Bacillus lentus, Bacillus licheniformis, Bacillus pumilus, Bacillus cereus, Pseudomonas stut-zeri, Micrococcus ssp., Staphylococcus xylosus, Staphylococcus saprophyticus, Staphylococillus haemillium, Aspergilus acridonium, Aspergilus haemillium Alternaria, Cladosporium, Epicocum, Mucor,* and *Penicillinium.*

Practically throughout the entire period of use, the CO_2 concentration exceeded 1000 ppm – an acceptable air quality standard used in many countries. There were overruns regardless of whether the windows were open or not.

In the summer season, the exceedances are smaller than in the heating season. The air temperature also exceeded the recommended values ensuring thermal comfort – 18 ÷ 20°C (winter) and 20–23°C (summer).

Referring to Krzysztofik's assumptions, according to which the concentration of bacteria in the classroom should not exceed $1.5*10^3$ CFU/m^3, and of fungi $2.0*10^2$ CFU/m^3, the concentration of bacteria was exceeded in almost all examined cases, regardless of the season. On the other hand, the concentration of fungi was within the recommended limit during the heating season (Figure 20.1).

Table 20.5 presents the average concentrations of bacterial and fungal aerosols collected in the indoor and outdoor air during the analyzed seasons. The samples were collected for each bioaerosol during both seasons, both inside and outside each location (outdoor—OUT; older children classrooms—O; younger children classrooms—Y). Generally, the bacteria levels were higher than the fungi levels. The results present that the bacterial, rather than fungal concentration depends on the occupants' activity. The small influence of the human activity on fungal levels was also confirmed by Shin et al. (Shin et al. 2015). The highest total concentration of bacterial (3670 CFU/m^3) and fungal (205 CFU/m^3) aerosols was found during non-heating season, respectively inside and outside the building. Similar results were obtained in two nursery schools situated about 10 km north of the city of Gliwice (Brągoszewska et al. 2016).

During the present study, the total concentrations of bacterial aerosols obtained indoors, were found levels of between 2310 and 3670 CFU/m^3 before thermo-modernization and 2150–2356 CFU/m^3 after thermo-modernization.

The research performed in Ankara, Turkey, underlined that among indoor urban environments, the highest concentrations of total bacteria aerosols were observed in kindergartens, at 649 and 1462

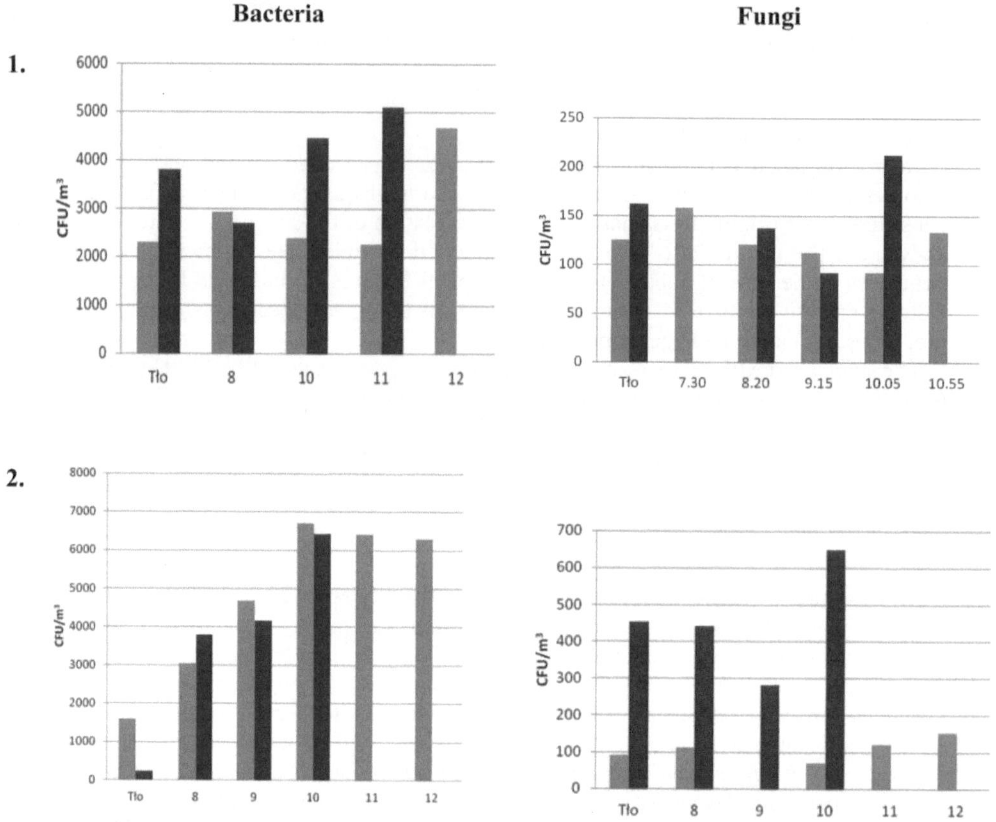

Figure 20.1. Total concentration of bacteria and fungi (CFU/m^3) – selected measurement day 1. before thermo-modernization; 2. after thermo-modernization in individual measuring hours, blue color – heating season, red color – without heating season.

CFU/m^3 in the winter and summer seasons, respectively (Mentese et al. 2012). Similar studies carried out in three schools in Portugal showed that the concentrations of bacteria in the indoor air of schools was between 500 and 1600 CFU/m^3 (Pegas et al 2010) Yang et al. (2009) studied South Korean schools between July and December, and reported the average concentration of bacterial aerosol to be 1300 CFU/m^3, while the maximum concentration reached 4700 CFU/m^3. The concentrations of airborne bacteria found in houses in Upper Silesian Province were on average 1021 CFU/m^3, while the same measure in offices was 300 CFU/m^3(Pastuszka et al 2000)

The indoor-to-outdoor (I/O) ratio indicates the source of bioaerosols. If this ratio is >1.0, there is a difference between the outdoor and indoor bioaerosol sources and the source exists in the indoor environment. The average I/O ratio calculated for all indoor and outdoor bacteria concentrations for the heating season before thermo-modernization was 3.2, while the average I/O ratio for the non-heating season was 2.9. After thermo-modernization, it amounted to 2.78 and 2.22, respectively.

Since the I/O ratio was >1 in both of these sampling site groups, it can be concluded that the major sources of these bioaerosols are likely internal, such as building occupants (in this case children and their activities) as well as building materials that host microbiological growth.

The I/O ratio for fungal aerosol levels varied between seasons, being <1 (0.25–0.82) in heating season and <1 (0.24–0.69) in non-heating season. Since the mean I/O ratio for fungi levels is below 1.0 the concentrations in indoor environments are relatively low (<489 CFU/m^3), it can be concluded that there is no significant mold source in these indoor environments. This result could demonstrate the significant role of outdoor infiltration on indoor fungi levels. Nevertheless, the I/O ratio <1.0

Table 20.5. Average concentrations of total bacterial and fungal colony-forming units per cubic meter of air in heating and non-heating season; (OUT) – outdoor; (I) – indoor; (Y) – younger children classroom; (O) – older children classroom before term-modernization.

Parameters			Classrooms No.			Outdoor
			1	2	3	
			Children's age			
			13/Older (O)	6–7/Younger (Y)	6–7/Younger (Y)	
Before termo-modernization						
Heating season	Bacterial aerosol (CFU/m^3)	AC	1896	1913	1727	576
		SD	130.1	284.3	113.5	163.5
		I/O Ratio	3.29	3.32	2.99	–
	Fungal aerosol (CFU/m^3)	AC	79	97	64	119
		SD	21,5	25,3	23,3	24.3
		I/O Ratio	0.66	0.82	0.54	–
Non-heating season	Bacterial aerosol (CFU/m^3)	AC	2310	3670	2590	996
		SD	74.3	115.8	211	154.5
		I/O Ratio	2.32	3.68	2.6	–
	Fungal aerosol	AC	439	305	489	710
	(CFU/m^3) SD	49.4	31.3	73.8	74.5	
		I/O Ratio	0.61	0.43	0.69	–
After termo-modernization						
Heating season	Bacterial aerosol (CFU/m^3)	AC	2356	2150	1997	779
		SD	115.9	259.3	89.5	196.7
I/O Ratio	3.02	2.76	2.56	–		
	Fungal aerosol (CFU/m^3)	AC	112	73	101	289
		SD	42.3	22.9	44.6	65.8
	I/O Ratio	0.39	0.25	0.35	–	
Non-heating season	Bacterial aerosol (CFU/m^3)	AC	3210	1862	2456	1129
		SD	256.1	125.9	233.2	289.2
		I/O Ratio	2.84	1.65	2.18	–
	Fungal aerosol (CFU/m^3)	AC	139	214	208	568
		SD	25.1	58.5	112.3	56.7
		I/O Ratio	0.24	0.38	0.37	–

during non-heating season suggests stronger deposition on solid surfaces, or decomposition in the indoor air, rather than differences in the filtering of the ventilation air when crossing the building threshold.

It is generally accepted that the microbiological concentration in the healthy buildings is similar to the corresponding outdoor values, which means that the I/O concentration ratio is close to 1. Comparing the obtained average values of the I/O ratios with the research of Stryjakowska-Sekulska et al. and the reference therein (2007) the indoor fungi levels in the tested rooms can be estimated to be relatively safe, in contrast to their poor air quality according to the bacterial aerosol levels (I/O $\leq$ 1.5: good indoor air conditions; I/O = 1.5–2.0: regular; I/O > 2: poor).

20.4 CONCLUSIONS

The data showed that the bacteria and fungi levels varied between seasons, with the non-heating season concentrations being higher than those in winter. The bacteria concentrations varied in the range from 1727 to 3670 CFU/m^3, while the fungi levels varied from 64 to 489 CFU/m^3. The indoor bacteria levels were between two and three times higher than the outdoor levels, while the I/O ratio for fungi was <1.

The highest concentrations of bacteria and fungi were obtained during tests at school before thermo-modernization.

Further research into the direct causes of the observed changes is recommended.

REFERENCES

Agranovski I.E., Agranowski V., Reponen T., Willeke K., Grinshupun S.A., 2002. Development and evaluation of a new personal sampler for culturable airborne microorganisms, *Atmospheric Environment* 36, 889–898.

An H.R., Mainelis G., Yao M., 2004. Evaluation of a high-volume portable bioaerosol sampler in laboratory and field environments, *Indoor Air* 14(6): 385–393.

Assar S.K., Block S.S., 2000. Survival of microorganisms in the environment. In: Disinfection, sterilization, and preservation. *Lippinkott-Williams*.

Barabasz W., Chmiel M.J., Albińska D., Mazur M.A., 2007. Składowiska odpadów jako źródła bioaerozolu i mikroorganizmów szkodliwych dla zdrowia. Składowiska odpadów komunalnych źródłem gazu, Konferencja Naukowo-Techniczna, Czarna 17–19.10.2007, *Instytut Nafty i Gazu*. 145 s. 143–152.

Bonetta S.A., Bonetta S.I., Mosso S., Sampo S., Carraro E., 2009. Assessment of microbiological indoor air quality in an Italina office building equipped with HAVAC system. *Environmental Monitoring Assessment* 161, 473–483.

Brągoszewska E., Mainka A., Pastuszka J. S., 2016. Bacterial and Fungal Aerosols in Rural Nursery Schools in Southern Poland. *Atmos.* 7(11), 142.

Breza-Boruta B., 2010. Ocena mikrobiologicznego zanieczyszczenia powietrza na terenie oczyszczalni ścieków. *Woda-Środowisko-Obszary Wiejskie*. T. 10. Z. 3(31) s. 49–57.

Cabral J.P.S., 2010. Can we use indoor fungi as bioindicators of indoor air quality? Historical perspectives and open questions. *Science of the Total Environment*. 408 s. 4285–4295.

Chmiel M., Frączek K., Grzyb J., 2015. Problemy monitoringu zanieczyszczeń mikrobiologicznych powietrza. *Water-Environment-Rural Areas* (I–III). ISSN 1642–8145 s. 17–27.

Dudzińska M.R., 2013. Aerozole w powietrzu wewnętrznym. *Monografia Komitetu Inżynierii Środowiska PAN* 112., pp.175

Dumała S., Dudzińska M.R., 2013. Microbiological indoor air quality in Polish schools. *Annual Set The Environment Protection* 15, 231 – 244.

Elazhary M.A., Derbyshire J.B., 1979. Effect of temperature, relative humidity and medium on the aerosol stability of infectious bovine rhinotracheitis virus. *Canadian Journal of Comparative Medicine* 43(2):158–67.

Fiegel J, Clarke R, Edwards DA., 2006. Airborne infectious disease and the suppression of pulmonary Bioaerosols. *Elsevier*, vol. 11(1/2). p. 51–7.

Flannigan B., Samson R.A., Miller J.D., 2011. Microorganisms in home and indoor work environments. Diversity, health impacts, investigation and control. Wyd. 2. *Londyn. CRC Press*. ISBN 9781420093346 ss. 539.

Gładysz J., Grzesiak A., Nieradko-Iwanicka B., Borzęcki A., 2010. Wpływ zanieczyszczeń powietrza na stan zdrowia i spodziewaną długość życia ludzi. *Problemy Higieny i Epidemiologii*. T. 91. Nr 2 s. 178–180.

Górny R.L., 2004. Biologiczne czynniki szkodliwe: normy, zalecenia i propozycje wartości dopuszczalnych. *Podstawy i Metody Oceny Środowiska Pracy* 3 (41): 17–39.

Górny R.L., 2004a. Biologiczne czynniki szkodliwe: normy, zalecenia i propozycje wartości dopuszczalnych. *Podstawy Metody Oceny Środowiska Pracy*. Nr 3 (41) s. 17–39. 26.

Górny R.L., 2004b. Cząstki grzybów i bakterii jako składniki aerozolu pomieszczeń: właściwości, mechanizmy emisji, detekcja. Sosnowiec. *Instytut Medycyny Pracy i Zdrowia Środowiskowego* s. 164.

Hameed A.A., Khoder M.I., I Brahim Y.H., Saeed Y., Osman M.E., Ghanem S., 2012. Study on some factors affecting survivability of airborne fungi. *Science of the Total Environment*. Vol. 414 s. 696–700.

Hatch M.T., Wolochow H., 1969. Bacterial survival: consequences of airborne state. An introduction to experimental aerobiology. New York: *John Wiley and Sons*, p. 267–95.

Hemmes J.H., Winkler K.C., Kool S.M., 1960. Virus survival as a seasonal factor in influenza and poliomyelitis. *Nature* 188:430–8.

Hinds W.C., 1999. Aerosol technology. Properties, behavior, and measurement of airborne particles. 11 respiratory deposition. 2nd ed. *John Willey & Sons*, Inc, p. 233–57.

Jong J.G., Winkler K.C., 1964. Survival of measles virus in air. *Nature* 201:1054–5.

Karim Y.G., Ijaz M.K., Sattar S.A., Johnson-Lussenburg C.M., 1985. Effect of relative humidity on the airborne survival of rhinovirus-14. *Canadian Journal of Microbiology* 31(11):1058–61.

Kiliszczyk A., Podlaska B., Sadowiec K., Zielińska-Polit B., Rytel M., Russel S., 2013. Ocena występowania grzybów oraz amoniaku i metanu w powietrzu w wybranym budynku inwentarskim. *Woda-Środowisko-Obszary Wiejskie*. T. 13. Z. 3(43) s. 79–89.

Krzyścko-Łupicka T., 2010. Zagrożenia mikologiczne w budownictwie – problem ogólnoświatowy. W: Problemy w ochronie środowiska w województwie opolskim w latach 2010–2020. Pr. zbior. Red. K. Oszańca. Opole. *Opolskie Ekoforum. Atmoterm S.A.* s. 203–222.

Krzysztofik B., 1992. Mikrobiologia powietrza. *Wyd. Politechniki Warszawskiej*, Warszawa.

Law A.K.Y., Chau C.K, Chang G.Y.S., 2001. Characteristics of bioaerosol profile in office buildings in Hong Kong, *Building and Environment* 36, 527–541.

Libudisz Z., Kowal K., Żakowska Z., 2009. Mikrobiologia techniczna. tom I, *Wyd. Nauk. PWN*. Warszawa.

Lidwell O.M., 1974. Aerial dispersal of micro-organisms from the human respiratory tract. *Society for Applied Bacteriology Symposium Series*, 3(0):135–54.

Loosli C.G., Lemon H.M., Robertson O.H., Appel E., 1943. Experimental airborne influenza infection: I. influence of humidity on survival of virus in air. *Proceedings of the Society for Experimental Biology and Medicine* 53:205–6.

Loudon R.G., Roberts R.M., 1967. Droplet expulsion from the respiratory tract. *American Review of Respiratory Disease* 95:435–42.

Lowen A.C., Mubareka S., Steel J., Palese P., 2007. Influenza virus transmission is dependent on relative humidity and temperature. *PLoS Pathogens* 3(10):1470–6. e151.

Menetrez M. Y., Foarde K.K., Dean T.R., Betancourt D.A., Moore S.A., 2007. An evaluation of the protein mass of particulate matter, *Atmospheric Environment* 41, 8264–8274.

Mentese, S., Rad, A.Y., Arisoy, M., Gullu, G. 2012. Seasonal and Spatial variations of bioaerosols in indoor urban environments, Ankara, Turkey. *Indoor Built Environ.* 21, 797–810.

Miller W.S., Artenstein M.S., 1967. Aerosol stability of three acute respiratory disease viruses. *Proceedings of the Society for Experimental Biology and Medicine* 125:222–7.

Morawska L., 2005. Droplet fate in indoor environments, or can we prevent the spread of infection? *Proceedings Indoor Air* 9–23.

Nicas M., Nazaroff W.W., Hubbard A., 2005. Toward understanding the risk of secondary airborne infection: emission of respirable pathogens. *Journal of Occupational and Environmental Hygiene* 2:143–54.

Pastuszka, J.S., Kyaw Tha Paw, U., Lis, D.O., Wlazło, A., Ulfig, K. 2000. Bacterial and fungal aerosol in indoor environment in Upper Silesia, Poland. *Atmos. Environ.* 34, 3833–3842.

Pegas, P.N., Evtyugina, M.G., Alves, C.A., Nunes, T., Cerqueira, M., Franchi, M., Pio, C., Almeida, S.M., Freitas, M.C. 2010. Outdoor/indoor air quality in primary schools in Lisbon: A preliminary study. *Quim. Nova* 33, 1145–1149.

Qian H., Li Y., Nielsen P.V., Hyldgaard C.E., Wong T.W., Chwang A.T.Y., 2006. Dispersion of exhaled droplet nuclei in a two-bed hospital ward with three different ventilation systems. *Indoor Air* 16:111–28.

Raczkowski A., Dumała S., Skwarczyński M., 2010. Układy wentylacji, klimatyzacji i chłodnictwa. *Monografie Komitetu Inżynierii Środowiska PAN* vol.77.

Shin, S.-K., Kim, J., Ha, S., Oh, H.-S., Chun, J., Sohn, J., Yi, H. 2015. Metagenomic insights into the Bioaerosols in the indoor and outdoor environments of childcare facilities. *PLoS ONE*,10, e0126960.

Stetzenbach L.D., Buttner M. P., Cruz P., 2004. Detection and enumeration of airborne biocontaminants. *Current Opinion in Biotechnology* 15, s. 170–174.

Stryjakowska-Sekulska, M., Piotraszewska-Pająk, A., Szyszka, A.; Nowicki, M., Filipiak, M. 2007. Microbiological quality of indoor air in university rooms. *Pol. J. Environ. Stud.*16, 623–632.

Tellier R., 2006. Review of aerosol transmission of influenza a virus. *Emerging Infectious Diseases* 12(11): 1657–62.

Wargocki P., 2005. Measurements of Perceived Indoor Air Quality. Energy Efficient Technologies in Indoor Environment – konferencja międzynarodowa, Politechnika Śląska, Gliwice.

Yang S., Lee G.W.M., Chen C.M., Wu C.C., Yu K.P., 2007. The size and concentration of droplets generated by Coughing in human subjects. *Journal of Aerosol Medicine* 20(4):484–94.

Yang, W., Sohn, J., Kim, J., Son, B., Park, J. 2009. Indoor air quality investigation according to age of the school buildings in Korea. *J. Environ. Manag.* 90, 348–354.

CHAPTER 21

Analysis of environmental hazards in soils located around cement plants

M. Widłak
Faculty of Environmental, Geomatic and Energy Engineering, Kielce University of Technology, Kielce, Poland

A. Widłak [a,b]
[a] *Kaufering, Hilti Entwicklungsgesellschaft mbH - Health, Safety and Environment Group manager*
[b] *Health, Safety and Environment Group manager, Hilti Entwicklungsgesellschaft mbH, Kaufering, Germany*

ABSTRACT: The research included the areas around three active cement plants located in the Swietokrzyskie province and one cement plant in Mazowieckie province which was shut down in 1997. The results of agricultural usable soil research on the granulometric composition and content of selected heavy metals (Cr, Cu, Ni, Pb, Zn) depending on pH value, humus content and soil sorption were presented. Test samples were taken at two distances from the emitters, from the east and west. The first distance is ten times the height of the emitter (about 1000 m), the second distance is forty times the height of the emitter (about 4000 m).

Keywords: soil, sorption, reaction, heavy metals, cement plants, danger

21.1 INTRODUCTION

The human activity and the pace of industrial development mean that the environment is increasingly polluted. A large amount of harmful substances infiltrates into the air, soil and water every day. The development of industry and the growing demand for a higher standard of living resulted in the degradation of the natural environment. Industry and agriculture – due to the use of fertilizers and plant protection products – as well as urbanization and communication have a negative impact on the environment. Pollution of agricultural soils, located at communication routes or within the range of influence of industrial plants, is particularly dangerous for the human and animal health. The content of heavy metals is one of the basic parameters determining the state of pollution and the degree of soil degradation.

Soil is a valuable renewable natural resource of the Earth, it is the basis for the functioning of all existing terrestrial ecosystems; it is a natural environment needed for the development of microorganisms and plants, which determines the distribution and production of biomass, energy flow and the circulation of matter in the ecosystem, it is also a filter in the groundwater circulation. As a result of the human activity, the natural properties of soil often undergo the changes which are difficult to reverse. Soil buffers, filters, and also accumulates harmful substances; hence, soil protection is a priority. Contaminated soil worsens the vegetation of crop plants as well as the aesthetic and ecological qualities of plants. The factors affecting the soil pollution are road traffic, population density, surface structure of a given area, waste and sewage management, agriculture chemization, microclimate and soil type. However, the greatest degradation of the natural environment is caused by industry – in particular metallurgy, mining and the electricity and cement industries. Soil pollution is mainly caused by the penetration of all chemical pollutants that cannot be neutralized by this environment (Dziubanek et al. 2012; Glinka & Wilk 2016).

The main source of soil contamination is the immission of atmospheric pollution, to a lesser extent they are caused by the direct introduction of pollution into the soil. The pollutants found in soils from the air depend on the strength and direction of the winds. The share of individual wind directions is not uniform throughout the year.

DOI 10.1201/9781003171669-21

In summer, the westerly and north-westerly winds prevail. In autumn, the share of the winds taking east and south-east directions increases. In winter, the winds from the south-west are predominant. Spring is characterized by a relatively even distribution of wind directions. In Poland, however, the dominant directions are always west and south-west. The height of the chimney plays a dominant role in the dispersion of exhaust gases. Industrial chimneys should have a height of at least 100 m (Oruba, 2019). The high height of the chimney reduces the concentration of pollutants in the immediate surroundings, but causes their movement over greater distances (Rup 2006).

It is a monitoring work the purpose of which was to assess the heavy metal content and physicochemical properties of the soils around the cement plants. The impact of blowing winds on the pollution of the studied soils used for agricultural crops was assessed.

21.2 PHYSICAL AND PHYSICOCHEMICAL PARAMETERS OF SOIL

21.2.1 *Grain size compostions*

The granulometric composition is the basic physical parameter of mineral soils, which constitutes the mutual proportions between the particles of different sizes in the soil. To a large extent, the share of individual granulometric fractions shapes the physical (density, porosity, water properties, etc.) as well as chemical properties of the soil (sorption properties, composition of organic matter, content of micro- and macroelements, etc.) (Stępień 2016).

The growth, development and yielding of plants depends not only on the grain size of the arable layer (usually from 0–20 to 0–30 cm), but also on the grain size of deeper layers, up to about 100 cm. The presence of clay or loam in the deeper layers of light soils reduces their sensitivity to drought, but at the same time increases the sensitivity to excess water, which can be retained by a more compact layer, especially after very heavy rainfall.

Under Polish conditions, the sensitivity to drought is more important than the sensitivity to excess water (Różański 2010).

21.2.2 *Organic matter*

Organic matter is one of the most important components of the soil solid phase, it undergoes mineralization and humification processes in soil. The organic substances in the substrate are subject to continuous quantitative and qualitative changes (Ukalska-Jaruga et al. 2015). The transformation of organic substances into soil humus is possible due to the high microbial activity in the soil and due to the continuous supply of fresh organic matter to the soil. The amount of soil humus is an indicator of soil fertility. In the mineral soils used for agriculture, humus accounts for 80–90% of all soil organic matter (Doran & Zeiss 2000). The content of soil humus, as the basic indicator of soil fertility, is mainly associated with the soil sorption properties, both in relation to plant nutrients, as well as unnecessary or harmful elements. The average humus content in the soils of Poland amount to 1.5–4.0%, whereas in the Swietokrzyskie province – 1.55% (Report 2012).

21.2.3 *Sorption*

Soil is characterized by the ability to retain and store the ions and particles from the soil solution, actively participates in the circulation of elements in the environment. The phenomenon known as sorption determines the bioavailability of chemical elements and performs protective functions against the excessive flow of undesirable substances in the environment (Adriano 2001). The sorption properties of soils play an important role as a factor regulating the leaching of nutrients from the soil, determine the efficiency of fertilization and regulate plant nutrition. The sorption properties of soils are determined, among others, by their granulometric composition, pH, as well as the content of mineral and organic colloids. The size of sorption capacity and the share of basic cations in the sorption complex is affected by the mineralogical composition of the clay fraction and pedogenic processes (Violante et al. 2010).

Metals differ in the strength with which they are bound by the sorption complex. The series of entries into the sorption complex for individual metals is as follows Pb> Cu> Zn ≥ Cr> Ni> Co> Cd – for sand soil (Sady & Smoleń 2004).

21.2.4 *Soil pH*

The soil pH is a feature that determines the course of many processes in the soil environment. It has a large impact on the development and settlement of soils by microorganisms, soil sorption, nutrient absorption, bioavailability and phytotoxicity of heavy metals, plant development, etc. The optimal range of soil pH for most plants is in the range of pH 6.0–7.0 (Sady & Smoleń 2004).

21.2.5 *Heavy metals*

21.2.5.1 *Chromium content*

Chromium is one of the elements necessary for the development and growth of plants. It is taken up by plants with the transpirational flow of water and stored in the vegetative parts of plants (Alloway 2012). The occurrence of chromium in soils is generally small, it is a derivative of the content in parent rocks, where the concentration range is 7 – 150 mg/kg, on average 30 mg/kg.

21.2.5.2 *Copper content*

Copper is a metal commonly found in the Earth's crust. It is strongly bound by organic matter and clay minerals in the top soil layer, creating low mobility forms. The mobility of this element increases under the influence of strong acidification, an increase in the soil pH usually reduces the availability of soil copper for plants. At high soil pH, copper immobilization may occur. Copper is an essential component of the metabolic processes, photosynthesis, respiration, and in transformations of the nitrogen compounds. Both excess and insufficiency of the element may cause a disturbance of these processes, and consequently cause a decrease in yielding (Alloway 2005; Yruela et al. 2000). The natural content of copper in soils varies and ranges between 1–140 mg/kg, on average 6.3 mg/kg.

21.2.5.3 *Nickel content*

The occurrence of nickel in soils is associated with the content in parent rocks, as well as the granulometric composition. The element is strongly associated with soil organic matter. The solubility increases with the acidity of the soil; however, due to the susceptibility of nickel to bonding with organic matter, in many soils there is a high mobility of nickel, even under neutral and alkaline conditions. The toxic effects of nickel are revealed in disorders of photosynthesis, transpiration and the process of nitrogen binding (Bai et al. 2006).

The natural nickel content in Polish soils at surface levels is between 40 – 50 mg/kg, and the average is 7.4 mg/kg.

21.2.5.4 *Lead content*

Lead belongs to the group of the most toxic elements for living organisms. In agricultural soils, the lead content is strongly related to the granulometric composition and organic matter. The acidic pH of the soil, low humus content, and poor sorption are conducive to the process of lead uptake by plants. The source of soil pollution is industry and motorization (Alkorta et al. 2004). The occurrence of lead in the surface soil layers is associated with the impact of anthropogenic factors. It is assumed that the natural content of this element is about 20 mg/kg, Polish average 18 mg/kg.

21.2.5.5 *Zinc content*

Zinc is an element found in nature in the form of easily soluble compounds, especially in an acidic environment. It creates permanent bonds with the organic substance, accumulates in the surface layers of soil, as well as gathers in the roots and leaves of plants.

Excess zinc can reduce the nitrification processes and hinder many metabolic processes. Deficiency may cause bone development disorders, skin diseases, atherosclerotic and allergic changes. The overall zinc content in soils is 10–200 mg/kg, the average concentration of zinc in Polish soils is 40 mg/kg (Sadeghzadeh & Rengel 2011).

21.3 RESEARCH SCOPE AND METHODOLOGY

The study covered soils in the Swietokrzyskie and Mazowieckie voivodships. The heavy metals content and physico-chemical properties of arable soils located on the east and west sides of the Dickerhoff Nowiny Cement Plant, Lafarge Małogoszcz Cement Plant, Grupa Ozarow Cement Plant in the Swietokrzyskie voivodship and Przyjazn Wierzbica Cement Plant closed in 1997 were examined. The soil samples (20 samples from each research point) for testing were taken at a distance of ten and forty times the height of the cement plant emitter, the soil sampling scheme was developed on the basis of PN-R-04031:1997 (Figure 21.1, Table 21.1).

The research areas are characterized by a very diverse geological structure, this is due to the large variety of parent rocks and terrain relief, as well as climatic conditions. The areas selected for testing are dominated by podzol, rendzinas and pararendzinas (Widłak 2015). These soils show the infiltration capacity during prolonged rainfall, while being the most susceptible to erosion. This is noticeable during spring thaws and after heavy rainfall (Widłak 2016).

The research area includes the arable land with a direct relationship between soil quality and the quality of food produced is observed. The test samples were taken from agricultural soils, from an accumulation level of 0–30 cm. The location of research points ensures the diversity of soil formations and soil types characteristic of the soil cover of the Swietokrzyskie and Mazowieckie regions (Report 2012).

The mixed soil samples were analyzed. Soil was collected from each point in the autumn of 2018, spring 2019 and summer 2019, with varying weather conditions (Gajec et al. 2018). In terms of the chemical properties, pH in 1M KCl was determined potentiometrically (PN-ISO 10390:1997).

The content of total forms of metals: Cr, Cu, Ni, Pb, Zn was determined by means of the ICP-AES technique after mineralization of samples with aqua regia, in accordance with the standard (EN-13346:2000). Organic matter – with the Tiurin method (Kabała & Karczewska 2019), sorption capacity – by determining the iodine value (IV) (PN-83C-97555.04) (Bezak-Mazur & Dańczuk 2013) and the granulometric composition by means of the laser diffraction method with use of the Mastersizer 3000 device (ISO 14887:2000, Mastersizer 3000).

Arithmetic mean, median, standard deviation were calculated for the results obtained. The size of the linear relationship between the tested parameters was determined. The statistical analysis of the results was carried out in Microsoft Excel 2016.

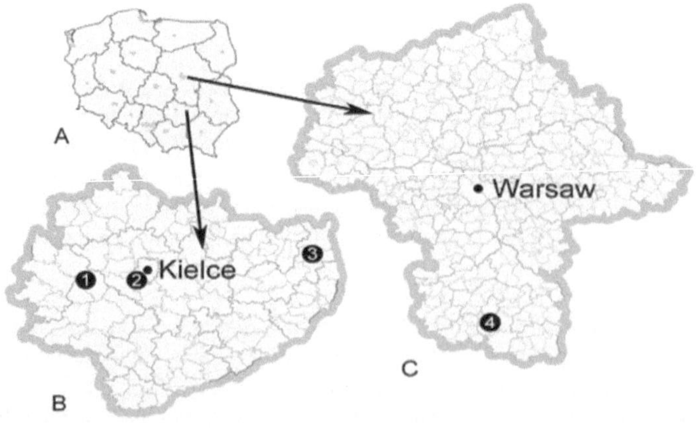

Figure 21.1. Location of sampling points: A – Poland, B – Swietokrzyskie voivodship, C – Mazowieckie voivodship (Map of Poland).

Table 21.1. The name and location number of the cement plant on the map of Poland and Swietokrzyskie and Mazowieckie voivodship

Position number and location on the map (Figure 21.1. Map of Poland)							
1		2		3		4	
Cement Plant Lafarge Małogoszcz		Cement Plant Dickerhoff Nowiny		Cement Plant Grupa Ozarow		Cement Plant Przyjazn Wierzbica	
Sampling place	Marking of the sampling place in text and drawings	Sampling place	Marking of the sampling place in text and drawings	Sampling place	Marking of the sampling place in text and drawings	Sampling place	Marking of the sampling place in text and drawings
Małogoszcz East 1000m	MW1	Nowiny East 1000m	NW1	Ozarow East 1000m	OW1	Wierzbica East 1000m	WW1
Małogoszcz East 4000m	MW2	Nowiny East 4000m	NW2	Ozarow East 4000m	OW2	Wierzbica East 4000m	WW2
Małogoszcz West 1000m	MZ1	Nowiny West 1000m	NZ1	Ozarow West 1000m	OZ1	Wierzbica West 1000m	WZ1
Małogoszcz West 4000m	MZ2	Nowiny West 4000m	NZ2	Ozarow West 4000m	OZ2	Wierzbica West 4000m	WZ2

21.4 RESULTS AND DISCUSSION

21.4.1 *Grain size compostions*

The studied soils were characterized by a diverse granulometric composition. At all soil sampling points, the lowest percentage corresponded to the clay fraction, in the average range of 0.04% - 1.25%. The silt fraction in the range of 11.00% - 29.45% was one of the higher percentages in the areas of the studied soils. However, the largest percentage was the fine sand and medium sand fraction. The average percentage share was 28.83% for fine sand, and 21.66% for medium sand (Table 21.1). The smallest clay and fractions shape the soil's sorption properties.

21.4.2 *Soil pH*

The pH of the soils in all research areas ranged from acidic to alkaline 5.39–8.48 (Table 21.3 and Table 21.4). The highest pH = 8.48 was recorded at the Małogoszcz cement plant area in a distance of 1000 m away from the emitter, in autumn. The lowest pH = 5.39, 5.50, 5.64 was recorded in the area of the closed Wierzbica Cement Plant from the west side, 4000 m away from the emitter in spring, summer and autumn. A similar pattern was observed in the research area within 1000 m of the emitter impact of the Ozarow Cement Plant from the east side (Table 21.4).

In the Małogoszcz and Nowiny research points, neutral and alkaline soil pH was dominant in the period of autumn 2018 - summer 2019, and ranged from 6.94 to 8.48 (Table 21.4, Figure 21.2).

The pH of the studied Ozarow soils ranged from 5.40 to 8.05. The lowest pH (5.40) was recorded in the summer of 2019 from the east (OW1), while the highest pH (8.05) was observed in the summer

Table 21.2. Average content [%] of all fractions of the studied soil in the area: Lafarge Małogoszcz Cement Plant (M), Dickerhoff Nowiny Cement Plant (N),Grupa Ozarow Cement Plant (O), Przyjazn Cement Plant Wierzbica (W).

Percentage content [%] of particular fractions in soil tested						
	[μm]	M	N	O	W	Average share of fractions from M, N, O, W locations [%]
Very coarse sand	(1000;2000)	1.61	0.79	1.74	5.28	2.36
Coarse sand	(500;1000)	9.89	7.64	6.39	14.03	9.49
Medium sand	(250;500)	20.39	21.50	21.02	23.73	21.66
Fine sand	(100;250)	23.23	33.99	25.12	32.96	28.83
Very fine sand	(50;100)	14.63	19.42	14.77	12.74	15.39
Silt	(2;50)	29.00	16.19	29.45	11.00	21.41
Clay	<2	1.25	0.47	1.31	0.04	0.77

Table 21.3. Statistical parameters at measuring points around the cement plants: Małogoszcz, Nowiny, Ozarow and Wierzbica.

Location	Małogoszcz			Nowiny			Ozarow			Wierzbica		
	pH [-]	Iodine value [mg/g]	Humus [%]	pH [-]	Iodine value [mg/g]	Humus [%]	pH [-]	Iodine value [mg/g]	Humus [%]	pH pH [-]	Iodine value [mg/g]	Humus [%]
Min	6.94	50.77	0.79	6.40	35.38	0.95	5.40	52.30	0.91	5.39	45.38	0.91
Max	8.48	101.54	2.45	8.42	104.23	2.74	8.05	101.54	2.36	8.32	96.53	2.93
Avg.	7.73	74.77	1.67	7.53	68.48	1.68	7.00	80.24	1.65	7.42	72.12	1.54
Med.	7.57	78.75	1.65	7.51	75.02	1.69	7.04	80.90	1.83	7.43	74.50	1.85
SD	0.44	13.66	0.63	0.28	26.18	0.41	0.26	14.03	0.46	0.16	18.79	0.58

of 2019 from the west (OZ1), 1000 m away from the emitter (Table 21.4, Figure 21.2). Slightly acidic pH of the soil tested was also noted in three cases: OW1 - fall 2018, spring 2019 and summer 2019 (Table 21.4, Figure 21.2).

The soil pH at measuring points Cement Plant Wierzbica showed slightly acidic values for all seasons for WZ2 point, alkaline for autumn 2018 and spring 2019 at all other research points (Figure 21.2 and Table 21.3).

Table 21.4. Soil pH during the research period at measuring points around the cement plant: Małogoszcz, Nowiny, Ozarow and Wierzbica.

Cement plant	pH [-]			
	Sampling location	Autumn 2018	Spring 2019	Summer 2019
Małogoszcz	MW1	8.48	7.93	7.94
	MW2	6.94	7.19	8.02
	MZ1	8.15	6.96	7.10
	MZ2	8,24	8.04	7.77
Nowiny	NW1	7.75	7.98	8.42
	NW2	7.62	8.09	8.11
	NZ1	7.49	7.35	7.44
	NZ2	7.23	6.52	6.40
Ozarow	OW1	5.94	5.40	5.70
	OW2	7.90	7.47	7.94
	OZ1	7.80	7.85	8.05
	OZ2	7.01	6.27	6.70
Wierzbica	WW1	8.21	8.07	8.23
	WW2	7.87	7.77	8.09
	WZ1	8.32	7.79	8.15
	WZ2	5.64	5.39	5.50

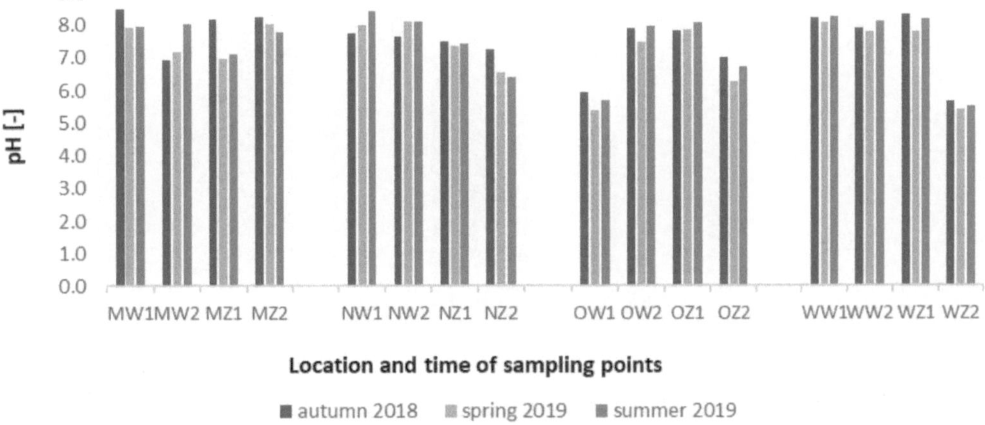

Figure 21.2. Values of pH of the tested soil during the sampling period in points from the east (W1, W2) and west (Z1, Z2) of the cement plant: Małogoszcz (M), Nowiny (N), Ozarow (O), Wierzbica (W).

21.4.3 *Iodine value of the soil*

The assessment of soil sorption capacity was determined using the method proposed at the Kielce University of Technology (Bezak-Mazur & Dańczuk 2013) using the determination of the iodine number (PN-83C-97555.04). The results of the studied soil in terms of the iodine value from the autumn of 2018, spring 2019 and summer 2019 show average sorption in the range of 35.38 mg/g - 104.23 mg/g. The lowest values were recorded in spring in the Nowiny town - East 1 (NW1) -35.38 mg/g and Wierzbica town west 1 (WZ1) - 45.38 mg/g. The soils in the Wierzbica Cement Plant area showed low values in autumn in the range of 45.38–96.53 mg/g. The highest value - 104.23 mg/g was recorded in the Nowiny Cement Plant on the east side at a distance of 4000 m from the emitter – in spring 2019 (NW2). A high value of 101.54 mg/g was recorded on the west side from the pollution emitter of the Małogoszcz Cement Plant and the Ozarow Cement Plant (Figure 21.3). The average iodine values in all research areas were in the range of 68.48–80.24 mg/g (Table 21.3).

The sorption properties of the tested soils depend on the season of the year. The soils showed the weakest sorption properties in autumn, which increased significantly during the spring vegetation and stabilized during the summer at an average level of 59.38–97.30 mg/g (Figure 21.3).

21.4.4 *Humus content in the soils*

The average humus content in the tested soils of all cement plants: Małogoszcz, Nowiny, Ozarow and Wierzbica in the autumn period ranges from 0.79% to 1.87%. The average humus content in all research areas remained at an even level in the range of 1.54–1.68 mg/g (Table 21.3). The percentage of humus content increases from 1.22% in autumn to 2.17% in spring and decreases to 1.73% in summer (Figure 21.4). The research was carried out on 20 samples from each research point, the samples were taken three times in each season of the year.

In spring 2019, there was an increase in soil humus content by 56% compared to autumn 2018, followed by a decrease of 20% in spring and summer 2019 (Figure 21.4).

21.4.5 *Heavy metals determination*

In the studied soils of all cement plants (Małogoszcz, Nowiny, Ozarow, Wierzbica) the content of metals such as Cr – chromium, Cu – copper, Ni – Nickel, Pb – lead, Zn – zinc was determined.

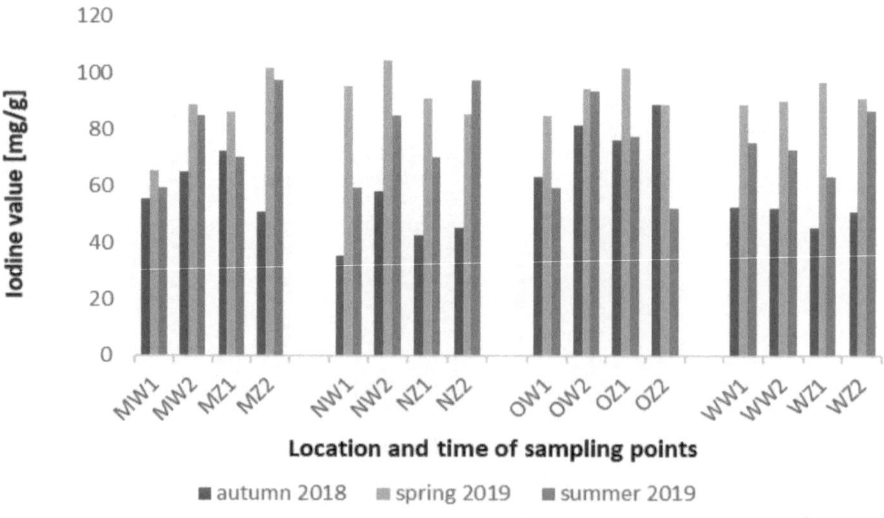

Figure 21.3. Iodine value of the tested soil during the sampling period in points, from the east (W1, W2) and western (Z1, Z2) cement plants: Małogoszcz (M), Nowiny (N), Ozarow (O), Wierzbica (W).

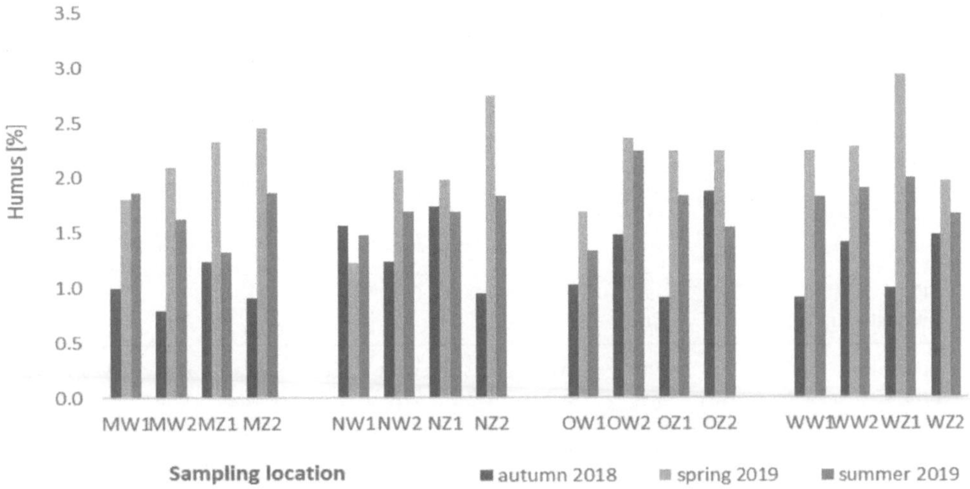

Figure 21.4. Average value of humus content in the tested soil during the sampling period at points from the east (W1, W2) and west (Z1, Z2) of the cement plants: Małogoszcz (M), Nowiny (N), Ozarow (O), Wierzbica (W).

Table 21.5 shows the minimum and maximum content of the tested metals at the measuring points and the values of statistical parameters.

Zinc had the highest content in all soil samples tested, but no excess was recorded at any test point (OJ. 1359, 2016). The zinc content in the samples from the whole research period, all areas ranged from 6 to 21 times lower than the limit value. Both from the east and west side, it was the same average value. The highest content of zinc, lead, nickel, copper and chromium occurred on the east and west sides of the Dickerhof Nowiny Cement Plant (OJ. 1359, 2016).

The highest mean zinc content in the studied soil of 196.28 mg/kg, was recorded from the eastern side of the Nowiny Cement Plant emitter, while from the west side of this emitter the zinc content was 86.6 mg/kg (Figure 21.5, Figure 21.6).

In the soil samples taken for testing from the east side of the cement plant emitters (Małogoszcz, Nowiny, Ozarow, Wierzbica), the average content of chromium, copper and nickel metals was slightly higher than from the west side.

The largest differences in the examined metals content, between the samples taken from different localizations, were observed in the cases of cooper and lead measured in the vicinity of Ozarow Cement Plant. Concentration of these metals were three and two times higher, respectively in soil taken from the west than east side of the emitter. (Figure 21.5, Figure 21.6).

In the cases of the lead and chromium in the soils from the east side of the Nowiny cement plant and the Wierzbica cement plant, the average contents determined during the study period were twice as high in relation to the values measured in the samples taken from west side of the studied plants (Figure 21.5, Figure 21.6).

Average values of the Cr, Cu, Ni content observed from the eastern side of the emitter were 30–35-fold smaller, while from the western side – 24–32-fold smaller than the current norms given in the regulation on the method of assessing the earth's surface pollution (OJ. 1359, 2016) (Table 21.6).

The increased content of the metals Cu, Ni, Pb, Zn occurred from the eastern side of the Nowiny Cement Plant, while in the case of the Ozarow emitter, the increased content of these metals occurred on the western side. This may indicate the impact of blowing winds

(Rup 2006) and the amount of raw material production in the Nowiny and Ozarow Cement Plant.

Assessment of the statistical parameters of the studied soil at the examined locations such as the median and arithmetic means for Cr, Cu, Ni showed that there is no, or there is very low differentiation, which indicates a correct interpretation of the average metal content. The SD values indicate a small variation in the average content of these metals in the studied areas. The parameters for zinc are

Table 21.5. Statistical parameters for heavy metal content (Cr, Cu, Ni, Pb, Zn) at measuring points, from the east and west, cement plants: Małogoszcz, Nowiny, Ozarow, Wierzbica.

Cement plant Małogoszcz	Cr mg/kg		Cu		Ni		Pb		Zn	
Location	East	West	East	West	East	West	East	West	East	West
Min	2.6	3.8	3.7	5.7	1.9	3.1	5.5	5.9	24.5	32.2
Max	11.0	8.7	6.5	8.4	5.5	5.3	22.9	15.8	60.5	72.6
Avg.	5.7	6.6	5.0	6.9	3.2	4.0	12.7	14.8	37.3	44.0
Med.	5.2	7.2	4.5	6.7	2.7	4.2	9.1	8.1	34.7	41.0
SD	2.8	2.0	1.3	0.9	1.3	0.8	7.5	15.6	13.2	15.0
Cement plant Nowiny	Cr		Cu		Ni		Pb		Zn	
Location	East	West	East	West	East	West	East	West	East	West
Min	4.4	5.2	1.3	7.1	2.4	2.1	19.9	11.6	44.3	43.8
Max	10.1	15.9	21.2	14.5	14.1	9.4	76.6	25.5	196.3	86.6
Avg.	6.4	9.0	8.2	9.3	6.0	5.9	45.3	19.5	98.0	61.9
Med.	5.9	7.8	6.6	8.3	5.0	5.9	32.8	19.7	71.6	61.1
SD	2.0	3.9	6.8	2.8	4.1	2.8	25.3	5.7	60.5	16.7
Cement plant Ozarow	Cr		Cu		Ni		Pb		Zn	
Location	East	West	East	West	East	West	East	West	East	West
Min	2.1	2.3	2.8	3.8	1.2	1.9	5.7	7.4	10.8	23.9
Max	9.3	10.4	9.0	59.0	6.0	8.2	7.7	26.8	30.7	42.0
Avg.	5.3	6.3	5.2	15.5	3.2	4.7	6.5	13.1	23.5	32.1
Med.	4.8	6.2	4.2	8.1	2.7	4.8	6.3	11.1	21.3	32.0
SD	2.6	2.7	2.6	21.5	1.7	2.1	0.7	7.0	8.9	6.3
Cement plant Wierzbica	Cr		Cu		Ni		Pb		Zn	
Location	East	West	East	West	East	West	East	West	East	West
Min	6.8	3.5	4.3	3.6	3.7	2.8	5.6	3.7	29.0	18.3
Max	17.4	14.9	14.0	9.9	10.6	10.8	18.8	17.8	66.1	22.4
Avg.	11.5	6.5	7.5	5.8	6.8	4.5	12.5	8.1	45.4	32.3
Med.	10.8	4.9	6.8	5.3	6.7	3.3	12.1	6.0	44.7	24.5
SD	3.9	4.3	3.4	2.5	2.6	3.1	4.7	5.1	12.9	22.6
Permissible content of selected metals in the soil [OJ. 1359, 2016]	200		200		150		200		500	

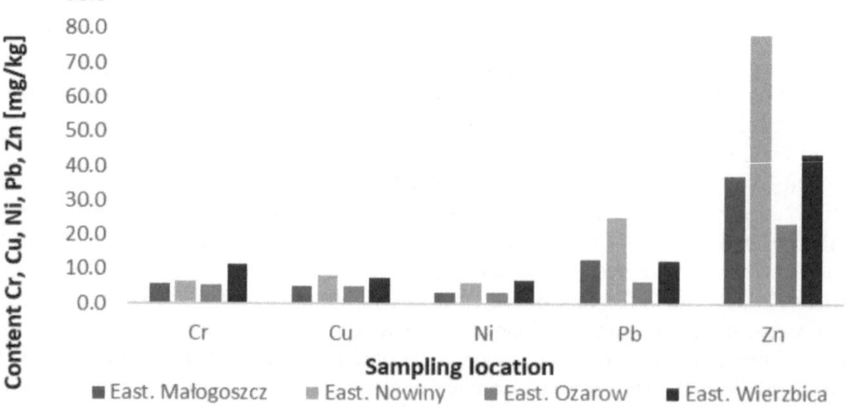

Figure 21.5. Average content of the Cr, Cu, Ni, Pb metals in the research area from the east of the emitter of the Małogoszcz, Nowiny, Ozarow, Wierzbica cement plants.

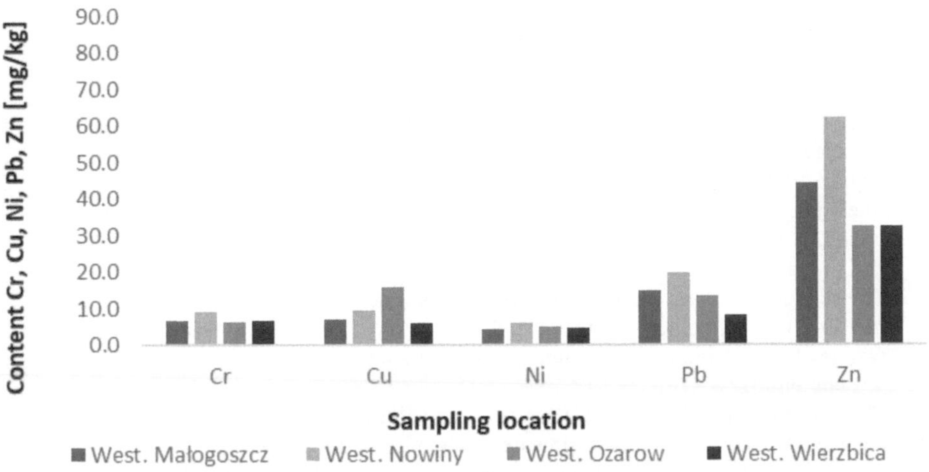

Figure 21.6. Average content of the Cr, Cu, Ni, Pb metals and the research area from the west of the emitter of the cement plant Małogoszcz, Nowiny, Ozarow, Wierzbica.

Table 21.6. Multiples of Cr, Cu, Ni, Pb, Zn metal content in the whole area and time of sampling for cement plants: Małogoszcz, Nowiny, Ozarow, Wierzbica, and average multiple of these metals content from the east and west of the emitter.

Sampling location	Multiple of metal contents [-]				
East of emitter of Cement Plant	Cr	Cu	Ni	Pb	Zn
Lafarge Małogoszcz	35	40	47	16	13
Dickerhof Nowiny	31	24	25	8	6
Grupa Ozarow	38	38	47	31	21
Przyjazn Wierzbica	17	27	22	16	12
Average for multiples of the east area	30	32	35	18	13
West of the cement of Cement Plant	Cr	Cu	Ni	Pb	Zn
Lafarge Małogoszcz	30	29	37	14	11
Dickerhof Nowiny	22	21	25	10	8
Grupa Ozarow	32	13	32	15	16
Przyjazn Wierzbica	31	34	34	25	15
Average for multiples of the west area	29	24	32	16	13

different, the median values and the arithmetic mean are comparable, the SD values indicate a greater variation in the minimum and maximum content of zinc in the studied soils (Table 21.5) (Manikandan 2011).

21.5 SUMMARY AND CONCLUSIONS

The locations of the sampling sites are within the impact range – from the east and west – of the emitter of the Lafarge Małogoszcz Cement Plant, Dickerhof Nowiny Cement Plant, Grupa Ozarow Cement Plant and Przyjazn Wierzbica Cement Plant closed in 1997r. Soil sorption was determined by identifying the iodine amount, which varied from 35.38 mg/g to 104.23 mg/g. The examined soils are characterized by average humus content typical for the soils of the Swietokrzyskie province. The humus content is in the range of 0.79% - 1.87%. These are low and medium class soils. The limit values for the metal concentrations in soil (Cr, Cu, Ni, Pb, Zn), from the east and west of the emitter, tenfold and fortyfold distances in relation to its height, were not exceeded in relation to the permissible values. The tested soils showed an increased content of zinc and lead in relation to other tested metals. The tests carried out including pH, iodine value, humus and heavy metal content indicate a different degree of soil degradation.

- The scope of tests carried out taking into account pH, iodine number, organic carbon, content of selected heavy metals and granulometric composition, indicates a different degree of soil degradation.
- The studied soils were characterized by a diverse granulometric composition. In the soil uptake points, the lowest content corresponded to the clay fraction – 0.77%, the silt fraction – 21.41%, and the largest – to fine sand – 28.83%.
- The soils in the research period showed an alkaline reaction pH in 74%, neutral in 15% and acidic in 13%. The acidic soils were located throughout the research period on the eastern side of the Ozarow Cement Plant emitter and on the west side of the Wierzbica Cement Plant emitter.
- The determined iodine value (IV) specifying soil sorption varied from 52.30 mg/g to 104.23 mg/g. An average IV of 74.89 mg/g indicates soils of medium sorptio.
- The tested soils were characterized by the humus content typical for the soils of the Świętokrzyskie voivodship 1.55%. The humus content was in the range of 0.79% - 2.93% for all areas of the studied soils and the time of the study. The soils with such humus content belong to the low and middle class.
- The tested soils were characterized by an increased content of zinc and lead in relation to other metals, i.e. Cr, Ni, Pb, and did not exceeded the limit value in Journal of Laws of 2016. For Zn, it was 13 times lower, for Pb – 18 times on the east side and 16 times on the west side of the emitter.
- The neutral and alkaline pH of the tested soils does not affect the heavy metal content. On acidic soils, the metal standards were not exceeded.
- No major differences were found in the content of heavy metals in the soils located on the eastern and western sides of the emitter, around the Małogoszcz Cement Plant and the Przyjazn Cement Plant in Wierzbica.
- Considering the use of soils located around the cement plants: Małogoszcz, Nowiny, Ozarow, Wierzbica, there is no hazard to the life and health of residents.

ACKNOWLEDGMENTS

The article was financed from the Program of the Minister of Science and Higher Education – Regional Initiative of Excellence, financed by the Ministry of Science and Higher Education under contract No. 025 / RID / 2018/19 of 28/12/2018, in the amount of PLN 12 million.

REFERENCES

Adriano, D.C. (2001). Trace elements in terrestrial environments: Biogeochemistry, Bioavailability and Risks of Metals, *Springer Verlag*, New York.

Alkorta, I., Hernandez-Allica, J., Becerril, J.M., Amezaga, I., Albizu, I, Gabisu, C. (2004). Recent findings on the phytoremediation of soils contaminated with environmentally toxic heavy metals and metalloids such as zinc, cadmium, lead and arsenic, *Environmental Science and Bio/Technology*, 3, 71–90.

Alloway, B.J. (2005). Copper-deficient soils in Europe. International *Copper association*, New York.

Alloway, B.J. (2012). Heavy Metals and Metalloids as Micronutrients for Plants and Animals. Oorts, K. Heavy Metals in Soils. Part of the Environmental Pollution book series (EPOL, 22, 367–394), *Chapter First Online*.

Bai,C., Reilly, C.C., Wood, B.W. (2006). Nickel deficiency disrupts metabolism of ureides, amino acids, and organic acids of young pecan foliage, *Plant Physiol, 140,* 433–443.

Bezak-Mazur, E., Dańczuk, M. (2013). Sorption capacity of conditioned sewage sludge in environmental conditions, *Archives of Waste Management and Environmental Protection*, 15(1), 87–92, ISSN: 1733–4381

Doran, J.W., Zeiss, M.R. (2000). Soil health and sustainability: managing the biotic, *Applied Soil Ecology*, 15, 3–11.

Dziubanek, G., Baranowska, R., Oleksiuk, K. (2012). Heavy metals in the soils of Upper Silesia – a problem from the past or a present hazard? *JEcolHealth,* 16(4), 169–176.

EN-13346:2000 — Characterization of sludges. Determination of trace elements and phosphorous. Aqua regia extraction methods.

Gajec, M., Król, A., Kukulska-Zając, E., Justyna Mostowska-Stąsiek, J. (2018). Soil sampling in the context of land surface pollution assessment, *Nafta-Gaz,* 74(3), 215–225, doi: 10.18668/NG.2018.03.05.

Glinka, K., Wilk, M. (2016). Analysis of soil contamination industrial establishments in Podkarpatie. *Budownictwo i Inżynieria Środowiska, JCEEA, 33*(63), 75–86, doi: 10.7862/rb.2016.111.

ISO 14887:2000 Sample Preparation—Dispersing procedures for powders in liquids.

Kabała, C., Karczewska, A. (2019). Methodology of laboratory analysis of soils and plants, *University of Life Sciences in Wroclaw,* Edition 8a.

Manikandan, S. (2011). Measures of central tendency: Median and mode, *J Pharmacol Pharmacother.* 2(3): 214–215, doi: 10.4103/0976-500X.83300

Map of Poland divided into voivodships, (access 28.01.2020), https://sites.google.com/site/maturamatematyka 20112012/mapa-polski-z-podzialem-na-wojewodztwa

Oruba, R., (2019) Structures of industrial chimneys discharging desulfurised flue gas, *Przeglad Budowlany,* 1(90), 19–23.

PN-83C-97555.04 – Węgle aktywne. Metodyka Badań. Oznaczenie liczby jodowej (Activated carbons. Research methodology. Determination of Adsorption Value of Iodine).

PN-ISO 10390:1997 – Jakość gleby. Oznaczenie pH (Soil quality. Determination of pH)

PN-R-04031:1997 – Instrukcja pobierania próbek glebowych z gruntów ornych i użytków zielonych (Instructions for taking soil samples from arable land and grassland)

Report, (2012). Monitoring of Chemistry of Arable Soils in Poland in 2010–2012, *IUNG* Puławy.

Różański, Sz. (2010). Texture of different types of soils with regard to their origin and changes in textural classification, *Roczniki Gleboznawcze Warszawa,* 61(3), 100–110.

Rozporządzenie Ministra Środowiska z dnia 1 września 2016 r. w sprawie sposobu prowadzenia oceny zanieczyszczenia powierzchni ziemi, (Regulation of the Minister of the Environment of 1 September 2016 on the conducting the assessment of land surface pollution (Journal of Laws, item 1395) in Poland

Rup, K. (2006). Pollution transfer processes in the environment, *WNT Warszawa.*

Sadeghzadeh, B., Rengel, Z. (2011). Zinc in Soils and Crop Nutrition In book: The Molecular and Physiological Basis of Nutrient Use Efficiency in Crops, *Oxford, UK: Wiley-Blackwell,* pp.499, doi: 10.1002/9780470960 707.ch16.

Sady, W., Smoleń, S. (2004). The influence of soil-fertilizing factors on the accumulation of heavy metals in plants, *X Polish National Symposium on the effectiveness of fertilizer application in horticultural crops Krakow*) 17–18 June.

Stępień, M. (2016). Soil grain size determines yielding, *Agropolska.*

Ukalska-Jaruga, A., Smreczak, B., Klimkowicz-Pawlas, A., Maliszewska-Kordybach, B. (2015). The role of organic matter in the processes of the accumulation of persistent organic pollutants (TZO) in soils, *Polish Journal of Agronomy,* 20, 15–23.

Violante, A., Cozzolino, V., Perelomov, L., A.G. Caporale, A.G., Pigna, M. (2010). Mobility and bioavailability of heavy metals and metalloids in soil environments. *J. Soil. Sci. Plant Nutr.* 10 (3), 268 – 292, doi: 10.4067/S0718-95162010000100005

Widłak, M. (2015). The occurrence assessment of selected heavy metals in soil of Swiętokrzyskie region, In the book: Health Work Environment - Contemporary Dilemmas, *WNIT-PIB.*

Widłak, M. (2016). The Variability of Selected Parameters of Arable Soils During the Vegetation Cycle of Plants, *Rocznik Ochrona Środowiska,* 18, 803–814, ISSN: 1506-218X.

Yruela, I., Alfonso, M., Baro'n, M., Picorel, R. (2000). Copper effect on the protein composition of photosynthesis, *Physiol Plant,* 110, 551–557.

Environmental aspects of heat and water sourcing

CHAPTER 22

Energetic, economic and environmental aspects of increasing the thickness of thermal insulation of district heating pipelines

M. Żukowski
Department of HVAC Engineering, Bialystok University of Technology, Bialystok, Poland

ABSTRACT: The purpose of this article was to determine the environmental, energetic, and economic impact assessment on modernization of DH networks depending on the use of the twin pipes insulation standard. Numerical modelling was a research method used to perform the energy balance of the DH pipelines buried in the ground. On the basis of the results of computer simulations performed in Ansys Workbench, it can be concluded that the highest class of thermal insulation can reduce the energy losses by more than 60%. The article also discusses the economic aspects of choosing the type of DH pipelines insulation. The analysis determining the payback period enabled to conclude that investing in pre-insulated pipes with increased insulation thickness is profitable. Another issue analysed in this paper was the impact of a series of pipe insulation on the reduction of pollutant emissions from the fuel combustion process.

22.1 INTRODUCTION

A simple analytical method for calculating the heat loss and the total conductivity of the twin pipe in district heating system was created by Nik and Adl-Zarrabi (2014). This model was based on estimating the parameters of heat transfer (conductive and convective) around the twin insulated pipe under steady state condition. An experimental research confirmed the accuracy of the proposed method.

A temperature independent conductance model for assessing the thermal performance of a twin heating pipe with vacuum insulation was tested by Berge et al. (2015). As demonstrated by the results of the analysis, the application of the new insulation concept allowed reducing the energy losses by 12% to 18% for return and supply pipes and by 29% to 39% only for the supply pipe.

Denarie et al. (2019) developed a new numerical model based on the method of characteristics for fast and accurate estimation of the heat transfer over pipes of DH networks. Proper validation of this model, constructed in MATLAB environment, was made by comparing it with the results from the node method and finite-volume method. The main advantage of the new approach for modelling long sections of heating networks was almost 1000 times faster calculation time compared to the finite volume method with the same accuracy.

As shown by this literature review, the topics related to modelling the heat transfer between DH pipelines and the surrounding soil are still popular issues and there are still new approaches to solving such complex problems. In addition, the author did not find a topic directly related to the subject of this article.

22.2 MODEL OF THE TWIN-PIPES BURIED IN THE GROUND

Twin steel pre-insulated pipe system can be used for both low and medium temperature hot water district heating and also cooling networks. The cross section throughout the pipeline is shown in Figure 22.1. Both steel pipes are insulated together using polyurethane (PUR) foam. The thermal insulation is surrounded by outer casing of high density polyethylene (PE-HD).

DOI 10.1201/9781003171669-22

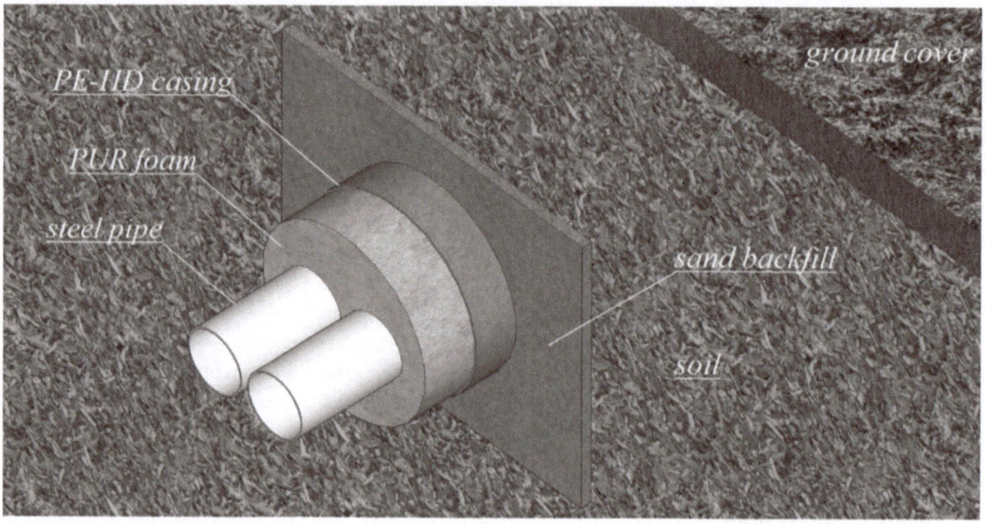

Figure 22.1. Cross section of the twin-pipe system showing the main elements included in the model of this research object.

Table 22.1. Margin settings for A4 size paper and letter size paper.

Steel pipe			Outer casing	
nominal diameter	outer diameter	pipe wall thickness	outer diameter	wall thickness
		insulation series 1		
100	114.3	3.6	315	4.1
		insulation series 2		
100	114.3	3.6	355	4.5
		insulation series 3		
100	114.3	3.6	400	4.5

This system is available in 3 different types of insulation thickness. A steel pipe with a nominal diameter of 100 mm was considered. The distance between the steel pipes in all three cases is the same and equals 25 mm. Other dimensions of the components of the pipeline for insulation series: 1, 2, and 3 are presented in Table 22.1.

In the analysis, the real average temperature of the heating medium, which are in the district heating network in Warsaw (Kręcielewska et al. 2011) was assumed. The supply and return temperature during the heating season was set to 82°C × 47°C and 73°C × 45°C from May to September.

The calculations were carried out in three locations: Szczecin, Warsaw, and Suwalki. These cities represent three Polish climate zones I, III, and V, which correspond to the following outside air design temperature: -16°C, −20°C, and −24°C. Due to the very good insulating properties of the piping system, the computational domain width was set to 5 m, i.e. 2.5 m on each side of the vertical pipe axis, assuming no effect of the hot pipe on the temperature field at this distance. This was confirmed by the simulations made by Danielewicz et al. (2017). The depth of the calculated area was set to 10 m below the ground level.

This value was adopted due to the fact that at this level there is almost constant soil temperature throughout the year. Figure 22.2 shows the heat transfer conditions at the domain boundaries. A two-dimensional model was developed for the calculation of heat losses due to the assumption of average operating temperature of the heating medium. The geometric model of the entire system was created in the SpaceClaim program, while a numerical mesh was developed in the Ansys Meshing

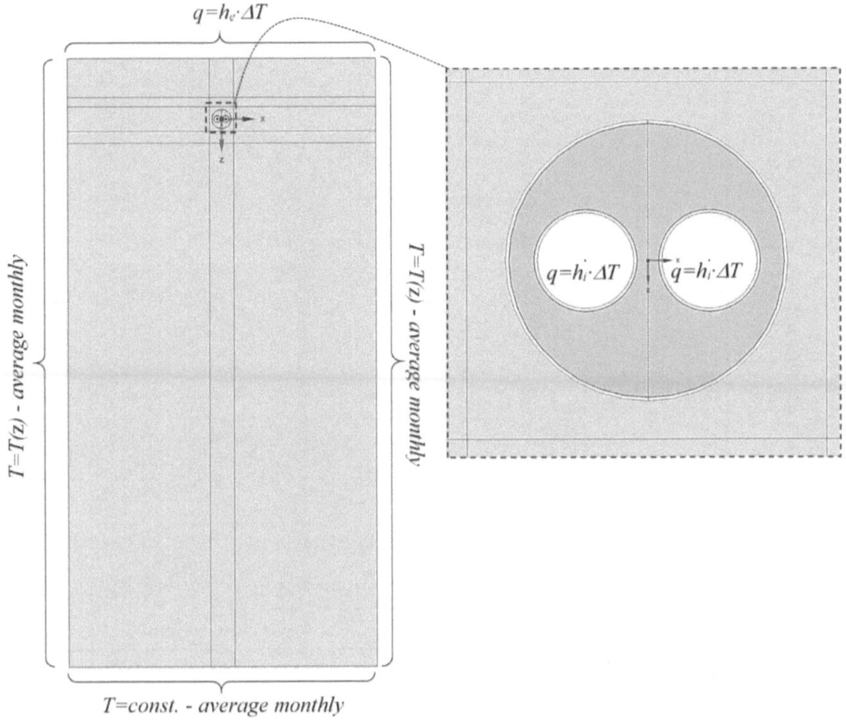

Figure 22.2. The heat transport domain with boundary conditions.

program. Fluent solver was used to perform the calculations and analysis of results. All these computer programs are included in the Ansys v. 19.2 package. As mentioned earlier, a 2D heat transfer model was developed. According to the author, the use of modelling in three dimensions in the case of large heating networks is pointless. This is due to the necessity of using very large mesh (tens of millions of nodes), which requires the use of very efficient computer arrays and a long calculation time. In addition, the results of calculations may have insufficient accuracy. A better solution is to use 2D modelling and, for example, apply the logarithmic mean temperature difference LMDT method to calculate the temperature change of the heating medium.

An element of novelty in the modelling of heat exchange in the ground is the use of vertical temperature profiles on the left and right boundaries of the computation domain. Until now, the most popular approach was to set a heat flux equal to 0. This caused a vertical linear change in temperature, which is not consistent with reality, especially below the surface of the ground. In the presented article, the vertical temperature profiles were determined using the method developed by Krarti et al. 1995. This model is used in the EnergyPlus software. The expression for calculating the soil temperature $T_{soil,t}$ at time t as a function of the depth z has the following form:

$$T_{soil,t} = T_{surf} - A_s \exp\left[-z\left(\frac{\pi}{365\alpha_s}\right)^{0.5}\right] \cos\left\{\frac{2\pi}{365}\left[t - t_0 - \frac{z}{2}\left(\frac{365}{\pi\alpha_s}\right)^{0.5}\right]\right\} [m] \quad (22.1)$$

where: T_{surf}– average temperature of soil surface [°C],
A_s– amplitude of the soil surface temperature change [°C],
α_s– soil thermal diffusivity [m^2/day],
t_0– phase constant of the soil surface [day].

Computer code CalcSoilSurfTemp, which is part of the EnergyPlus software package, was used to calculate three parameters (different for each location) occurring in Equation 22.1: T_{surf}, A_s, and t_0.

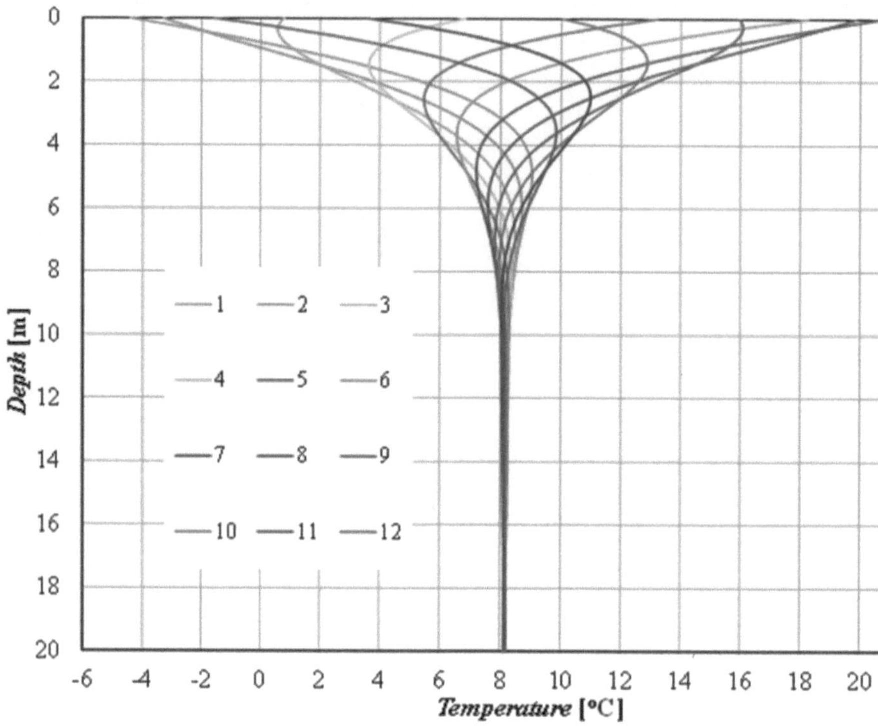

Figure 22.3. Average monthly ground temperature profile for Warsaw.

Variation in the ground-temperature profile determined for each month separately based on the data of a typical meteorological year for the Warsaw-Okecie station is presented in Figure 22.3.

As can be seen in Figure 22.3, the temperature of the soil changed at a depth less than 10 meters only to a limited extent, which is why this value was taken as the vertical dimension of the computational domain. The following values of the thermal conductivity coefficient of materials in W/m/K were assumed in the computer simulations: 1.6 for soil, 0.42 for polyethylene casing, and 50 for steel pipe.

A quite complex and, at the same time, very important issue is the dependence of thermal conductivity of insulation materials k on the operating temperature. Most manufacturers provide only one value for the average temperature of 50°C. Experimental studies show (Kristjansson & Bøhm 2006) that the temperature of the medium significantly affects the thermal properties of the insulation. Figure 22.4 shows thermal conductivity as a function of temperature for polyurethane foam. As can be determined from the graph in Figure 22.3, the thermal conductivity increases by 25.9%, when the insulation temperature changes from 50°C to 90°C. This is an undesirable physical phenomenon due to the increase in heat loss. Thus, this is another argument that supports the use of low temperature district heating (LTDH) networks of the fourth generation.

User Defined Function (UDF) was used to take into account the change in the thermal conductivity of the insulation material. UDF is a function created in the C language that is dynamically linked to the Fluent solver during calculations.

22.3 RESULTS OF NUMERICAL SIMULATIONS

In total, 108 series of energy balance calculations were performed for the model presented here. The heat flow emitted from 1 m^2 of the surface of the outer casing was the main result of the calculations. Due to the fact that the insulation classes have different outer surface area (series 1 – 0.989 m^2, series 2 – 1.115 m^2, series 3 – 1.257 m^2), the value of the heat flux was referred to 1 m of the pipeline

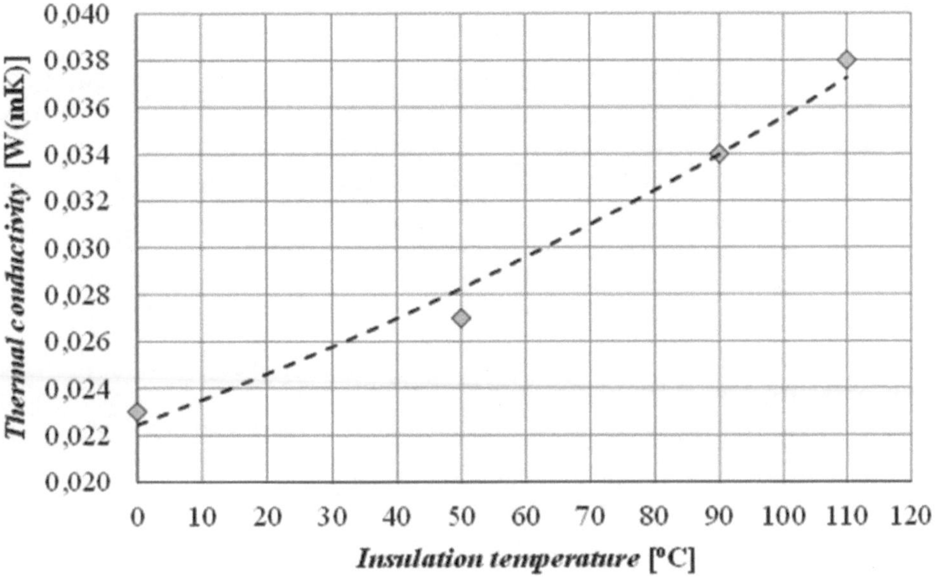

Figure 22.4. The dependence of the coefficient of thermal conductivity on temperature – based on Kristjansson and Bøhm (2006).

Table 22.2. Margin settings for A4 size paper and letter size paper.

Month	Szczecin			Warsaw			Suwalki		
	Insulation series			Insulation series			Insulation series		
	1	2	3	1	2	3	1	2	3
1	15.59	11.67	9.66	16.12	11.98	9.99	17.32	13.05	10.74
2	14.24	10.73	8.83	14.50	10.89	8.99	15.51	11.69	9.62
3	15.48	11.67	9.61	15.44	11.62	9.57	16.43	12.31	10.13
4	14.30	10.77	8.87	13.97	10.53	8.66	14.50	10.93	8.99
5	12.46	9.39	7.73	11.93	8.98	7.40	12.13	9.13	7.52
6	11.14	8.40	6.91	10.60	8.00	6.59	10.52	7.93	6.53
7	10.81	8.15	6.71	10.41	7.85	6.47	10.18	7.68	6.32
8	10.55	7.96	6.55	10.42	7.88	6.47	10.24	7.72	6.36
9	10.45	7.88	6.48	10.64	8.02	6.60	10.68	8.05	6.63
10	12.82	9.67	7.96	13.25	10.05	8.23	13.72	10.34	8.51
11	13.34	10.05	8.27	14.00	10.54	8.68	14.72	11.10	9.13
12	14.74	11.11	9.14	15.46	11.65	9.58	16.50	12.44	10.23
Average	12.99	9.79	8.06	13.06	9.83	8.10	13.54	10.20	8.39

length -q_l. Owing to this, we can objectively compare the results of calculations of heat losses as a function of the length of DH network (Table 22.2). Figure 22.5 presents a graphical illustration of exemplary results as the temperature field around the twin pipes with insulation series 1 in January in Suwalki.

On the basis of the analysis of the results, two main conclusions can be formulated. Location has a small influence on the heat loss of pipelines, because the greatest difference is about 4.2% compared to the results obtained for Szczecin and Suwalki. The main difference arises when the class of applied insulation is compared. The reduction in energy losses is about 21% when we compare the insulation

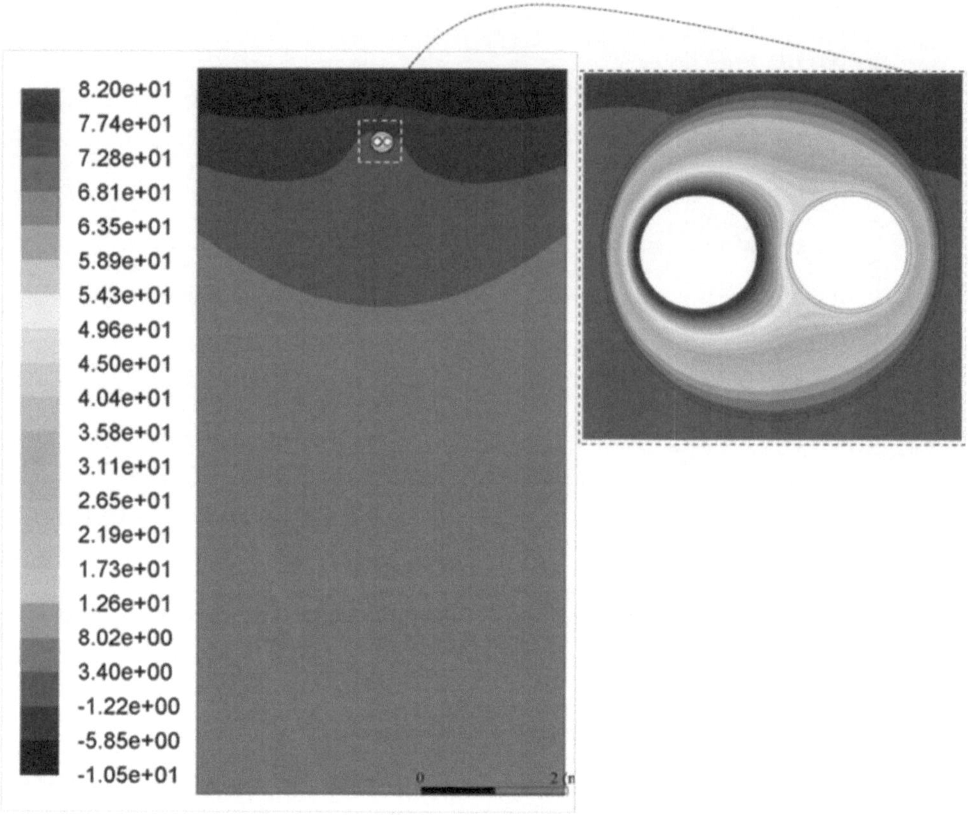

Figure 22.5. Temperature distribution in the analysed section of the ground.

of series 2 instead of the series 1. An even greater reduction, which is about 61%, we can obtained using the insulation of 3 series.

Energy losses in the whole year can be calculated from the following dependence:

$$E_{loss} = \sum_{i=1}^{i=12} q_{l,i} \cdot l \cdot 24 \cdot n \cdot 10^{-3} \quad \text{[kWh/year]} \tag{22.2}$$

where:
l – pipeline length [m],
n – number of days in each month [-]

Figure 22.6 shows the change in heat loss from the pipeline of the heating network located in Suwalki for three isolation series. The E_{loss} values were calculated for each month with the assumption of a unit length (1m) of twin-pipeline. Total annual energy loss from a 1 m pipeline located in Suwalki is 162.45 kWh (series 1), 122.36 kWh (series 2), and 100.71 kWh (series 3). In the other two locations, the energy losses are about 4% lower, and their distribution is similar.

22.4 COMPARATIVE ANALYSIS OF INVESTMENT AND OPERATING COSTS

As one would expect, lower heat losses through pipelines with greater insulation thickness are obvious. However, another important issue are the investment costs, the change of which is not easy to predict.

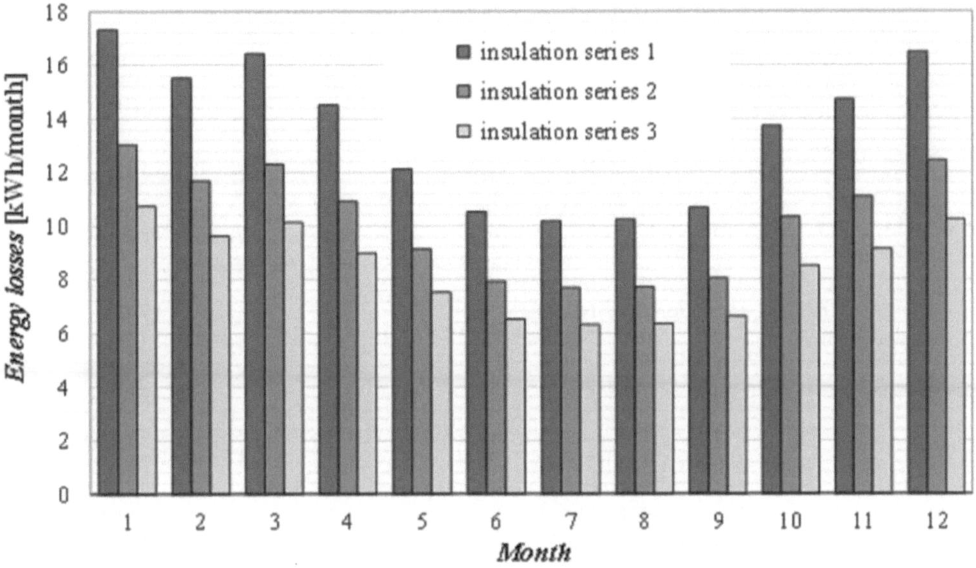

Figure 22.6. Average monthly energy loss in Suwalki for different insulation standards.

They are not directly proportional only to the prices of pipelines because additional costs arise. A comparative analysis was performed for the DN 100 pipeline with a length of 100m. It should be noted that the costs associated with the installation of a district heating network, regardless of the thickness of insulation, are identical. Therefore, it was not necessary to include them in this analysis. The component results for determining the total costs of a fragment of the heating network are presented in Table 22.3. All component prices are net values.

As can be seen in Figure 22.7, the differences in capital expenditure are not high and they change almost linearly. This increase is only about 14% when comparing insulation series 2 with series 3 and about 23% when comparing insulation series 1 with series 3. The differences related to the heat losses costs are more noticeable and are about 25% and 38%, comparing insulation series 1 to series 2 and 3, respectively.

Simple Pay Back Time (*SPBT*) was used to determine real time return on investment:

$$SPBT = \frac{\Delta CI}{\Delta E_{loss} \cdot E_{cost}} [\text{year}] \tag{22.3}$$

where:

ΔCI – difference in investment costs [Eur],
ΔE_{loss} – difference in the energy losses using different types of the insulation in the whole year [kWh],
E_{cost} – price of energy from the district heating system [Eur/kWh]

Calculations of the *SPBT* value were performed assuming an average purchase price of energy from the district heating system of 0.05 Eur/kWh for Poland in 2020. It turned out that the time at which the return on investment reaches a break-even point is about 6 and 9.5 years for the pipelines with increased insulation thickness, i.e. series 2 and series 3, respectively. The increase in net financial expenses and energy savings was determined in comparison to the base value determined for the pipeline with insulation series 1.

To summarise, it can be stated that the use of twin pipe system with increased thickness of thermal insulation allows for a relatively quick return on investment.

Table 22.3. List of investment costs – divided into individual elements of the system.

No.	pc.	Item name	Unit price	Net value
-	-	-	[EUR]	[EUR]
Series 10,0,00				
1	8	2x114,3/315 Pre-insulated pipe 12m TWIN-PIPE	948.86	7 590.88
2	13	315 SXWP Muff D315 L=650	59.54	774.01
3	2	2x114,3/315 Horizontal pre-insulated elbow 90 degrees, TWIN-PIPE L=1m	340.57	681.14
4	26	Insulation foam No. 11	19.74	513.14
Total amount:				9 559.17
Series 2				
1	8	2x114,3/355 Pre-insulated pipe 12m TWIN-PIPE PLUS	1 080.84	8 646.75
2	13	355 SXWP Muff D355 L=750	77.25	1 004.28
3	2	2x114,3/355 Horizontal pre-insulated elbow 90 degrees, TWIN-PIPE L=1m	364.63	729.26
4	26	Insulation foam No. 9	14.09	366.27
Total amount:				10 746.56
Series 30.00				
1	8	2x114,3/400 Pre-insulated pipe 12m TWIN-PIPE	1 254.62	10 036.94
2	13	400 SXWP Muff D400 L=750	90.59	1 177.72
3	2	2x114,3/400 Horizontal pre-insulated elbow 90 degrees, TWIN-PIPE L=1m	404.61	809.21
4	13	Insulation foam No. 10	16.45	213.88
5	13	Insulation foam No. 11	19.74	256.57
Total amount:			0,00	12 494.31

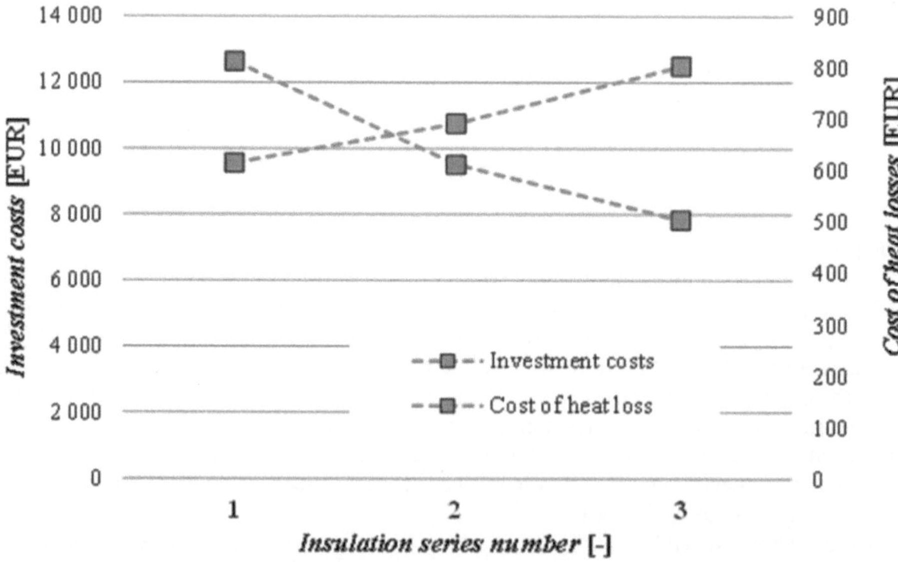

Figure 22.7. Comparison of investment and energy losses costs for different types of insulation.

Table 22.4. Air pollutant emission factors in kg/kWh based on (PGE 2017).

PF_{SO2}	PF_{NOx}	PF_{PM10}	PF_{CO}	PF_{CO2}
0.000234	0.000173	0.000007	0.000151	0.241007

Table 22.5. Annual reduction of air pollutants [kg].

	SO_2	NO_x	PM10	CO	CO_2
Case I	1.34	0.99	0.04	0.87	1380.28
Case II	2.06	1.52	0.06	1.33	2125.68

22.5 THE ECOLOGICAL ASPECT RELATED TO THE USE OF THICKER THERMAL INSULATION

The reduction of air pollutant emissions is an important issue that involves the use of thick thermal insulation. A smaller amount of fuel is burned in heating and cogeneration plants as a result of reducing the heat loss from direct heating networks. In order to determine the amount of dust and polluting gases released into the ambient air, the following general relationship (EMEP/EEA 2019) should be used:

$$\Delta P_{poll} = \Delta E \cdot PF_{pol} \quad \text{[kg/year]} \tag{22.4}$$

where:
ΔE – reduction of energy produced in the power plant [kWh/year],
PF_{pol} – emission factor of the pollutant [kg/kWh].

The difference in pipeline heat loss should be divided by the efficiency of the heating plant η_q to calculate the energy reduction ΔE.

$$\Delta E = \Delta E_{loss}/\eta_q \quad \text{[kWh/year]} \tag{22.5}$$

The emission factor of the pollutants (Table 22.4) was determined on the basis of measurements carried out by Polish Energy Group S.A. in 2017 (PGE 2017).

The reduction in annual heat losses E_{loss} of the 100 m pipeline was determined based on the analysis presented in Chapter 3. The following two cases were selected:

- *Case I* – 4009 kWh/year – consider the use of series 2 insulation instead of series 1,
- *Case II* – 6174 kWh/year – consider the use of series 3 insulation instead of series 1.

Table 22.5 presents the results of calculations for reducing the pollutant emissions, taking into account 70% efficiency of heat production in a cogeneration power plant.

While analysing the results of the calculations presented above, it can be concluded that increasing the insulation thickness has a noticeable effect on the amount of harmful substances released into the atmosphere, especially in the case of carbon dioxide. It should be noted that the analysis above was conducted only for a 100 m of DN 100 twin pipes. In large cities, the length of DH networks reaches up to several dozen kilometres and the main pipelines can have a diameter larger than one meter. Therefore, in the scale of the whole city or even a single district, the reduction of the impact of power plants on environmental pollution can be significant.

22.6 CONCLUSION

The modernization of heating systems is one of the most important projects that can significantly affect the quality of the environment in a short time.

Twin steel pre-insulated pipes are an increasingly often used system for replacing old DH networks due to many advantages. Perhaps the most important one corresponds to the low energy losses associated with the flow of a heating or cooling medium. Twin pipes are manufactured using different thicknesses of insulation casing. This work concerns the analysis of the thermal insulation class impact on the energy losses. In order to solve this problem, numerical simulations using the Ansys v. 19.2 software package were applied. The calculations of the energy balance were performed for the district heating pre-insulated twin pipes buried in the ground. In order to increase the accuracy of numerical modelling, temperature profiles on the vertical boundaries of the model were used. Three pipeline locations corresponding to various climate zones in Poland were selected, including: Szczecin (I zone), Warsaw (III zone) and Suwalki (V zone). The results of multi-variant calculations showed that there is a small impact of the pipeline' location on energy losses, amounting to about 4%. Three standards of thermal insulation were subjected to analysis. In these cases, the differences in energy losses were significant. The twin pipes insulation made in series 3 allowed limiting the heat flux passing through the outer casing by 61%, and in the case of series 2 by about 21%, in relation to the standard thermal insulation of series 1. As indicated by the simplified economic analysis, the use of increased insulation is economically justified. The payback period was only 6 and 9.5 years when using insulation series 2 and 3, respectively.

An additional effect of using a larger thickness of insulation is the reduction in heat demand and thus the reduction of environmental pollution. Table 22.5 shows the values of the potential avoided emissions of pollutants released into the ambient air as a result of using different types of thermal insulation of a twin pipeline with a length of 100 m. In the case of carbon dioxide, its emissions can be reduced by 1380 and 2126 kg/year, when thermal insulation of series 2 and 3 are applied instead of the basic insulation thickness, respectively.

This type of multi-faceted analysis has not yet been carried out. The results presented here will allow designers and heat supply companies to choose the optimal solution for DH networks in terms of reducing the heat loss and minimising environmental impact.

Finally, it should be mentioned that many research teams are working on optimising the shape of the twin pipes insulation. One example of such solution was developed by Teleszewski and Zukowski (2020) and their protection right for a utility model (W.127874 – Polish Patent Office) which was obtained in September of this year.

ACKNOWLEDGEMENTS

This work was performed within the framework of Grant of the Bialystok University of Technology and WZ/WBiIS/9/2019 financed by the Ministry of Science and Higher Education of the Republic of Poland. The calculations were carried out with the use of ANSYS 19.2 software, which is provided to BUT based on an agreement between BUT and ANSYS Inc. (Canonsburg, USA) and MESco Sp. z o.o. (Poland).

Special thanks to Mr. Ireneusz Iwko from the Logstor company, who provided the necessary data for the analysis of investment costs.

REFERENCES

Berge, A., Adl-Zarrabi, B., Hagentoft, C.E. 2015 Assessing the Thermal Performance of District Heating Twin Pipes with Vacuum Insulation Panels. *Energy Procedia* 78: 382 – 387.

Danielewicz, J., Śniechowska, B., Sayegh, M.A., Fidorów, N., Jouhara H. 2016 Three-dimensional numerical model of heat losses from district heating network pre-insulated pipes buried in the ground. *Energy* 108: 172–184.

Denarie, A., Aprile, M., Motta, M. 2019 Heat transmission over long pipes: New model for fast and accurate district heating simulations. *Energy* 166: 267–276.

EMEP/EEA air pollutant emission inventory guidebook 2019 *Technical guidance to prepare national emission inventories*. European Environment Agency Report No 13/2019, Denmark.

Krarti, M., Lopez-Alonzo, C., Claridge, D. E. & Kreider, J. F. 1995 Analytical model to predict annual soil surface temperature variation. *Journal of Solar Energy Engineering* 117: 91–99.

Kręcielewska, E., Łebek, A., Smyk, A. 2011 Analysis of Possible Reduction of Costs and Losses Concerning Heat Transfer trough Integration of Traditional Pre-insulated Double and Flexible Pipelines. *Ciepłownictwo, Ogrzewnictwo, Wentylacja* 42: 7–8.

Kristjansson, H. & Bøhm, B. 2006 *Optimum design of distribution and service pipes*. Euroheat and Power vol. 3 (English Edition) 34–42.

Modelica® and the Modelica Standard Library© 2019 – https://www.modelica.org/

Nik, V.M., Adl-Zarrabi, B. 2014 *An analytical method for calculating the thermal conductivity of a twin pipe in district heating system*. 10th Nordic Symposium in Building Physics: 517–524.

PGE (2017) https://elbelchatow.pgegiek.pl/Ochrona-srodowiska/Wskazniki-emisji – access to the website 2.04.2020.

Teleszewski, T.J., Zukowski, M. 2019 *Modification of the Shape of Thermal Insulation of a Twin- Pipe Pre-Insulated Network*. AIP Conference Proceedings 2078, 020030; https://doi.org/10.1063/1.5092033.

Van der Heijde, B., Aertgeerts, A., Helsen, L. 2017 Modelling steady-state thermal behaviour of double thermal network pipes. *International Journal of Thermal Sciences* 117: 316–327.

Van der Heijde, B., Fuchs, M., Ribas, Tugores, C., Schweiger, G., Sartor, K., Basciotti, D., Müller, D., Nytsch-Geusen, C., Wetter, M., Helsen, L. 2017 Dynamic equation-based thermo-hydraulic pipe model for district heating and cooling systems. *Energy Conversion and Management* 151: 158–169.

CHAPTER 23

Effect of using a multi-pipe thermal insulation in heating installations on reduction of heat loss and pollutant emission

T.J. Teleszewski
Department of HVAC Engineering, Faculty of Civil Engineering and Environmental Sciences, Bialystok, Poland

ABSTRACT: The publication presents the influence of multi-pipe insulation on heat losses in heating installations and the effect of reducing the pollution emitted into the atmosphere in relation to thermal insulation. Currently, single round thermal insulation is the most commonly used solution in heating pipelines. The article proposes the thermal insulation in which six heating pipes in were placed common thermal insulation. This solution was compared with classic heating pipelines located in unheated corridors in the building of the Bialystok University of Technology, where the average outside temperature was 8°C. The calculations of heat losses in heating pipelines were performed using the author's calculation program written in Fortran with the boundary elements method (BEM). The emissions of pollutants into the atmosphere by a heat source in relation to heat losses in the analyzed heating installation were determined based on the emission factors dedicated for a given heat source. The adopted shape of flat multi-pipe thermal insulation with three pairs of heating pipes allowed to reduce the emission of pollutants into the atmosphere by about 43%, compared to the existing six heating pipes equipped with single thermal insulation. Flat multi-tube thermal insulation with three pairs of heating pipes is also characterized by about 20% smaller amount of thermal insulation material, compared to six single pipes.

23.1 INTRODUCTION

The pollutants emitted into the atmosphere by heating systems in buildings can be reduced by using renewable energy sources, thermomodernization of buildings and modernization of the heating installations inside buildings (Guelpa et al. 2018; Ravina et al. 2017). Most publications from recent years have focused on reducing the emissions to the atmosphere from heat sources by changing the combustion technique (Alphones et al. 2020; Yan et al. 2014) or by using renewable energy sources (Đurđević 2018; Günkaya 2020). Improving the thermal insulation of central heating installations, domestic hot water pipes and domestic hot water circulation allows reducing heat losses, which contribute to decreasing the consumption of fuels in heat sources, which in turn results in lower emissions of pollutants into the atmosphere. Improvement of thermal insulation of heating pipes can be achieved by placing a few heating pipes in common thermal insulation. An example of such a solution are twin pipes, used in the underground and above-ground heating networks, as well as central heating installations in which the supply and return pipes are in common thermal insulation (Bøhm & Kristjansson 2005; Khosravi & Arabkoohsar 2019; Nowak-Ocłoń & Ocłoń 2020; Ocłoń et al. 2019). The second way to reduce the heat loss is to modify the shape of the thermal insulation of the twin pipe from round to elliptical (Teleszewski & Zukowski 2019), oval (Krawczyk & Teleszewski 2019a) and egg-like (Kristjansson & Bøhm 2006; Krawczyk & Teleszewski 2019b) shape.

The presented paper demonstrates a solution of a flat shape of thermal insulation, in which three pairs (Figure 23.1) of heating pipes composed of supply and return pipes are placed in unheated rooms. The purpose of the work was to determine the reduction of heat losses and emissions of pollutants emitted by heat sources in relation to the heat losses in the heating installations by using multi-pipe thermal insulation on a selected existing example of a heating installation. The presented solution is an alternative to single standard round thermal insulation and enables not only to reduce heat loss, but

DOI 10.1201/9781003171669-23

Figure 23.1. General view of six heating pipe with single round thermal insulation (three supply pipes and three return pipes).

also to reduce the amount of material used for the production of thermal insulation. The heat losses were determined for the actual operating conditions of the heating system using a computer program written in Fortran using the boundary element method (BEM). After determining the heat loss for existing (Figure 23.1) and modified thermal insulation (Figure 23.3b), for the existing heat source using the emission factors of pollutants into the atmosphere, the reduction of pollutant emissions such as nitrogen oxides (NOX), carbon dioxide (CO_2), carbon monoxide (CO), non-methane volatile organic compounds (NMVOC), sulfur oxides (SOX), total suspended particles (TSP), particulate matter (PM10 and PM2.5), carbon dioxide (CO_2) and methane (CH_4) emitted into the atmosphere by a heat source after applying the proposed multi-tube thermal insulation.

23.2 MATERIALS AND METHODS

23.2.1 *Description of the selected section of the heating installation*

The subject of the research was a heating installation located in unheated rooms of the Bialystok University of Technology, in which the average air temperature amounted to 8°C. The heating installation consisted of three heating supply pipes with a medium supply temperature of 70°C and three return pipes with a medium return temperature of 50°C (Figure 23.1). The tested cable section was made of steel pipes with a diameter of 48.3 mm and an external diameter of thermal insulation equal to 110 mm. The length of heating pipes was 50 m. Heating pipes with single round thermal insulation were made in accordance with PN-B-02421: 2000. The thermal insulation was made of polyurethane foam with a thermal conductivity coefficient of λ=0.0265 W/m/K (Jarfelt & Ramnas 2006; BING 2016). The heating installation was used during the heating season for 255 days.

23.2.2 *The concept of multi-pipe thermal insulation in heating installations*

In the case of the district heating pipes located inside a building, circular thermal insulation with a thickness selected according to PN-B-02421: 2000 is used most often. This solution is characterized by simplicity of implementation. The main disadvantage of this thermal insulation solution are longitudinal thermal bridges (Figure 23.2b), which appear at the joints of thermal insulation (Figure 23.2a). Figure 23.2a shows an example of a general view of a heating pipe with longitudinal connection of thermal insulation, while Figure 23.2b shows a view of a linear thermal bridge resulting from the connection of thermal insulation. The temperature difference on the surface of the thermal insulation connection and the surface of the thermal insulation outside the connection for this example (Figure 23.2b) is about 1°C ($T_{sp1} - T_{sp2}$=1°C).

It should be noted that the heat losses through the thermal bridges created by the thermal insulation connections depend primarily on the quality of the thermal insulation connections. The number of longitudinal thermal bridges is equal to the number of single pipes, i.e. six linear thermal bridges per six pipes. The number of thermal bridges resulting from longitudinal joints of thermal insulation can be limited by using common thermal insulation for a group of heating pipes. The following example shows the use of one thermal insulation common to six heating pipes. The cross-section of the thermal insulation consists of a rectangle and two semi-circles, which are located on the left and right of the first and last heating pipe (Figure 23.3b). The geometry of multi-pipe thermal insulation limits the number of external walls on the left and right of a single heating pipe, which reduces the heat loss. The adopted multi-tube thermal insulation assumes the maximum thickness of the thermal insulation equal to the thickness of the thermal insulation of a single pipe.

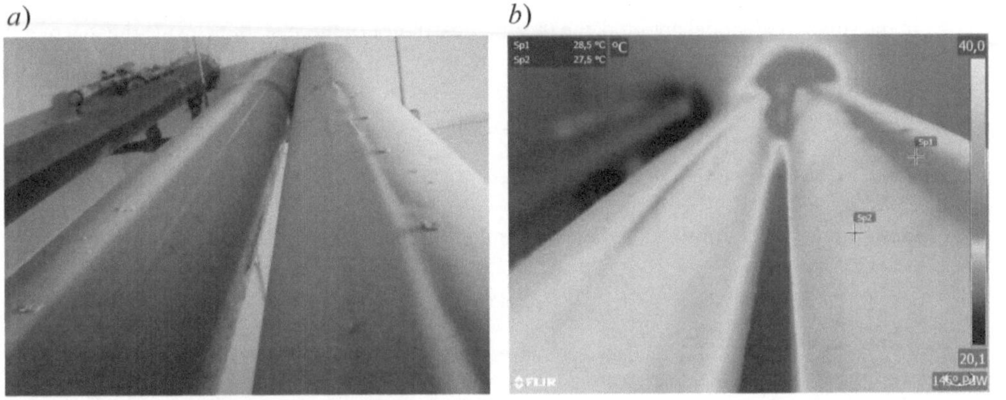

Figure 23.2. Linear bridges on the longitudinal joints of the thermal insulation of two heating pipes: a) photograph of the joint of thermal insulation, b) thermogram.

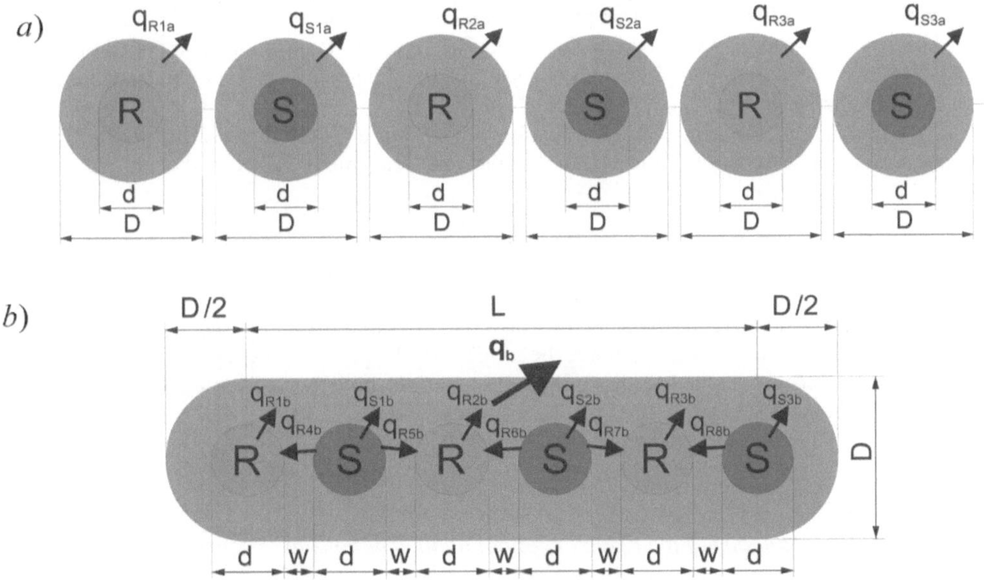

Figure 23.3. Thermal insulation of heating pipes (three supply S and three return R pipes): a) calculation diagram of six heating pipes with single insulation, b) calculation diagram of multi-pipe thermal insulation with six heating pipes.

23.2.3 *A simplified two-dimensional model for determining the unit heat loss in heating installations*

In order to determine the heat loss through multi-tube thermal insulation, a simplified stationary two-dimensional model of heat conduction in thermal insulation was applied, described by the following differential equation (Figure 23.3b):

$$\frac{\partial^2 T}{\partial x^2}\frac{\partial^2 T}{\partial y^2}=0, \quad q_x=-\lambda\frac{\partial T}{\partial x}, \quad q_y=-\lambda\frac{\partial T}{\partial y} \tag{23.1}$$

where T is the temperature (°C), x and y are the coordinates of the Cartesian system in the cross-section of multi-tube thermal insulation (m), q_x and q_y are components of the heat flux density (W/m^2), while λ is the thermal conductivity of the thermal insulation (W/m/K).

In the simplified model adopted, the thermal resistance of the steel pipe inside the thermal insulation was neglected, and the Neumann boundary condition described by the following relationship was used in the calculations:

$$q'=-\lambda\frac{\partial T}{\partial n}=h(T-T_o) \quad [\mathrm{W/m^2}], \tag{23.2}$$

where T_o is the external temperature in the unheated room of 8°C, while h is the heat transfer coefficient of 25 W/m^2/K.

Unit heat losses in multi-tube thermal insulation result from the heat flux balance (Figure 23.3b):

$$q_b=q_{R1b}-q_{R4b}+q_{S1b}+q_{R2b}-q_{R5b}-q_{R6b}+q_{S2b}+q_{R3b}-q_{R7b}-q_{R8b}+q_{S3b} \quad [\mathrm{W/m}], \tag{23.3}$$

Negative heat flux q_{R4b}, q_{R5b}, q_{R6b}, q_{R7b}, q_{R8b} means the heat flowing from the supply pipe S to the return pipe R (Figure 23.3b).

The calculations were performed using the boundary element method (BEM) using the author's calculation program written in Fortran. The boundary element method is often used in thermal calculations. This method involves solving the boundary integral equation on the boundary of the cross-section of thermal insulation. In order to solve the differential equation (1) no internal mesh is needed, while the boundary line of the thermal insulation requires discretization for constant elements. In order to perform calculations, an edge line of 3000 elements was adopted. The verification of the method was carried out at work (Teleszewski & Zukowski 2019), where for the edge composed of 3000 elements, the relative temperature error did not exceed 0.001%. The details of the boundary element method in heat conduction can be found in (Brebbia 1984).

In the case of the heat losses in heating pipes with a single thermal insulation, the heat losses through a single supply or return heating pipe were determined based on the analytical formula:

$$q_{Ria}, qSia=\frac{(T_s-T_a)\pi}{\left(\frac{1}{2\pi}\ln\left(\frac{D_i}{d_i}\right)+\frac{1}{hD_i}\right)} \quad [\mathrm{W/m}], \tag{23.4}$$

where d_i is the diameter of the steel pipe inside the thermal insulation, D_i is the outer diameter of the thermal insulation, whereas i denotes the number of the pipe (Figure 23.3a).

In order to compare the solution of heat loss in multi-pipe thermal insulation with single pipes, the sum of heat losses in single pipes was determined (Figure 23.3b):

$$q_a=q_{R1a}+q_{S1a}+q_{R2a}+q_{S2a}+q_{R3a}+q_{S3a} \quad [\mathrm{W/m}], \tag{23.5}$$

23.2.4 *Determination of the reduction of pollutant emissions to air by using multi-pipe insulation*

The heating installations are supplied from the heat center, to which heat is supplied by the district heating network. The municipal heating network is connected to the municipal combined heat and power plant (CHP), in which hard coal is burned. In order to determine the emission of pollutants generated for the heat losses in the heating installation, the coefficients of pollutants emitted by heating plants presented in the guidelines were used (Gómez et al. 2006; EMEP / EEA 2016). The calculations were performed for the following pollutants emitted to the atmosphere by CHP plants: nitrogen oxides (NOX), carbon monoxide (CO), non-methane volatile organic compounds (NMVOC), sulfur oxides (SOX), total suspended particles (TSP), particulate matter (PM10 and PM2.5), carbon dioxide (CO_2) and methane (CH_4). Reductions of the pollutants emitted to the atmosphere were determined from the following formula:

$$\Delta E = (E_a - E_b)^* E_f \quad \text{[kg/year]}, \tag{23.6}$$

where E_a is the annual energy consumed for heat losses by six separate single heating pipes, E_b is the annual energy for the heat losses through six pipes placed in a common multi-pipe thermal insulation, while E_f is the emission factor based on the work of the authors Gómez et al. (2006) and EMEP / EEA guidelines (2016).

Annual energy for the heat losses through six separate single heating pipes and six pipes in a common multi-pipe thermal insulation is determined by the following formula:

$$E_i = q_i L_i t, \quad i = a, b \quad \text{[GJ]}, \tag{23.7}$$

where L_i is the length of heating pipes, t is time, q_i are the unit heat losses determined from formulas (3) and (5) for six heating pipes in common multi-pipe thermal insulation ($i = b$) and for six individual heating pipes respectively ($i = a$).

23.3 RESULTS AND DISCUSSION

This section presents the results of the calculations of unit heat losses, energy losses and reduction of selected pollutants emitted to the atmosphere by heat sources in relation to the heat losses in heating installations. It should be noted that the calculations did not include the effect of reduction of linear thermal bridges in single heating pipes (Figure 23.2b) on the reduction of the heat losses through the use of multi-pipe insulation. The influence of linear thermal bridges on the reduction of heat loss can be significant. The circular thermal insulation of a standard single heating pipe consists of two half-round parts, resulting in the formation of two thermal bridges. In the case of six pipes with round thermal insulation, twelve linear thermal bridges are obtained. In the proposed multi-pipe thermal insulation, assuming that it consists of two parts, only six linear connections of insulation, i.e. two linear thermal bridges, were obtained for six heating pipes.

Figure 23.4 presents a comparison of unit heat losses for six single pipes in single-round thermal insulation and six heating pipes placed in a common multi-pipe thermal insulation as a function of external temperature at actual operating parameters of the 70/50°C installation. As the temperature decreases, the unit heat losses decrease linearly for both single and multi-pipe thermal insulation. The unit heat losses in the case of multi-tube thermal insulation with six heating pipes for the outdoor temperature range from 0 to 20°C are on average 43% smaller than in the case of six pipes with single round thermal insulation. The reduction of heat loss in twin pipe heating networks compared to single pipes ranges from 31 to 35% (Kristjanssom & Bohm 2006; Babiarz & Zieba 2012; Krawczyk & Teleszewski 2019b; Teleszewski et al. 2019). Adding more pipes to the common thermal insulation can contribute to a further reduction of heat losses provided that the appropriate thickness of thermal insulation around the pipes is maintained. The annual energy consumption for heat loss by an installation consisting of six single pipes is equal to 56 GJ, while for the pipes in common thermal insulation, it is equal to 31 GJ. The use of multi-tube thermal insulation reduced the annual energy for heat losses by 43%.

Figure 23.5 shows the temperature field and heat lines in multi-pipe thermal insulation. A characteristic feature of multi-tube thermal insulation is the heat exchange between the supply and return pipes. This problem has been thoroughly investigated in the work (Kristjanssom & Bohm 2006). As the distance between the supply and return pipes decreases, the heat exchange between the supply and return pipes is reduced. In the presented calculations, the standard distance between the pipes used in the twin pipe was adopted and it was equal to 0.02 m for the diameter of 48.3 mm (Kristjanssom & Bohm 2006; Kristjanssom & Bohm 2006). The cross-sectional area of multi-pipe thermal insulation

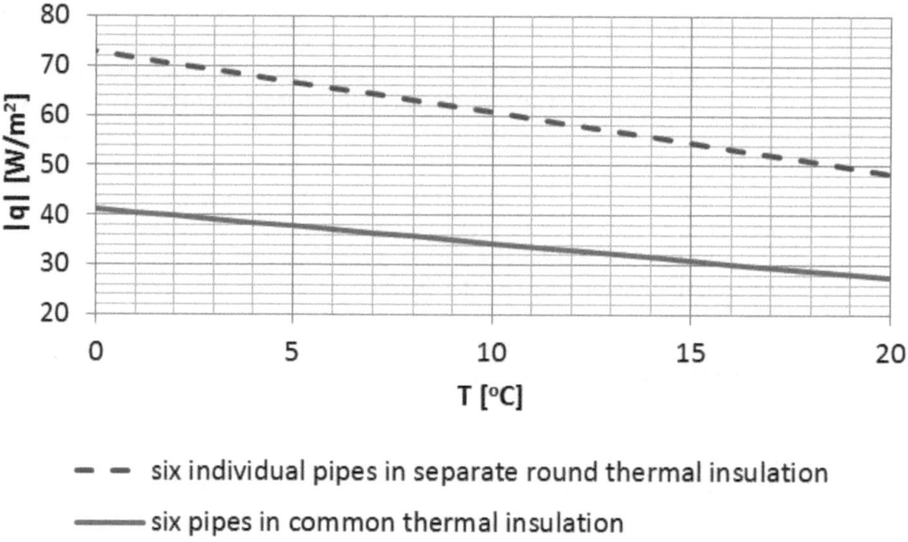

Figure 23.4. Comparison of unit heat losses as a function of the external temperature of the analyzed example of six single pipes with round thermal insulation and the proposed solution of multi-pipe thermal insulation.

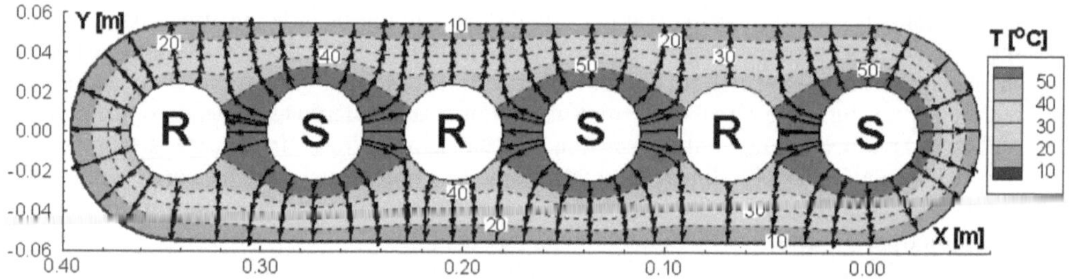

Figure 23.5. Temperature field and heatlines in multi-tube thermal insulation (S-supply pipe, R-return pipe).

was 0.036075 m^2, while the area thermal insulation of six single pipes equaled 0.046026 m^2. The consumption of polyurethane foam for multi-pipe insulation with six pipes is about 22% lower than for six single pipes in round thermal insulation. The reduction of material consumption for insulation can also contribute to reducing the emissions from the production of polyurethane foam.

Table 23.1 presents the emissions of pollutants by a heat source in relation to heat losses at the analyzed heating installation. The emission of pollutants into the atmosphere depends on the fuel burned in the heat source, and so in the case of coal, the largest reductions were obtained for CO_2 (by about 2372 kg / year) and NOx (by about 5 kg / year). It should be noted that these values have been determined only for a fragment of the installation with a length of 40 m. In the case of the presented multi-pipe insulation, the emission of pollutants into the atmosphere turned out to be 1.7 times lower compared to the existing installation with single heating pipes, in relation to the heat losses through thermal insulation. The last column of Table 23.1 calculates the reduction of pollution per one meter of the installation, so that other authors in the future can compare their test results with the results presented in this paper.

Table 23.1. Annual emissions of selected pollutants into the atmosphere for six single pipes with round insulation and multi-pipe thermal insulation.

Pollutant	E_f [g/GJ]	E_aE_f [kg/year]	E_bE_f [kg/year]	ΔE_{a-b} [kg/year]	$\Delta E_{a-b}/L$ [kg/year/m]
NO_x	209.0	11.616	6.573	5.0	0.126
CO	8.7	0.484	0.274	0.210	0.005
NMVOC	1.0	0.056	0.031	0.024	0.001
SO_x	820	45.6	25.8	19.8	0.49
TSP	11.4	0.634	0.359	0.275	0.007
PM10	7.7	0.428	0.242	0.186	0.005
PM2.5	3.4	0.189	0.107	0.082	0.002
CO_2	98300	5463.5	3091.5	2372.1	59
CH_4	1	0.056	0.031	0.024	0.001

23.4 CONCLUSIONS

Modernization of heating networks and installations transporting heat from a heat source to radiators can significantly contribute to the ecological effect of heating systems. The paper proposes multi-pipe thermal insulation, i.e. thermal insulation of many pipes in common thermal insulation instead of using individual thermal insulation for each heating pipe. The solution helps to reduce the heat loss in heating installations. The reduction of the heat losses in heating installations causes lower consumption of fuels in heat sources, and thus contributes to the decrease of the pollutants released into the atmosphere. In the presented example, for six heating pipes, a reduction of 43% in relation to heat losses was obtained compared to the classic solution of separate circular thermal insulation. The presented multi-tube thermal insulation can be used both in a vertical arrangement of cables, e.g. under the ceiling, as well as in a horizontal configuration in the form of installation risers. The use of the proposed thermal also causes lesser consumption of materials for the purpose of making such thermal insulation compared to individual round insulation pipes, which in turn reduces the pollution in the production of a smaller amount of polyurethane foam.

23.4.1 *Author disclosure statement*

The authors declare that the research is in compliance with ethical standards, and they have no conflict of interest. This article does not contain any studies with human participants or animals performed by any of the authors.

ACKNOWLEDGMENTS

This work was performed within the framework of Grant No. WZ/WBiIS/9/2019 of Bialystok University of Technology and financed by the Ministry of Science and Higher Education of the Republic of Poland.

REFERENCES

Grove, A.T. 1980. Geomorphic evolution of the Sahara and the Nile. In M.A.J. Williams & H. Faure (eds), *The Sahara and the Nile*: 21–35. Rotterdam: Balkema.

Jappelli, R. & Marconi, N. 1997. Recommendations and prejudices in the realm of foundation engineering in Italy: A historical review. In Carlo Viggiani (ed.), *Geotechnical engineering for the preservation of monuments and historical sites*; *Proc. intern. symp., Napoli, 3–4 October 1996*. Rotterdam: Balkema.

Johnson, H.L. 1965. Artistic development in autistic children. *Child Development* 65(1): 13–16.

Polhill, R.M. 1982. *Crotalaria in Africa and Madagascar*. Rotterdam: Balkema.

Alphones, A., Damodara, V., Wang, A., Lou, H., Li, X., Martin, Ch.B., Chen, D.H. and Johnson, M.R. 2020. Response Surface Modeling and Setpoint Determination of Steam- and Air-Assisted Flares. *Environmental Engineering Science*. 37(4): 246–262. http://doi.org/10.1089/ees.2019.0089

Babiarz, B. & Zięba, B. 2012. Heat losses in the preinsulated district heating systems. *Civ. Environ. Eng.* 283: 5–19.

BING Federation of European Rigid Polyurethane Foam Associations. Thermal Insulation Materials Made of Rigid Polyurethane Foam (PUR/PIR); Report N1; BING Federation of European Rigid Polyurethane Foam Associations: Brussels, Belgium, 2006; Available online: http://highperformanceinsulation.eu/wp-content/uploads/2016/08/Thermal_insulation_materials_made_of_rigid_polyurethane_foam.pdf (accessed on 2 April 2020)

Bøhm, B., Kristjansson, H. 2005. Single, twin and triple buried heating pipes: On potential savings in heat losses and costs. *Int. J. Energy Res.* 29: 1301–1312. https://doi.org/10.1002/er.1118

Đurđević, D., Blecich, P. and Lenić, K. 2018. Energy Potential of Digestate Produced by Anaerobic Digestion in Biogas Power Plants: The Case Study of Croatia. *Environmental Engineering Science*. 35(12): 1286–1293. http://doi.org/10.1089/ees.2018.0123

European Environment Agency, Publications Office of the European Union. EMEP/EEA Air Pollutant Emission Inventory Guidebook; European Environment Agency, Publications Office of the European Union: Luxemburg, 2016.

Gómez, D.R., Watterson, J.D., Americano, B.B., Ha, C., Marland, G., Matsika, E., Namayanga, L.N., Osman-Elasha, B., Kalenga Saka, J.D., Treanton, K., et al. 2006. Intergovernmental Panel on Climate Change (IPCC) Guidelines for National Greenhouse Gas Inventories. Volume 2: Energy, Stationary Combustion. Institute for Global Environmental Strategies; Institute for Global Environmental Strategies (IGES) on behalf of the IPCC: Hayama, Kanagawa, Japan, Chapter 2.

Guelpa, E.; Mutani, G.; Todeschi, V.V. 2018. Reduction of CO2 emissions in urban areas through optimal expansion of existing district heating networks. *J. Clean. Prod.* 204:117–129.

Günkaya, Z. 2020. *Environmental Engineering Science*. 37(3): 214. http://doi.org/10.1089/ees.2019.0272

Jarfelt, U., Ramnas, O. Thermal conductivity of polyurethane foam—Best performances. Heat distribution—pipe properties. Sektion 6a. In Proceedings of the 10th International Symposium of District Heating and Cooling, Hanover, Germany, 3–5 September 2006; pp. 1–11.

Khosravi, M. and Arabkoohsar, A. 2019. Thermal-hydraulic performance analysis of twin-pipes for various future district heating schemes. *Energies*. 12: 1299. https://doi.org/10.3390/en12071299

Krawczyk, D.A. and Teleszewski T.J. 2019b. Reduction of heat losses in a pre-insulated network located in central Poland by lowering the operating temperature of the water and the use of egg-shaped thermal insulation: A case study. *Energies* 12: 2104. https://doi.org/10.3390/en12112104

Krawczyk, D.A. and Teleszewski, T.J. 2019a. Optimization of geometric parameters of thermal insulation of pre-insulated double pipes. *Energies* 12: 1012. https://doi.org/10.3390/en12061012

Kristjansson, H. and Bøhm, B. 2006. Advanced and Traditional Pipe Systems Optimum Design of Distribution and Service Pipes; Technical Paper. Euroheat and Power: Brussels, Belgium. 34–42.

Nowak-Ocłoń, M., Ocłoń, P. 2020. Thermal and economic analysis of preinsulated and twin-pipe heat network operation. *Energy*. 193, 116619. https://doi.org/10.1016/j.energy.2019.116619

Ocłoń, P., Nowak-Ocłoń, M., Vallati A., Quintino, A., Corcione, M. 2019. Numerical determination of temperature distribution in heating network. *Energy*. 183: 880–891. https://doi.org/10.1016/j.energy.2019.06.163

PN-B-02421:2000 Heating and district heating – Heat insulation of pipelines, valves and equipment – Specification and reception tests.

Ravina, M.; Panepinto, D.; Zanetti, M.C.; Genon, G. 2017. Environmental analysis of a potential district heating network powered by a large-scale cogeneration plant. *Environ. Sci. Pollut. Res.* 24: 13424–13436.

Teleszewski, T.J. and Zukowski M. 2019. Modification of the shape of thermal insulation of a twin-pipe pre-insulated network. *AIP Conf. Proc.* 2078, 020030. https://doi.org/10.1063/1.5092033

Teleszewski, T.J.; Krawczyk, D.A.; Rodero, A. 2019. Reduction of Heat Losses Using Quadruple Heating Pre-Insulated Networks: A Case Study. *Energies* 12, 4699.

Yan, L., Lu, X., Wang, Q., and Guo, Q. 2014. Recovery of SO2 and MgO from By-Products of MgO Wet Flue Gas Desulfurization. *Environmental Engineering Science*. 31(11): 621. http://doi.org/10.1089/ees.2014.0004

CHAPTER 24

Preliminary studies of PM10 and PM2.5 particles concentration in the atmospheric air in a city located in Central Europe – a case study

D.A. Krawczyk, T.J. Teleszewski, M. Żukowski & P. Rynkowski
Department of HVAC Engineering, Bialystok University of Technology, Bialystok, Poland

ABSTRACT: The paper presents the results of preliminary studies on the PM_{10} and $PM_{2.5}$ aerosols in the city of Bialystok, located in the north-eastern part of Poland. The studied area is characterized by the diversity of heat sources, which is why the research was carried out in a few points with different proportions of heat sources occurrence. The main heat sources in Bialystok are district heating substations supplied from the district heating network, and local boiler rooms (natural gas, oil, coal, wood and multi-fuel supply). The official data on the concentrations of suspended aerosols in the city originate from two measurement stations of the Inspectorate of Environmental Protection (IEP) located in the area where the main heat source comprises local heating centers connected to the district heating network. The measurements of PM_{10} and $PM_{2.5}$ were conducted in four other locations. The results showed that the concentrations of PM_{10} and $PM_{2.5}$ in these locations were higher than the official data of the IEP, where in extreme cases, they reach near 400%. It was caused by the occurrence of various heat sources in the studied areas. The official data on PM_{10} and $PM_{2.5}$ of the IEP are not representative for the entire city; therefore, it is recommended to perform tests in different areas of the city taking into account different types of heat sources. All measurements were conducted in the heating season at evening time, when the PM_{10} and $PM_{2.5}$ emissions in the city of Bialystok is the maximum.

24.1 INTRODUCTION

Monitoring the concentrations of suspended particles in the air is one of the important tasks of the air quality assessment in cities. The main marker of atmospheric, aerosol in the air were PM_{10} and $PM_{2.5}$.

Particulate matter $PM_{2.5}$ constitutes one of the most harmful and complex pollutants that influence both the human and animal health as well as the environmental quality (Wu & Zhang 2018). An exposure to PM2.5 was indicated as a source of several health problems, particularly asthma symptoms (Fan et al. 2016; Gleason et al. 2014; Zora et. al 2013). Kuo et al. (2019) found that particulate matters PM_{10} and $PM_{2.5}$ were positively associated with childhood asthma hospitalization in Taiwan. Xu et al. (2018) analyzed the results of a long-term study of trends in respiratory mortality in the urban area of Shenyang (China) and found 4.7% increase; they also noted that males were more susceptible to the effect of $PM_{2.5}$ exposure. Liu et al. (2018) studied the short-term effects of ambient air pollution ($PM_{2.5}$, PM_{10}, SO_2, NO_2) on chronic obstructive pulmonary disease hospital admissions in Shandong province (China) for different age groups and gender, and concluded that the air pollution was significantly associated with adverse health outcomes and stronger effects were observed for females. Zhang et al. (2018) concluded that PM_{10} and $PM_{2.5}$ were associated with an increasing risk of death events in residential cardio-cerebrovascular disease in Yanbian, (China). Pothirat et al. (2019) found the negative impact of a short-term increase in outdoor particulate matter in which the particles are less than 10 microns in diameter (PM10) during a seasonal smog period on the quality of life (QoL) of chronic obstructive pulmonary disease (COPD) patients. This review of literature that aimed at the factors influencing the association between the exposure to environmental pollutants and health effects in the population living area of the Northern Italy showed the connection between the pollutants present in air matrix and soil as well as the health effects (Alias et al. 2019).

DOI 10.1201/9781003171669-24

Most of the published papers report the results from the studies conducted in selected single sampling stations that were assumed as the representative points for a test areas. Rajšić et al. (2004) studied the PM_{10} and $PM_{2.5}$ mass concentrations in the Belgrade urban area. They found it as high average values (77 μg and 61 $\mu g\ m^{-3}$) in comparison with other European cities and indicated a traffic emission, road dust resuspension, and individual heating emissions as main sources of pollutants. Their conclusions were drawn after measurements in three points of the city. Tositti et al. (2014) developed models and discussed seasonal variation of the PM_{10} and $PM_{2.5}$ concentration, based on the measurements in a single measurement point, located at university campus. Along similar lines, a research of Villanueva et al. (2015) was conducted to examine the PM_{10} seasonal and diurnal variations, and their emission sources in the urban area of Ciudad Real, in central-southern of Spain. The results of 1-month campaign (Siciliano et al. 2018) conducted in a rural coastal area in South Italy showed the PM10 concentrations in a range from 6.8 to 62.6 $\mu g\ m^{-3}$ (a mean value of 22 $\pm$ 14 $\mu g\ m^{-3}$), whereas the $PM_{2.5}$ concentrations varied from 2.8 to 28.0 $\mu g\ m^{-3}$ (a mean value of 11 $\pm$ 6 $\mu g\ m^{-3}$). Kim et al. (2013) found the mean concentration of PM10 around 7 tested schools located in different parts of Seoul (Korea) as 54.6 $\mu g/m^3$, with minimum 13 $\mu g/m^3$ and maximum 103 $\mu g/m^3$, both values recorded for the same location (industrial area). In turn, Pandolfi et al. (2011) conducted the research on ambient PM_{10} and $PM_{2.5}$ at four strategic sampling locations around the Bay of Algeciras (southern Spain), and found the lowest levels of PM as 31.9 $\mu g\ PM_{10}/m^3$ and 21.8 μg PM2.5/m^3 in a small urban agglomeration.

As shown by Widziewicz et al. (2018) during a 12-year period, 348 large-scale smog episodes occurred in Poland, mostly in the winter season, which is characterized by increased emissions from residential heating.

The concentration of aerosols suspended in the air in cities during the heating season depends primarily on the type of heat source. Municipal research stations (air quality monitoring stations (AQMSs)) of particles suspended in the air are often located in the places where the emission of pollutants is not adequate for the whole city. The aim of the research was to estimate the concentration of PM_{10} and $PM_{2.5}$ concentration in the Bialystok agglomeration and to compare the obtained results with the Inspectorate of Environmental Protection (IEP) AQMSs located in Białystok, the results of which are officially reported to the local media.

24.2 EXPERIMENTAL SETUP

While analyzing the available data of the Inspectorate of Environmental Protection in Bialystok, during the heating season, it was noted that the highest concentrations of PM_{10} and $PM_{2.5}$ occur in the evening hours 19–24. Figure 24.1 shows a typical distribution of PM_{10} and $PM_{2.5}$ during the selected days of January. A significant increase in the PM_{10} and $PM_{2.5}$ concentrations in the evening is caused by the activation of coal and wood boilers by the inhabitants of Bialystok. The results of the tests for the peak period of PM_{10} and $PM_{2.5}$ particulate matter were presented in the further part of this research.

The measuring points are shown in satellite images presented in Figs. 24.2a–e. The research was carried out simultaneously outside the buildings and inside the rooms. It should be noted here that the measurements performed by the Inspectorate of Environmental Protection in Bialystok are performed only outside the building. All measurements were performed in single-family buildings equipped with stack ventilation. The windows were not opened during the measurement. The average outside temperature were fluctuated between -6.8°C and 9.8°C, while the average indoor temperature was maintained at the level of 20°C.

The first point is located about 4.5 km from the city center and is characterized by single-family housing. The second measuring point is located 2.3 km from the city center. It is a mix of a multi-family and single-family house, the third measuring point is located in the suburbs of the city at a distance about 7.7 km from the city center and the fourth point is located outside the urban area at a distance of about 9.3 km from the city center. Two AQMSs of the Inspectorate of Environmental Protection are located at a distance of 2.1 km and 1.4 km from the center of Bialystok. The results of PM_{10} and $PM_{2.5}$ measurement taken from these stations were used for comparison with the measurements obtained by the authors of this article.

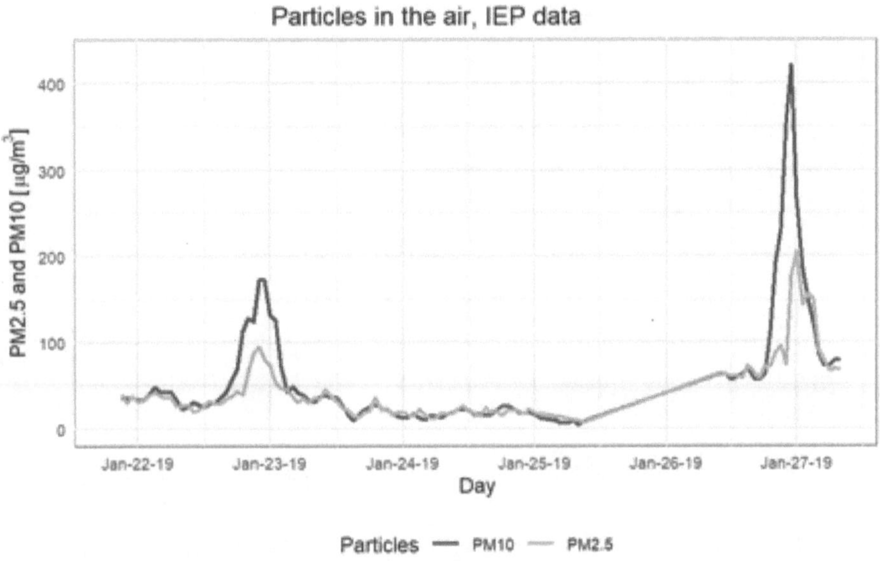

Figure 24.1. Daily changes of PM10 and PM2.5 on a typical days during January in the heating season in Bialystok (powietrze.gios.gov.pl).

24.3 METHODOLOGY

The measurements were performed at four locations within the Bialystok agglomeration in the north-eastern part of Poland. The measurements were carried out with the aid of a particle recorder suspended in air type DT-9880 with the following measurement parameters: counting efficiency: 100% for particles > 0.45 μm, 50% for particles=0.3 μm, accuracy of temperature measurement $\pm$ 0.5°C.

The heat sources occurring within the measurement points have the greatest influence on the PM_{10} and $PM_{2.5}$. On the basis of the inventory, the estimated area is 62 500 m^2, the percentage of individual heating energy sources were estimated in the range of measuring points and are presented in Table 24.1.

24.4 MEASUREMENT RESULTS AND DISCUSSION

An exemplary diagram of daily changes in the PM_{10} and $PM_{2.5}$ concentrations (IEP data) during one day in a cold day is shown in Figure 24.3. It may suggest other causes of pollutant growth, like the energy sources mentioned above. The graph below shows a significant increase in PM_{10} and $PM_{2.5}$ concentrations in the evening hours, which is probably caused by the activation of solid fuel heat sources with manual fuel loading (wood or coal).

Results of measurements PM_{10} and $PM_{2.5}$ from three days (22–24.01.2019) during the heating season are presented in Figures 24.4–24.5. Additionally, the distribution of the PM_{10} concentration from point 5 and $PM_{2.5}$ from point No. 6, both from the municipal AQMSs (IEP) were plotted.

The highest increases in the PM_{10} and $PM_{2.5}$ concentrations were recorded in the places where coal and wood boilers have the highest quantitative share. In the case of municipal AQMSs located in the area of multi-family housing, the concentration of PM_{10} and $PM_{2.5}$ remained at a nearly constant level (Figure 24.4). The measurement point No. 4, located outside the urban area, the concentration of PM_{10} and $PM_{2.5}$ is clearly lower. At measurement point No. 2, the concentration of PM_{10} and $PM_{2.5}$ on 22.01.2019 was 5 times and 7 times higher, respectively, than the concentrations recorded by AQMSs, at the point of the municipal measuring station of the Inspectorate of Environmental Protection.

Figure 24.2. Daily changes Satellite images of measuring points (www.google.pl/maps): a) No. 1, b) No. 2, c) No. 3, d) No. 4, e) No. 5 (measuring point of the first municipal stations of the Inspectorate of Environmental Protection), No. 6 (measuring point of the second municipal stations of the Inspectorate of Environmental Protection).

Table 24.1. Basic percentage share of heat sources, the estimated area is 62500 m^2, the tested measurement points.

Measuring point	Type of heat source			
	Boiler rooms for coal and wood [%]	Gas boiler rooms [%]	District heating substations [%]	Heat pump [%]
No.1	55	45	0	0
No.2	40	30	30	0
No.3	40	60	0	0
No.4	60	38	0	2
No.5	10	0	90	0
No.6	20	0	80	0

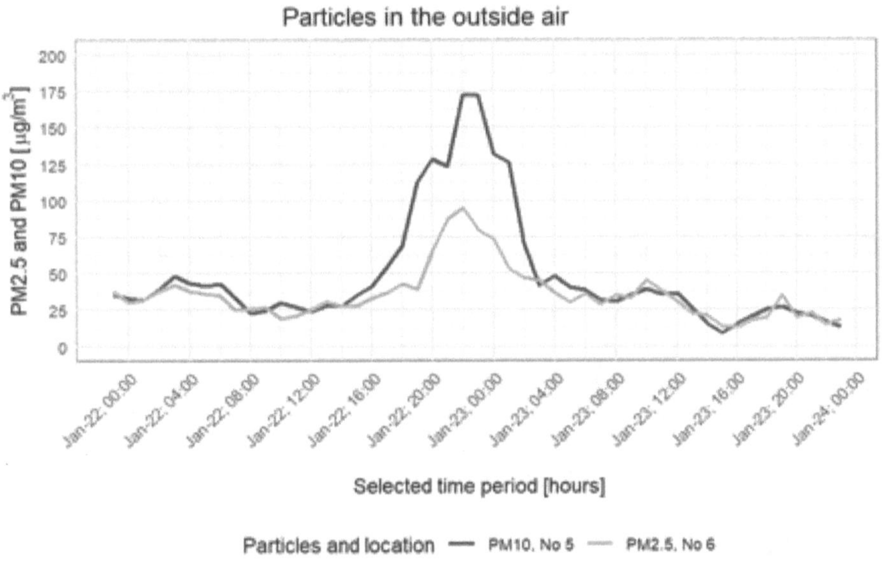

Figure 24.3. Sample graph of changes in PM_{10} and $PM_{2.5}$ concentrations during a cold day.

In the case of measuring points no. 1 and 3, where the concentration of PM_{10} and $PM_{2.5}$ is the highest and an increase in the concentration of aerosol up to 20 and 21 hours was observed, followed by a decrease.

The change in the pollution concentration trend is probably caused by intensive exploitation of wood- and coal-fired boilers plants and wood with an open combustion chamber in the evening and spontaneous extinction of furnaces at night. Similar trends in concentrations in winter PM_{10} and $PM_{2.5}$ were found in the works of Wernecke et al. 2015 and Haynes et al. 2019, whose maximum value of PM10 and PM2.5 outside the building reached 19 and 22, respectively. Point 4 located outside the urban area was characterized by much lower concentrations of PM_{10} and $PM_{2.5}$ from points 1–3. The air in non-urban or rural areas is usually less polluted than in urban areas (Clemens et al. 2016).

In the analyzed time, during the first five hours of the evenings, the increase in the PM_{10} and $PM_{2.5}$ concentrations outside the building (Figure 24.6), affect the concentrations of these pollutants inside the residential building. It should be emphasized that the tests were carried out in buildings equipped with stack ventilation, which is characterized by low efficiency and the windows were not opened. Only in the case of the building in point 3 (Figure 24.5) there was a significant increase in PM_{10} and $PM_{2.5}$, which was caused by the efficient operation of stack ventilation and leaky external doors. In the case of airing buildings by opening the windows, the impact of PM_{10} and $PM_{2.5}$ from the outside

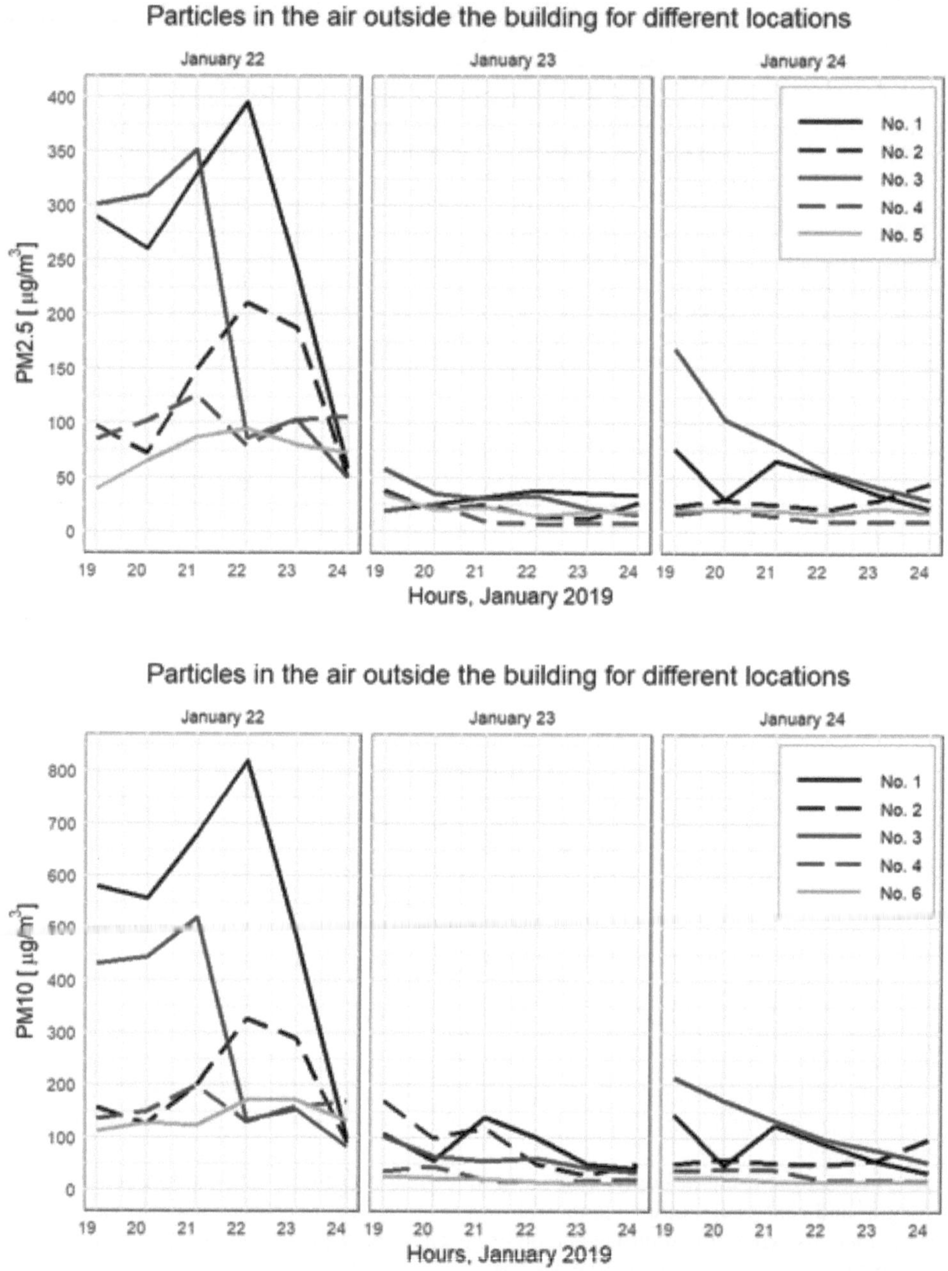

Figure 24.4. Changes the PM2.5 and PM10 concentration outside the building (no. 1–4 –measurement points, no.5–6 - municipal AQMSs) for selected days and hours, from 19 to 24 p.m.

of the building on the indoor air quality could be greater. It should be noted that the preliminary tests carried out concern only the short measurement period in the evening hours, when the concentration of smog is the highest. The concentrations of PM_{10} and $PM_{2.5}$ in rooms inside the building depend primarily on the activity of residents such as cooking, cleaning, etc., as described in the works of Nasir et al. 2013 and Nasir and Colbeck 2013.

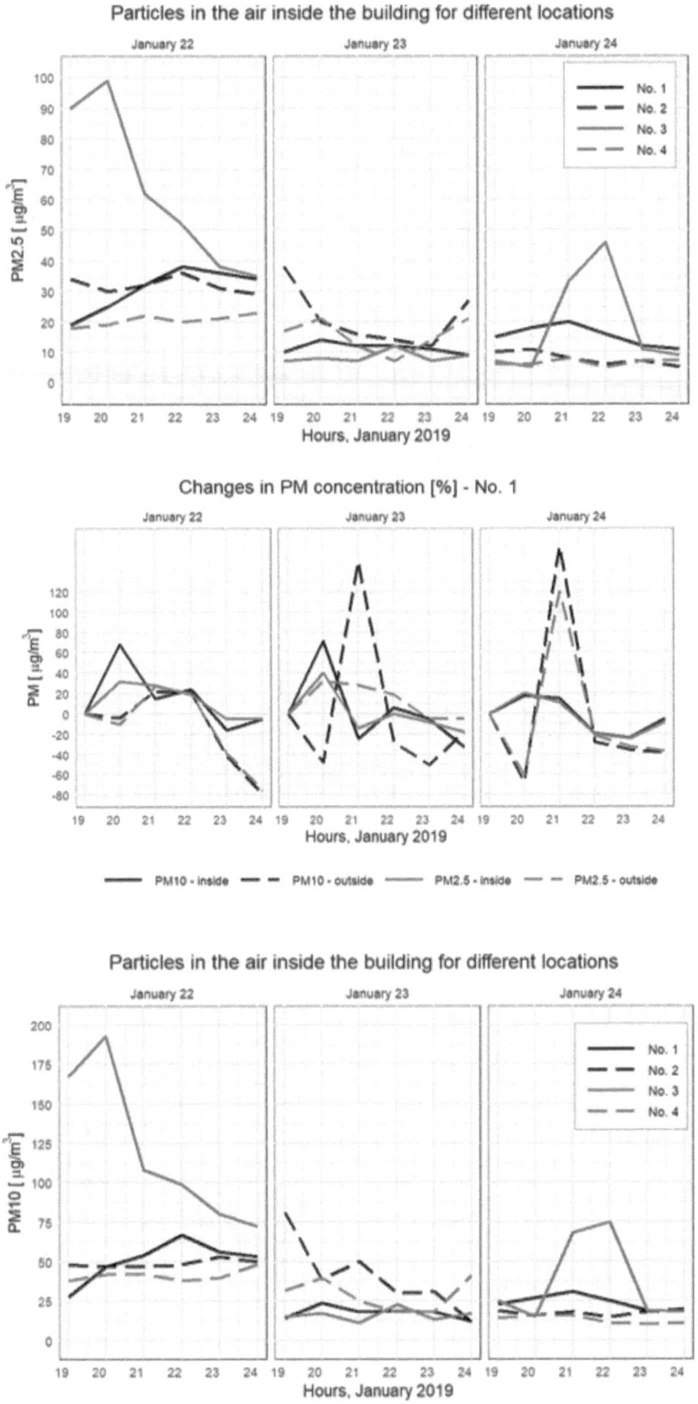

Figure 24.5. Changes the PM2.5 and PM10 concentration inside the building (no. 1–4 –measurement points) for selected days and hours, from 19 to 24 p.m.

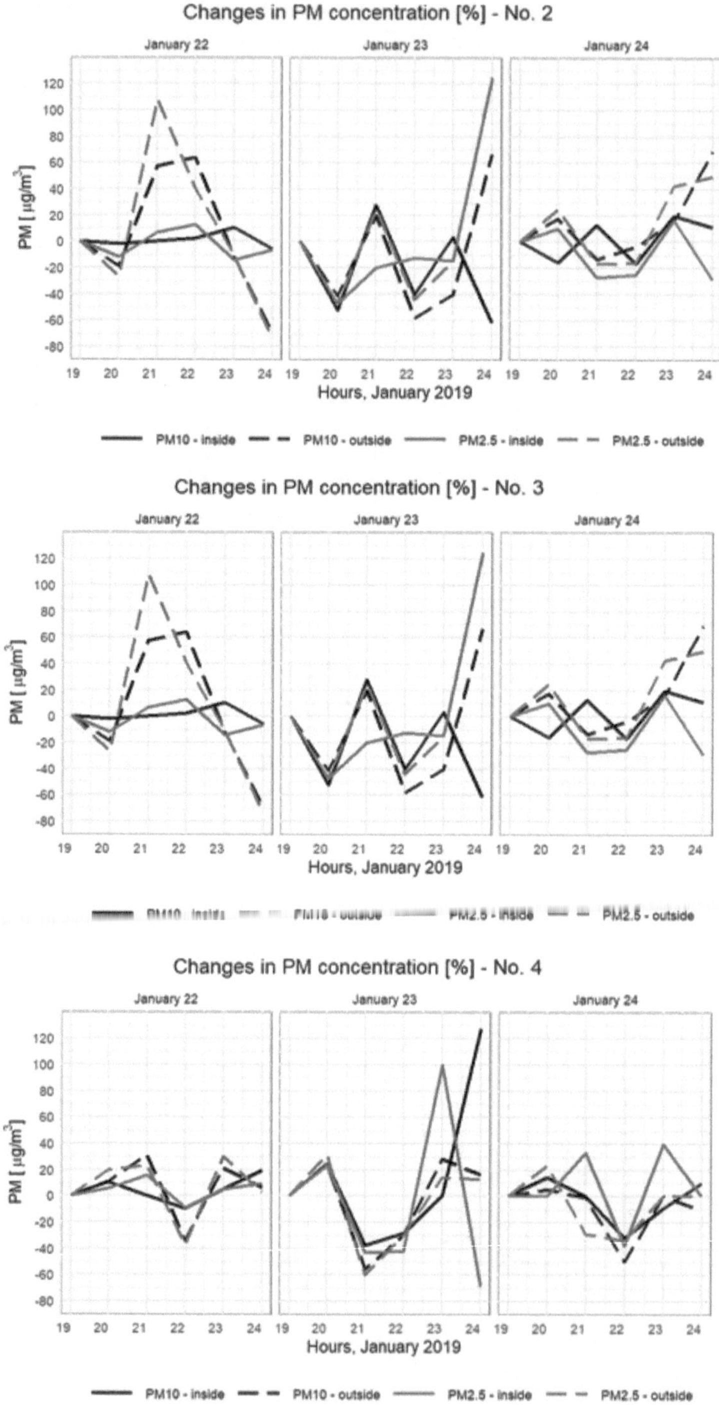

Figure 24.6. Changes in PM2.5 and PM10 concentration at measuring points no. 1–4 from 19 to 24, inside and outside the building.

The median and the quartiles are calculated directly from the data. The median is represented by the line in the box. Comparing medians, there is a discrepancy in the measurements between the measuring points by authors and the municipal AQMSs. It shows a relatively high unevenness of the PM_{10} and $PM_{2.5}$ concentrations in the air. For the PM10 (Figure 24.7) values are 104 μg/m^3 for no. 1 and 23 μg/m^3 for no. 6. Treating the AQMSs as the reference value, it is a 352% difference, which is significant. Similarly for the other values of minimum, first quartile, third quartile and maximum (Table 24.2). This indicates a large impact of the existing heat sources and the atmospheric factors, as described earlier.

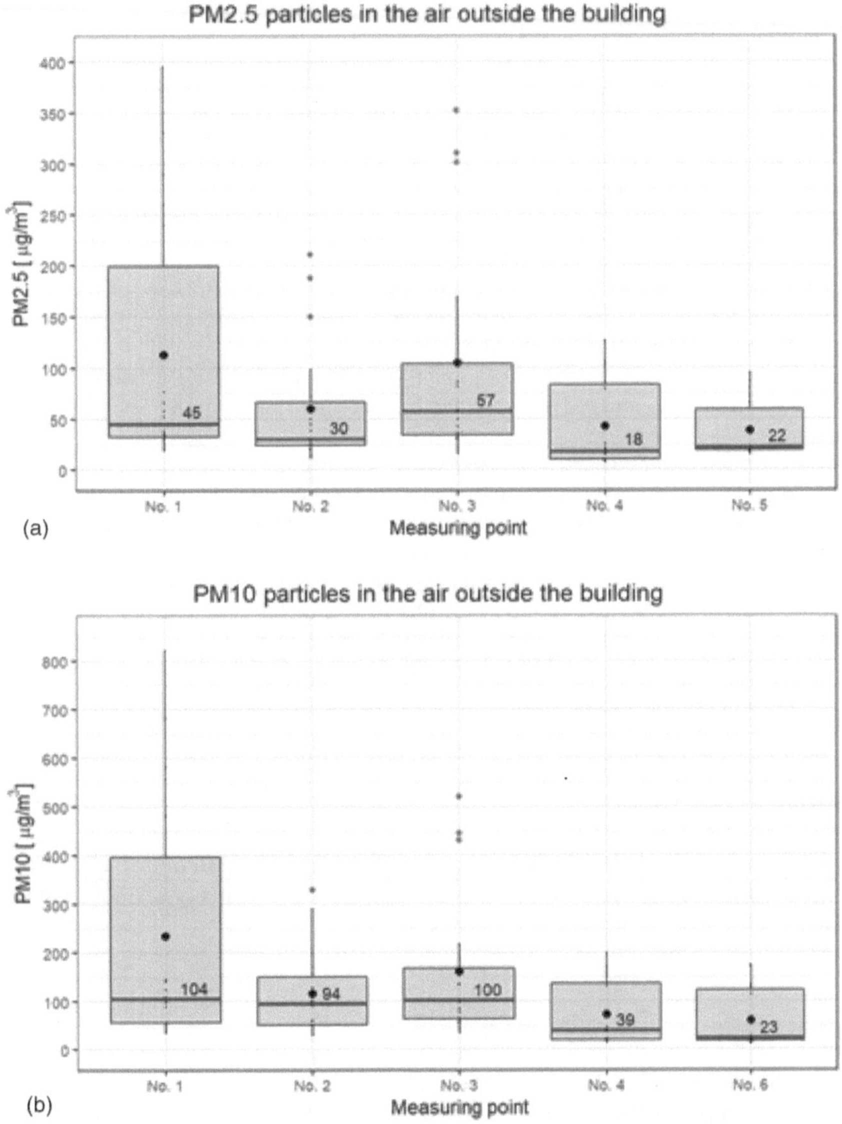

Figure 24.7. Changes in PM2.5 and PM10 concentration at measuring points no. 1–4 from 19 to 24, inside and outside the building.

Table 24.2. Results of box plot charts for particles PM_{10} and $PM_{2.5}$ in the air outside the building.

Results for particles $PM_{2.5}$					
No.	minimum	first quartile	median	third quartile	Maximum
1	19	32.5	45.0	199.25	395
2	12	24.25	30.0	66.50	98
3	16	33.75	57.0	104.00	169
4	7	10.00	18.0	83.50	126
5	15	19.25	21.5	58.50	95
Results for particles PM_{10}					
No.	minimum	first quartile	median	third quartile	Maximum
1	35	56.25	104.50	396.00	820
2	30	51.25	94.00	149.50	289
3	35	62.75	100.50	166.50	216
4	14	20.00	38.85	134.75	198
6	12	17.25	23.25	120.88	172

24.5 CONCLUSIONS

The obtained results showed significant differences in the concentrations of PM_{10} and $PM_{2.5}$ between the municipal pollution measurement stations of the Inspectorate of Environmental Protection (No. 5–6) and measurement points No. 1–3.

The measured concentrations of PM_{10} and $PM_{2.5}$ obtained from AQMSs are lower than the measured concentrations of PM_{10} and $PM_{2.5}$ at selected measurement points no. 1–3, where coal and wood boilers predominate.

The maximum difference between the AQMS station and measurement points was as high as 450% for PM10 and 644% for $PM_{2.5}$ measurements. The local concentrations of PM_{10} and $PM_{2.5}$ are mainly influenced by the heating energy sources located in the vicinity of the examined points. Municipal measuring points No. 5 and 6 are located within multi-family buildings, where the main source of energy are district heating stations and substations; therefore, the measurement results in these points are lower than in selected points no. 1–3. The results of the research indicate how important is the location of the measurement stations. While analyzing Figure 24.7, it is recommended to use a larger number of AQMSs located in different parts of the city, so that the variation in the occurrence of different energy sources of heat could be taken into account.

ACKNOWLEDGEMENTS

This work was performed within the framework of Grant of the Bialystok University of Technology and WZ/WBiIS/9/2019 financed by the Ministry of Science and Higher Education of the Republic of Poland.

REFERENCES

Alias, C., Benassi, L., Bertazzi, L., Sorlini, S., Volta, M., Gelatti, U. 2019. Environmental exposure and health effects in a highly polluted area of Northern Italy: a narrative review. *Environ Sci Pollut Res* 26: 4555. https://doi.org/10.1007/s11356-018-4040-5

Amara, S., Baghdadli, T, Knapp, S., Nordell, B. 2017. Legionella disinfection by solar concentrator system. *Renew Sust Energ Rev* 70: 786–792. https://doi.org/10.1016/j.rser.2016.11.259

Clements N, Hannigan M P, Miller S L, Peel J L, Milford J B. 2016. Comparisons of urban and rural PM10–2.5 and PM2.5 mass concentrations and semi-volatile fractions in northeastern Colorado. *Atmos. Chem. Phys.* 16, 7469–7484. https://doi.org/10.5194/acp-16-7469-2016

Fan, J., Li, S., Fan, C., Bai, Z, Tang, K. 2016. The impact of PM2.5 on asthma emergency department visits: a systematic review and meta-analysis. *Environ Sci Pollut* Res 23: 843. https://doi.org/10.1007/s11356-015-5321-x

Gleason, J.A, Bielory, L, Fagliano, J.A. 2014. Associations between ozone, PM2.5, and four pollen types on emergency department pediatric asthma events during the warm season in New Jersey: a case-crossover study. *Environ Res* 132:42. https://doi.org/10.1016/j.envres.2014.03.035

Haynes A, Popek R, Boles M, Paton-Walsh C, Robinson S.A. 2019. Roadside Moss Turfs in South East Australia Capture More Particulate Matter Along an Urban Gradient than a Common Native Tree Species. *Atmosphere* 10, 224. https://doi.org/10.3390/atmos10040224

Kim, H-H., Lee, C-D., Jeon, J-M., Yu, C-W., Park, J-H., Shin, D-C., Lim, Y-W. 2013. Analysis of the association between air pollution and allergic diseases exposure from nearby sources of ambient air pollution within elementary school zones in four Korean cities. *Environ Sci Pollut Res* 20: 4831. https://doi.org/10.1007/s11356-012-1358-2

Kuo, Ch.-Y., Chan, Ch-K., Wu, Ch-Y., Phan D-V., Chan, Ch-L. 2019. The Short-Term Effects of Ambient Air Pollutants on Childhood Asthma Hospitalization in Taiwan: A National Study. *Int. J. Environ. Res. Public Health* 16(2), 203. https://doi.org/10.3390/ijerph16020203

Liu, Y., Sun, J., Gou, Y., Sun, X., Li, X., Yuan, Z., Kong, L., Xue, F. 2018. A Multicity Analysis of the Short-Term Effects of Air Pollution on the Chronic Obstructive Pulmonary Disease Hospital Admissions in Shandong, China. *Int. J. Environ. Res. Public* Health 15(4), 774. https://doi.org/10.3390/ijerph15040774

Nasir ZA, Colbeck I, Ali Z, Ahmad S. 2013. Indoor particulate matter in developing countries: a case study in Pakistan and potential intervention strategies, *Environ. Res. Lett.* 8 024002. https://doi.org/10.1088/1748-9326/8/2/024002

Pandolfi, M., Gonzalez-Castanedo, Y., Alastuey, A., de la Rosa, J.D., Mantilla, E., Sanchez de la Campa, A., Querol, X., Pey, J., Amato, F., Moreno. T. 2011. Source apportionment of PM10 and PM2.5 at multiple sites in the strait of Gibraltar by PMF: impact of shipping emissions. *Environ Sci Pollut Res* 18: 260. https://doi.org/10.1007/s11356-010-0373-4

Pothirat, Ch., Chaiwong, W., Liwsrisakun, Ch., Bumroongkit, Ch., Deesomchok, A., Theerakittikul, T., Limsukon, A., Tajaroenmuang P., Phetsuk, N. 2019. Influence of Particulate Matter during Seasonal Smog on Quality of Life and Lung Function in Patients with Chronic Obstructive Pulmonary Disease. *Int. J. Environ. Res. Public Health* 16(1), 106. https://doi.org/10.3390/ijerph16010106

Rajšić, S.F., Tasić, M.D., Novaković, V.T., Tomašević, M.N. (2004) First assessment of the PM10 and PM2.5 particulate level in the ambient air of Belgrade city. *Environ Sci & Pollut Res* 11: 158. https://doi.org/10.1007/BF02979670

Siciliano, T., Siciliano, M., Malitesta, C. et al. 2018. Carbonaceous PM10 and PM2.5 and secondary organic aerosol in a coastal rural site near Brindisi (Southern Italy) *Environ Sci Pollut Res* 25: 23929 https://doi.org/10.1007/s11356-018-2237-2

Tositti, L., Brattich, E., Masiol, M., Baldacci, D., Ceccato D., Parmeggiani, S., Stracquadanio, M., Zappoli S. 2014. Source apportionment of particulate matter in a large city of southeastern Po Valley (Bologna, Italy) *Environ Sci Pollut Res* 21: 872. https://doi.org/10.1007/s11356-013-1911-7

Villanueva, F., Tapia, A., Cabañas, B., Martinez, E., Albaladejo, J. 2015. Characterization of particulate polycyclic aromatic hydrocarbons in an urban atmosphere of central-southern Spain. Environ Sci Pollut Res 22: 18814. https://doi.org/10.1007/s11356-015-5061-y

Wernecke, B, Language B, Piketh S, Burger, RJ. 2015. Indoor and outdoor particulate matter concentrations on the Mpumalanga highveld – a case study. *Clean Air Journal.* 25. 12–16. https://doi.org/10.17159/2410-972X/2015/v25n2a1

Widziewicz, K., Rogula-Kozłowska, W., Loska, K., Kociszewska, K., Majewski, G. 2018. Health Risk Impacts of Exposure to Airborne Metals and Benzo(a)Pyrene during Episodes of High PM10 Concentrations in Poland. *Biomedical and Environmental Sciences.* 31, 1, 23–36. https://doi.org/10.3967/bes2018.003

Wu, W., Zhang, Y. 2018. Effects of particulate matter (PM2.5) and associated acidity on ecosystem functioning: response of leaf litter breakdown *Environ Sci Pollut Res* 25: 30720. https://doi.org/10.1007/s11356-018-2922-1

Xue, X., Chen, J., Sun, B., Li, X. 2018. Temporal trends in respiratory mortality and short-term effects of air pollutants in Shenyang, *China Environ Sci Pollut Res* 25: 11468. https://doi.org/10.1007/s11356-018-1270-5

Zora, J.E, Sarnat, S.E., Raysoni, A.U., Johnson, B.A., Li, W.W., Greenwald, R., Holguin, F., Stock, T.H., Sarnat, J.A. (2013) Associations between urban air pollution and pediatric asthma control in El Paso, Texas. *Sci Total Environ* 448:56–65. https://doi.org/10.1016/j.scitotenv.2012.11.067

CHAPTER 25

Development of the method of heat demands determination in residential buildings

D. Leciej-Pirczewska
West Pomeranian University of Technology, Szczecin, Poland

ABSTRACT: The paper presents the method of building thermal power demands determination on the basis of the building heat consumption measurements. Firstly, the heat balance for building was worked out. It takes varying external temperature, wind velocity, solar radiation, heat accumulation in the walls and internal heating gains into consideration. In order to determine the building daily heat consumption and to verify the method, the measurements of heat consumption were carried out in three apartment buildings located in Szczecin. The measurements were investigated day by day during two heating seasons. Secondly, using the heat balance and the measurements results, the real heat demands for the buildings and the parameters affecting heat consumption were determined. The calculations were carried out using the regression analysis method. Finally, the influence of duration and frequency of measurements on obtained results was determined.

25.1 INTRODUCTION

The demand for thermal power in a building is a very important parameter, significant particularly when heat is supplied from central sources. Ordered power that crucially influences the global heating costs is determined on a basis of this parameter. The thermal power demand is determined at building design process. The building can be changed, rebuilt, insulated during exploitation, its windows can be replaced. Its thermal needs vary as well. All modifications can be expressed analytically in the thermal heating demand calculations. This process is relatively simple when a building owner has a detailed project. Otherwise, the matter gets complicated and reliability of findings may be questionable. The paper presents the method of building thermal power demands determination on the basis of the building heat consumption measurements. The point of the method is to determine the thermal characteristic of the building; the heat demand can be derived from this characteristic.

In order to verify the method at real objects, and to determine the range of its application, the accuracy of building thermal characteristic and frequency of the heat consumption measurements, three apartment buildings were investigated day by day during two heating seasons.

In order to specify the thermal characteristics of the building the heat balance for building was worked out. This balance, in its most detailed form, considers varying external temperature, wind velocity, solar radiation, heat accumulation in the walls and internal heating gains. In order to verify the influence of particular parameters on heat balance, their values were identified for data sets using the regression analysis method. The building thermal characteristics were determined this way.

Carrying out heat consumption measurements each day for all heated objects during the heating season (seven or eight months per year) is practically impossible. Therefore, the influence of time interval and measurements frequency on the obtained results were to be considered as the next step.

The possibility of shortening the period of daily measurements execution and the possibility of extending the time intervals between measurements were checked out and the analysis of the measurement frequency influence on the accuracy of buildings thermal characteristic determination was performed.

DOI 10.1201/9781003171669-25

25.2 HEAT BALANCE OF THE BUILDING

During exploitation of the building the continuous energy flow between the building and surroundings takes place. General energy balance for the building has the form:

$$\sum Q = U \tag{25.1}$$

where U = change of the internal energy of the system, kJ; $\sum Q$ = sum of energy being supplied to and taken off from the building, kJ.

Considering this balance for the time period $\Delta\tau$ and assuming that the thermal fluxes within this period are constant, the following relationship is obtained:

$$\sum \dot{Q}\Delta\tau = \dot{U}\Delta\tau \tag{25.2}$$

hence:

$$\sum \dot{Q} = \dot{U} \tag{25.3}$$

Sum of supplied and taken off thermal fluxes that are supplied equals to:

$$\sum \dot{Q} = \dot{Q}_{co} - \dot{Q}_p - \dot{Q}_w + \dot{Q}_s + \dot{Q}_i \tag{25.4}$$

where $\dot{Q}_{co}$ = thermal flux absorbed from central heating system, kW; $\dot{Q}_p$ = transmission heat losses, kW; $\dot{Q}_w$ = ventilation heat losses, kW; $\dot{Q}_s$ = solar heat gains, kW; $\dot{Q}_i$=internal heat gains, kW.

By substituting the relation (25.4) into (25.3), the general form of heat balance for the building is obtained:

$$\dot{Q}_{co} = \dot{Q}_p + \dot{Q}_w - \dot{Q}_s - \dot{Q}_i + \dot{U} \tag{25.5}$$

Depending on the method of determination of elements in equation (25.5) one can obtain the formulae with different minuteness of detail.

25.2.1 *Heat transmission losses*

The heat transmission losses are usually determined (Salakij et al. 2016) as a sum of heat losses by particular walls or their parts with the same value of the designed heat transfer coefficient:

$$\dot{Q}_p = \sum_{i=1}^{n} F_i k_i (t_{ii} - t_{ei}) \tag{25.6}$$

where F = area of wall or its part, m^2; k = heat transfer coefficient, $kW/(m^2K)$; t_i = inner temperature, °C; t_e = outer temperature, °C; i = wall index; n = number of walls.

It can be assumed that the temperature outside building partition is equal to the temperature of environment. The temperatures inside wall (temperatures in apartments) can be different. Assumption of their average value gives small error during heat balancing. Then, the temperature difference can be extracted from the sum operator, and the heat losses for the building are determined as:

$$\dot{Q}_p = (t_i - t_e)\sum_{i=1}^{n} F_i k_i \tag{25.7}$$

where t_i = average internal temperature, °C; t_e = average external temperature, °C.

The heat transfer coefficient depends on the wall structure and surface conductance. The wall is made of different materials which have various coefficients of thermal conductivity. The values of thermal conductivity coefficients of porous materials that are used in the building industry depend on moisture, and the values of surface conductance coefficients depend on the temperatures of wall and environment as well as on the air flow velocity. During exploitation, the parameters mentioned above may vary. Due to the relatively small influence of the changes of specified parameters on heat transfer coefficient k during exploitation, their values can be assumed as constants in relation (25.7). Therefore, for the given building:

$$\sum_{i=1}^{n} F_i k_i = \text{const} = A \tag{25.8}$$

while penetrating heat flux is defined as:

$$\dot{Q}_p = A(t_i - t_e) \tag{25.9}$$

25.2.2 *Heat for inflowing air heating*

Heat for ventilation air heating is defined as [1,2]:

$$\dot{Q}_w = \sum_{i=1}^{m} \dot{V}_i \rho c_p (t_{ii} - t_{ei}) \tag{25.10}$$

where $\dot{V}_i$ = average volume of infiltrating air flow, m^3/s; ρ = air density, kg/m^3; c_p = specific heat of air, kJ/(kgK); i = ventilated room index; m = number of ventilated rooms.

Under real conditions a volume of the infiltrating air flow is variable. Air infiltration in the building is caused by pressure difference on both sides of the building wall. It is produced by wind action and air temperature difference between inner and outer side of the building. The pressure difference produced by wind depends on the building shape and its height. The pressure difference produced by the temperature difference depends on storey on which the room is situated as well as on the room fittings with additional heat sources.

Blowing wind generates the pressure difference at external surfaces of walls. The values of under- and overpressure depend on the wind direction and velocity towards the surface.

If air temperatures are different at both sides of building external wall and if the building is not equipped with any device producing additional air flow, the pressure difference in a given point of the building can be defined as:

$$\Delta p = p_{st} + p_w - P_i \tag{25.11}$$

where Δp = pressure difference inside and outside the building in given point; p_{st} = external static pressure at undisturbed air flow at the building height; p_w = pressure caused by the wind action in the point; p_i = internal static pressure in the point.

The pressure caused by the wind action can be defined on the basis of its velocity as:

$$p_w = C\frac{\rho w^2}{2} \tag{25.12}$$

where C = aerodynamic resistance coefficient; w = wind velocity, m/s.

Coefficient C depends on the position of the room in the building and wind direction.

Static pressure difference between the outer and inner side of the building is caused by air density difference resulting from the temperature differences. During the heating season, the warm air from the room is removed by natural ventilation ducts to the building surroundings and is replaced by cold air from outside flowing throughout the supply air outlets. The pressure differences caused by temperature difference between outer and inner side of the building Δp_t can be specified in the following way (ASHRAE 2005; Chmielnicki 1996; Shaw & Tamura 1997):

$$\Delta p_t = p_{st} - p_i = (\rho_{st} - \rho_i)g\Delta h \cong \rho g \Delta h \frac{\Delta T}{T_e} \tag{25.13}$$

where Δh = difference between height of the point in the building and height of neutral level, m; ΔT = internal and external air temperature difference, K; T_e = external air temperature, K.

The pressure difference between outside and inside the room causes air infiltration into the room. This can be determined from relation:

$$\Delta p = D\dot{V}^2 \tag{25.14}$$

where D = flow factor.

Relation (25.14) was used in calculation. After substituting equations (25.12–25.14) into (25.11) and its transformation the following relation for volume infiltrating air flow rate is obtained (ASHRAE 2005; Chmielnicki 1996; Shaw & Tamura 1997):

$$\dot{V} = L\sqrt{A\Delta t + Bw^2} \tag{25.15}$$

where L = effective gap surface, m^2; A = coefficient including the chimney effect, $(m^3/s)^2/(m^4\ K)$; B = wind factor, $(m^3/s)^2/(m^4\ (m/s)^2)$.

Coefficient A depends on the building height and varies within the range from 0.0145 to 0.04352 $(m^3/s)^2/(m^4\ K)$ (ASHRAE 2005; Shaw & Tamura 1997). Factor B depends on the number of storeys in the building and from building shield ratio, and varies from 0.00316 to 0.04946 $(m^3/s)^2/(m^4(m/s)^2)$ (ASHRAE 2005).

Accepting simplification that the building can be substituted by one room, relation (25.15) can be inserted into equation (25.10):

$$\dot{Q}_w = L\rho c_p\sqrt{A\Delta t + Bw^2}(t_i - t_e) \tag{25.16}$$

Assuming that air density and specific heat changes in connection with change of external temperature are small the following relation is acceptable $L\rho c_p = \text{const} = E$, the relation (25.16) is expressed as:

$$\dot{Q}_w = E\sqrt{A\Delta t + Bw^2}\Delta t \tag{25.17}$$

Received function can be solved by expanding in power series. It can be done in the following two ways:

- assuming that the wind action has larger influence on infiltration than the temperature difference, the following relation for heat losses related to infiltration is obtained:

$$\dot{Q}_w = a_1\Delta t w + a_2(\Delta t)^2 w^{-1} + a_3(\Delta t)^3 w^{-3} + a_4(\Delta t)^4 w^{-5} + a_5(\Delta t)^5 w^{-7} + \cdots \tag{25.18}$$

Assuming that the importance of successive series elements diminishes, series may be reduced to:

$$\dot{Q}_w = a_1\Delta t w + a_2(\Delta t)^2 w^{-1} \tag{25.19}$$

where $a_1 = E\sqrt{A}$, $a_2 = 0{,}5E\sqrt{B}A/B$

- assuming that the internal and external temperature difference has larger influence on infiltration than the wind action, the heat losses related to infiltration can be determined as:

$$\dot{Q}_w = a_1(\Delta t)^{3/2} + a_2(\Delta t)^{1/2}w^2 + a_3(\Delta t)^{-1/2}w^4 + a_4(\Delta t)^{-3/2}w^6 + \cdots \tag{25.20}$$

Assuming that the importance of successive series elements diminishes, series may be reduced to:

$$\dot{Q}_w = a_1(\Delta t)^{3/2} + a_2(\Delta t)^{1/2}w^2 \tag{25.21}$$

where $a_1 = E\sqrt{A}$, $a_2 = 0{,}5E\sqrt{A}B/A$

25.2.3 *Heat gains*

The heat gains in heated rooms can be classified in two groups – external gains caused by solar radiation and internal ones.

The solar and internal heat gains form the total heat gain:

$$\dot{Q}_z = \dot{Q}_s + \dot{Q}_i \tag{25.22}$$

The solar heat gains in heated building arise due to solar power passive utilization. The short-wave radiation is transmitted through the building walls, absorbed on wall surfaces and converted into heat that is emitted as long-wave radiation (thermal). This radiation is not transmitted back outside. The absorbed heat is scattered in the building, accumulated in massive barriers and by internal air in rooms.

The solar heat gains depend on the geographical latitude, season, building location and location of transparent barriers towards North, surfaces of these barriers, kind of glazing, building shading and solar radiation value. Solar radiation is absorbed through surfaces exposed to sunlight. The value of solar heat gains differs for various building rooms and depends on their location. The largest part of external heat gains in total heat gains occurs at the beginning and at the end of heating season, particularly in spring as a result of growing solar radiation.

The solar heat gains can be determined as:

$$Q_s = A_a KF \tag{25.23}$$

where A_a = absorption coefficient for partition surface; F = partition surface, m^2; K = average total incident radiation upon surface unit area, W/m^2.

The average solar incident radiation on the surface unit area can be obtained as follows [1, 6]:

$$K = \frac{p}{\exp(R/\sin\alpha)}(\sin\alpha + G)\left(p + r\frac{s}{s_0}\right) \tag{25.24}$$

where A_a = absorption coefficient for partition surface; F = partition surface, m^2; K = average total incident radiation upon surface unit area, W/m^2.

P, R = constants depending on the Sun – Earth distance variations during the year, the moisture content in the atmosphere and its pollution; α = sun rise head (latitude-, date- and hour-dependent); G = constant depending on the year season; s = real sunshine, h; s_0 = astronomically possible sunshine, h; p, r = regression coefficients.

The internal heat gains ($\dot{Q}_i$) for housing are the sum of heat gains from humans, cooking, warm water devices, electric devices, artificial lighting, heat transmission through walls from neighbouring rooms and heat gains resulting from differences of supplied and exhausted air temperatures (ASHRAE 2005). All components of internal heat gain values change during a day, but the average values for twenty-four hours can be accepted as constant. The values of twenty-four hours average internal heat gains are almost constant during a year, except the lighting heat gains.

25.2.4 *Change of internal energy*

The temperature distribution in a partition depends on the wall construction, the air temperatures on both sides of barrier and incident radiation. Any change of these values may cause the heat energy accumulation in partition or the energy dissipation.

Using the principle of superposition, the flux of accumulated heat or heat taken from barrier resulting from temperature changes in diurnal and long-wave cycles can be determined as follows:

$$\dot{U} = \dot{Q}_d + \dot{Q}_\tau \tag{25.25}$$

where $\dot{Q}_d$ = heat flux resulting from the temperature changes in a 24-hour cycle, W; $\dot{Q}_\tau$ = heat flux resulting from the temperature changes in a long-term cycle, W.

Assuming repeatability of the diurnal cycle, a quasi-steady state occurs in a wall after some time. At certain points of a wall, the temperature values repeat periodically each twenty-four hours and the diurnal heat balance is closed and equal to zero. The accumulated heat flux depends on the temperature difference resulting from the long-term temperature changes.

It is possible to assume that the internal temperature is constant while the external temperature changes linearly at long-term variations. The analysis of the boundary value problem solution for the above boundary conditions shows that (Carslaw & Jaeger 1959):

$$\dot{Q}_z(\tau) = [t_e(\tau = 0) - t_e(\tau)]f(\tau, z) \tag{25.26}$$

where z = temperature equalization coefficient, m^2/s; τ = time, s.

Table 25.1. Buildings basic parameters

		Building 1	Building 2	Building 3
Cubic volume	m^3	21,063	19,663	24,689
Design volumetric heating requirement factor	$W/(m^3K)$	0.38	0.57	0.60
Number of floors		6	5	5
Number of occupants		189	249	290
Number of flats		56	70	90
Number of occupants to heating cubature ratio	m^{-3}	0.0245	0.0228	0.0211
Additionally insulated gable walls		no	yes	yes

Considering the temperature changes and assuming that time is given in days, the heat flux dissipated or adsorbed by wall for "i" day can be determined by the relation:

$$\dot{Q}_\tau = \dot{U}_j(t_{e,i} - t_{e,i-1}) \tag{25.27}$$

where $\dot{U}_i$ = building accumulation coefficient, W/°C; $t_{e,I}$ = external temperature in "i" day, °C; $t_{e,i-1}$ = external temperature in "i−1" day (a day before), °C.

The coefficient of accumulation characterizes the building thermal inertia. It can be interpreted as an average heat flux that is accumulated in the external barriers of a given building at constant internal temperature and an increase of the external temperature by about one degree during a 24-hour period.

25.2.5 *Heat balance equation*

The parameters described above are included in the equation (25.5) that presents the heat balance for the building. Depending on the way of its determination, the following relation regarding the influence of all factors on building heat demand is obtained:

$$\dot{q}_{co} = \dot{q}_j(t_i = t_e) + \dot{q}_w(t_i - t_e)w + \dot{q}_i(t_i - t_e)^2w^{-1} + \dot{u}_{j,1}(t_e - t_{e,i-1}) - aK - \dot{q}_{zw} \tag{25.28}$$

where $\dot{q}_{co}$ = heat flux taken in from heating system for 1 m^3 of cubage, W/m^3; $\dot{q}_i$ = coefficient of cubage heat demand for building heating, W/(m^3K); $\dot{q}_w$ = coefficient of the influence of wind on the amount of air infiltrating into the room; $\dot{q}_i$ = coefficient of the influence of temperatures difference on both sides of the partition on the amount of air infiltrating into the room; $\dot{q}_{zw}$ = average internal heating gains, W/m^3; $\dot{u}_{j,1}$ = accumulation coefficient of the influence of external temperature that occurred a day before, W/(m^3K).

25.3 CHARACTERISTIC OF TESTED OBJECTS

The heat consumption measurements (indispensable for thermal power demand determination) were carried out in three apartment buildings located in Szczecin. The buildings were constructed in the 1970s and 1980s. Two of them, were built in the Wk–70 system, with five floors and additional insulation of the gable walls. The third six floors building was built in so-called co-operative system. Objects are supplied with heat from the urban heat distribution network. The systems are connected with the heat distribution network by heat exchangers in thermal stations. Those buildings are large heat capacity objects. The characteristics of these buildings are shown in Table 25.1.

25.4 HEAT CONSUMPTION MEASUREMENTS

The aims of measurements were as follows:

- step one – to determine the daily heat consumption of a building,
- step two – to analyze the determined consumption variations with regard to the weather conditions, in order to identify the parameters characterizing the real thermal needs of chosen objects.

Therefore, the daily readings of heat meters were performed in building thermal stations during two heating seasons. The readings were performed every 24 – hours using electronic heat meters installed in thermal stations. The accuracy of the heat meters was 2.5%. On this basis, the unit heat consumption was determined [W/m^3] for particular days. The weather conditions data, i.e. average diurnal external temperature, average diurnal wind velocity and daily sunshine, originating from weather station in Szczecin – Dąbie, were received from the thermal emergency service. The accuracy of the external temperature measurement was about ± 1°C.

The building internal temperature was additionally measured over the second year of tests. The continuous internal temperature records were carried out using thermographs in randomly chosen flats of each building. In order to find the mean diurnal internal temperature, the value the hourly readings were performed and averaged.

25.5 REAL BUILDING'S HEAT DEMANDS

Using the aforementioned heat balance and the measurements results, the real heat demands for the buildings as well as the parameters affecting the heat consumption were determined. The calculations of the parameters characterizing the heat demands of objects were carried out using multiple regression analysis. The STATISTICATM software package was used for calculations.

Different forms of heat balance can be analysed. These balances considered the influence of the following elements on the heat consumption:

a) variable external temperature,

$$\dot{q}_{co} = \dot{q}_j(t_i - t_e) - \dot{q}_z \tag{25.29}$$

b) variable external temperature and variable wind velocity,

$$\dot{q}_{co} = \dot{q}_j(t_i - t_e) + \dot{q}_w(t_i - t_e)w + \dot{q}_i(t_i - t_e)^2 w^{-1} - \dot{q}_z \tag{25.30}$$

c) variable external temperature and heat accumulation in partitions,

$$\dot{q}_{co} = \dot{q}_j(t_i - t_e) + \dot{u}_{j,1}(t_e - t_{e,i-1}) - \dot{q}_z \tag{25.31}$$

d) ariable external temperature, variable wind velocity and heat accumulation in partitions,

$$\dot{q}_{co} = \dot{q}_j(t_i - t_e) + \dot{q}_w(t_i - t_e)w + \dot{q}_i(t_i - t_e)^2 w^{-1}\dot{u}_{j,1}(t_e - t_{e,i-1}) - \dot{q}_z \tag{25.32}$$

e) variable external temperature and variable solar radiation,

$$\dot{q}_{co} = \dot{q}_j(t_i - t_e) - aK - \dot{q}_z \tag{25.33}$$

f) all considered elements – heat accumulation in partitions, variable external temperature, wind velocity and solar radiation (25.28).

Table 25.2 shows some equations and values of correlation coefficients received for one of the buildings (building 2) for the 2nd heating season for all examined relations. For the next two objects and for the 1st heating season the received correlation coefficients were similar.

The analysis shows that these factors have various influence on the heat demand. The crucial one is external temperature. The other parameters cause a small increase of the correlation coefficient or are statistically insignificant. Hence, in order to determine the heat characteristics of a building, the

Table 25.2. Equations and values of correlation coefficients obtained for examined forms of heat balance for the building 2 and 2nd heating season.

Equation	Correlation coefficient
$\dot{q}_{co} = 0{,}479\,(t_i - t_e) - 1{,}692$	0.9317
$\dot{q}_{co} = 0{,}481\,(t_i - t_e) + 0{,}006\,(t_i - t_e)\,w - 0{,}001\,(t_i - t_e)^2\,w^{-1} - 2{,}079$	0.9396
$\dot{q}_{co} = 0{,}399\,(t_i - t_e) + 0{,}094\,(t_e - t_{e,i-1}) - 1{,}927$	0.9376
$\dot{q}_{co} = 0{,}501\,(t_i - t_e) + 0{,}006\,(t_i - t_e)\,w - 0{,}002\,(t_i - t_e)^2\,w^{-1} + 0{,}093\,(t_e - t_{e,i-1}) - 2{,}333$	0.9452
$\dot{q}_{co} = 0{,}467\,(t_i - t_e) - 0{,}002K - 1{,}107$	0.9398
$\dot{q}_{co} = 0{,}481\,(t_i - t_e) + 0{,}006\,(t_i - t_e)\,w - 0{,}001\,(t_i - t_e)^2\,w^{-1} + 0{,}094\,(t_e - t_{e,i-1}) - 0{,}002K - 1{,}717$	0.9524

Table 25.3. Values of heat balance (25.29) coefficients obtained for the buildings

	$\dot{q}_j$ W/m^{3}K	$\dot{q}_z$ W/m^3	Correlation coefficient
	Season 1st		
BUILDING 1	0.306	−1.696	0.9064
BUILDING 2	0.448	−0.940	0.9646
BUILDING 3	0.450	−1.540	0.9505
	Season 2nd		
BUILDING 1	0.301	−0.532	0.9055
BUILDING 2	0.479	1.692	0.9317
BUILDING 3	0.465	1.185	0.9269

simplest form of heat balance (25.29) considering only the difference between internal and external temperatures was adopted.

The values of individual coefficients from equation (25.29) are shown in Table 25.3.

The values of correlation coefficients for the received relations are satisfactory. They are higher for 1st season. This likely results from the fact that in this season the external temperatures were lower and the relation between heat consumption and external temperature is more distinct. The values of t statistics for particular coefficients are higher than critical values $t_{\alpha/2,f}$ which proves their statistical significance.

The values of correlation coefficients for the received relations are satisfactory. They are higher for 1st season. This likely results from the fact that in this season the external temperatures were lower and the relation between heat consumption and external temperature is more distinct. The values of t statistics for particular coefficients are higher than critical values $t_{\alpha/2,f}$ which proves their statistical significance.

The values of heat gains received for 2nd season for Building 2 and Building 3 are somewhat smaller than the theoretical values presented in literature and concerning the Polish conditions (Leciej 1997). In other cases, the values of these gains were negative. At derivation, the formula for 1st season the internal temperature was assumed as equal to 20°C and negative heat gains indicate that the temperature in rooms was higher. The negative value for 2nd season for Building 1 can be caused by the lack of thermostatic valves in the installation and the rooms, where the internal temperature measurements were not performed, could be overheated.

Figure 25.1 shows unit heat demand for 2nd season in the external and internal temperature difference function for all buildings with 95% confidence interval.

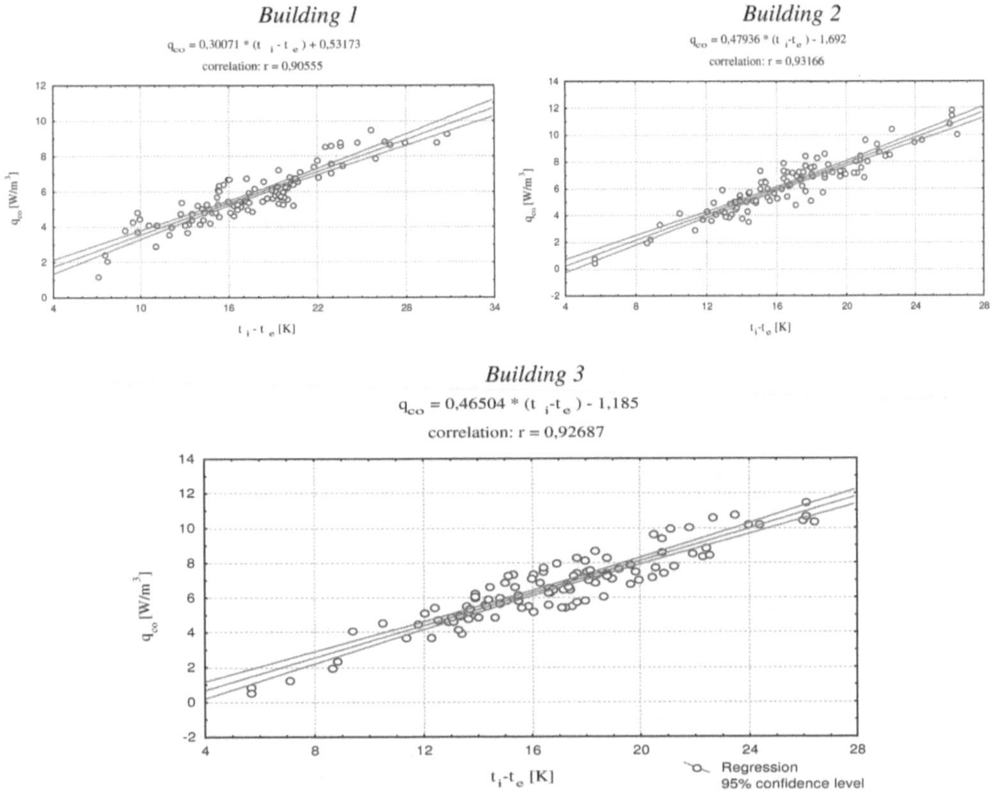

Figure 25.1. Unit heat demand in external and internal temperature difference function.

25.6 THE INFLUENCE OF DURATION AND FREQUENCY OF MEASUREMENTS ON THE OBTAINED RESULTS

Performing daily heat consumption measurements and data analysis for every heated object over seven or eight months during a year is impossible under real conditions (Alzetto et al. 2018). Therefore, the next stage of investigations was to determine the influence of measurements duration on the accuracy of the obtained results and to define its optimal frequency.

The heat balance coefficients value analysis was performed for all three buildings at various number of records. It turned out that for short periods (e.g. a week, fortnight) the values of the analyzed coefficients differ even by about 100%. For longer periods those deviations are smaller (20–30%), but those results are unsatisfactory. This may appear from the temperature changes in rooms, accuracy of measurements and building heat storage, the more so as the relation of heat capacity on 1 m^2 of external barriers to average heat-transfer coefficient k for examined objects is rather high (about 600 000 s) which means that at 1K difference between external and internal temperature and lack of heating, it takes about 7 days for the heat dissipation from building till the moment of temperatures equalization.

The Table 25.4 presents values of mean cubage heat demand coefficient obtained for one, up to eight following weeks in 2nd season.

Next, the parameters describing heat demands were determined assuming heat consumption during different number of days for averaged weather conditions. The calculations were based on the simplest form of heat balance (25.29). The analysis was carried out for weekly periods, up to eight weeks periods. For this purpose, the average internal and external temperatures for consecutive weeks (up to eight weeks) as well as the average heat consumption in these periods were determined. Then, the values of

Table 25.4. Values of average cubage heat demand coefficient and their mean deviation for different measuring periods in 2nd season for particular buildings.

Measuring period	Building 1 average cubage heat demand coefficient W/m³K	Building 1 mean deviation	Building 2 average cubage heat demand coefficient, W/m³K	Building 2 mean deviation	Building 3 average cubage heat demand coefficient, W/m³K	Building 3 mean deviation
1 Week	0.164	0.210	0.306	0.070	0.267	0.140
2 Weeks	0.192	0.063	0.355	0.060	0.316	0.078
3 Weeks	0.222	0.059	0.393	0.039	0.359	0.053
4 Weeks	0.238	0.050	0.422	0.032	0.386	0.045
5 Weeks	0.257	0.046	0.439	0.023	0.406	0.039
6 Weeks	0.270	0.036	0.445	0.025	0.416	0.032
7 Weeks	0.280	0.029	0.451	0.019	0.421	0.027
8 Weeks	0.292	0.013	0.455	0.009	0.423	0.022

Table 25.5. Values of cubage heat demand index and coefficient and heat gains for Building 1 for average weather conditions in 2nd season.

Average values for	Average total heat gains $\dot{q}_z$ W/m³	Cubage heat demand coefficient $\dot{q}_j$ W/m³K	Cubage heat demand index in 2nd season: Using measurement periods W/m³	Cubage heat demand index in 2nd season: Average annual W/m³
Day	−0.532	0.300	5.88	
Week	0.052	0.335	5.92	
2 Weeks	0.110	0.339	5.93	
3 Weeks	0.469	0.360	5.95	
4 Weeks	0.722	0.373	5.92	5.92
5 Weeks	0.860	0.380	5.91	
6 Weeks	0.697	0.372	5.93	
7 Weeks	0.679	0.370	5.91	
8 Weeks	0.581	0.365	5.92	

heat balance coefficients were determined using the regression analysis. These values were used for the calculation of cubage heat demand indices. The obtained results are presented in Tables 25.5–25.7.

The differences between the values of cubage heat demand coefficient and heat gains determined for average temperatures decrease as the period between measurements lengthens.

In the next phase of calculations, the thermal characteristics of the buildings were determined once again. This time, the heat consumption during different number of days for averaging weather conditions was taken into account for this purpose and the temperatures for calculations were taken as equal to the average temperatures for heating season. For 2nd season, average external temperature was 4.2°C and average internal temperature −21.68°C and for Building 1–22.02°C.

After the analysis of the obtained results, it can be observed that the values of cubage heat demand indices for different periods of measurements differ from the average in less than by 1%, which may prove the correctness of applied method.

Reading heat meters records every 4 weeks seems to be an optimal period for determining the thermal characteristics of buildings. The difference between average year value and value determinate

Table 25.6. Values of cubage heat demand index and coefficient and heat gains for Building 2 for average weather conditions in 2nd season.

Average values for	Average total heat gains $\dot{q}_z$ W/m^3	Cubage heat demand coefficient $\dot{q}_j$ W/m^{3}K	Cubage heat demand index in 2nd season: Using measurement periods W/m^3	Cubage heat demand index in 2nd season: Average annual W/m^3
Day	1.691	0.479	6.68	
Week	3.218	0.569	6.73	
2 Weeks	3.884	0.609	6.76	
3 Weeks	4.039	0.619	6.78	
4 Weeks	4.763	0.659	6.76	6.78
5 Weeks	5.935	0.725	6.74	
6 Weeks	6.094	0.737	6.79	
7 Weeks	4.947	0.671	6.78	
8 Weeks	4.858	0.666	6.78	

Table 25.7. Values of cubage heat demand index and coefficient and heat gains for Building 3 for average weather conditions in 2nd season.

Average values for	Average total heat gains $\dot{q}_z$ W/m^3	Cubage heat demand coefficient $\dot{q}_j$ W/m^{3}K	Cubage heat demand index in 2nd season: Using measurement periods W/m^3	Cubage heat demand index in 2nd season: Average annual W/m^3
Day	1.185	0.465	6.94	
Week	2.388	0.537	7.00	
2 Weeks	2.879	0.566	7.01	
3 Weeks	2.852	0.563	6.99	
4 Weeks	3.499	0.601	7.01	7.02
5 Weeks	3.999	0.629	7.00	
6 Weeks	4.069	0.634	7.01	
7 Weeks	3.589	0.607	7.02	
8 Weeks	3.460	0.599	7.01	

for 4 weeks period is about 0.3%. At such small difference, carrying out the readings every 4 weeks it is not troublesome yet.

25.7 BUILDINGS HEAT DEMANDS DETERMINATION FOR DESIGN EXTERNAL TEMPERATURES

The presented method of cubage heat demand index determination can also be used for design cubage heat demand index evaluation This is widely applied coefficient enabling determination of the thermal needs for heated objects in a simple way. It defines the amount of heat [W] that should be supplied in unit time to 1 m^3 of object cubature at difference of temperatures: internal - suitable for building destination and external – adopted due to climatic zone. The values of design cubage heat demand index obtained for the investigated buildings are shown in Table 25.8.

Table 25.8 also presents the average value of design cubage heat demand index obtained after rejecting the one-week values – deviating the most from the rest. The values obtained from readings

Table 25.8. Values of design cubage heat demand index obtained for investigated buildings for averaging weather conditions in 2nd season.

Averaging values for	Building 1 W/m^3	Building 2 W/m^3	Building 3 W/m^3
Week	12.008	17.266	16.944
2 Weeks	12.094	18.040	17.497
3 Weeks	12.464	18.245	17.416
4 Weeks	12.706	18.961	18.137
5 Weeks	12.820	20.165	18.645
6 Weeks	12.692	20.438	18.755
7 Weeks	12.641	19.209	18.263
8 Weeks	12.559	19.118	18.104
Average from 2-8 weeks	12.568	19.168	18.117

performed every 4 weeks are close to average value. The maximum deviation equals approximately 1%, what which confirm the correct choice of the measurements period.

25.8 CONCLUSIONS

The paper presents a method of determining the heat demand in apartment buildings based upon the heat consumption measurements. In order to develop this method, the values of everyday measurements were used. The influence of the length of measurement period and the size of intervals between measurements on the results was analyzed.

Short periods of everyday measurements cause a large value of deviations from the average annual values. If a period of seven or eight weeks is considered, the deviation from average annual value is about 20–30%. Performing measurements over more than 50 days seems to be too troublesome. The results that are based on this laborious collected data, can be biased with relatively large error.

The second proposed way, based on lengthening the period between measurements and averaging external temperatures during this period, seems to be better. The thermal characteristics for three buildings were determined using this method. The differences between the characteristics for the building, for the periods between measurements that lasted longer than a week, were small (approximately 0.5 %). The performed analysis shows that the intervals between measurements lasting 4 weeks (1 month) produce results similar to average values. This period corresponds with standards of measurements for the purpose of settlement with heat receivers. When reading the heat meters every 4 weeks (which gives about 7–10 readings during a season), the maximum deviations from the average value are approximately 1%.

REFERENCES

Alzetto F., Pandraud G., Fitton R., Heusler I., Sinnesbichter H. 2018. QUB: A fast dynamic method for in-situ measurement of the whole building heat loss. *Energy and Buildings* vol.174, 124–133.

ASHRAE 2005. *Handbook, Fundamentals, SI edition, American Society of Heating, Refrigerating and Air Conditioning Engineers*.

Carslaw H. S. & Jaeger J. C. 1959. Conduction of heat solid. *At the Clereadon Press*, Oxford.

Chmielnicki W.J. 1996. Sterowanie mocą w budynkach zasilanych z centralnych źródeł ciepła. *Instytut Podstawowych Problemów Techniki PAN*, Warszawa.

Leciej D. 1997. Heat gains in apartment buildings. *Inf. Istal*, vol. 12, in polish.

Salakij S., Yu N., Paducci S., Antsaklis P. 2016. Model-based predictive control for building energy management. *Energy and Buildings* vol.133, 345–358.

Shaw C. Y. & Tamura G. T. 1997. The calculation of air infiltration rates caused by wind and stack action for tall buildings, *ASHRAE Transactions* 83 (2), 145.

CHAPTER 26

Real temperature field and numerical calculations in a solar assisted ground source heat pump system

P. Rynkowski
Bialystok University of Technology, Poland

ABSTRACT: In the paper, the numerical and experimental studies were performed for a solar assisted ground source heat pump system (SGSHPS) with vertical ground heat exchanger (VGHE). The objective of this study was to show the change in temperature distribution over the time in the vicinity of the borehole in SGSHPS. One of the VGHE boreholes and five control probes were used in experiment during the summer season in 2018. The heat extraction rate from solar collectors, by the water tank to the ground were monitored. The boundary element method (BEM) and line source model (LSM) were used to calculate the disturbed temperature profile. The simulating of SGSHPS operation and the numerical results were analysed. The results indicate relatively high convergence between the numerical calculation and experimental data. The experimental studies showed that the ground was effectively regenerated by the solar radiation. The temperature of the control boreholes in the vicinity of VGHE increased markedly, up to 0.5°C over undisturbed temperature in the soil, after the summer season.

26.1 INTRODUCTION

Heat pumps are nowadays used for heating and cooling in many kinds of building. In the countries with a cool climate, the ground source heat pump system (GSHPS) is well-known and popular. Against the background of saving fossil fuels and developing a variety of new energy sources, it is regarded as a new highly-efficient and energy saving air-conditioning technology, which employs the underground energy storage (Deng et al. 2012). In order to use these GSHPS, it is necessary to accurately predict the heat extraction and injection rates of the ground heat exchangers (GHEs) (Nam et al. 2008). In Poland, the vertical ground heat exchangers (VGHEs) are very popular for extracting the heat from the ground. Ground is a suitable source of heat for heating system based on heat pump (America Society of Heating Refrigerating and Air-Conditioning Engineers 2015). VGHE is a very important part of GSHPS. In order to ensure the proper operation of GSHPS, it is necessary to find the heat transfer efficiency of the VGHE. The boreholes with the heat exchangers must meet the performance and environmental requirements. For this reason, many VGHE studies have appeared in recent years. Most of them concentrate on the heat transfer between the U pipe and the surrounding soil. Understanding this heat transfer is essential for calculating the required length of the GHE, both vertical and horizontal. The vertical ground heat exchanger has near constant flow temperature in short period with small changeable flow temperature during the heating season. During the heat pump operation, the temperature of the ground fall or rise around the heat exchanger as the boreholes might be used for both heating and cooling. The radial ground cooling occurs in the distance of a few meters around the borehole (Fidorow et al. 2015). The active soil regeneration during summer time is crucial for the coefficient of performance (COP) of GSHPS. For this reason the integration of solar system and geothermal energy, especially in cold climate is being analyzed. The engineering project the integration of solar system and geothermal energy in cold climate was presented in (Wang et al. 2011).

The heating systems with solar energy and ground source heat pump system may reduce the conventional energy sources and, what is important, improve the efficiency of SGSHPS. According to

DOI 10.1201/9781003171669-26

the Emmi et al. (2015) solar thermal collectors can balance the ground loads over a yearly cycle and assist in maintaining more efficient heat pumps. Experimental studies on a solar assisted heat pump water heating system were presented by Yang et al. (2015). Pärisch et al. (2014) investigated the heat pump behavior under non-standard conditions for the operation of SGSHPS in combination with solar collectors. The higher temperature and varying flow rates were analysed in comparison to non-solar systems. Some of the parameters influencing the energy efficiency of a novel solar-assisted absorption ground-coupled heat pump in its cooling mode operation were presented by Rad et al. (2013). Experimental investigation of solar assisted ground source heat pump system was carried out from morning to evening for the winter conditions in (Verma et al. 2018). With the heat input from solar collector and ground heat exchanger, the heat pump was able to absorb heat at temperature 5°C higher than the ambient temperature. Samii et al. (2016) indicate the effect of solar heat energy on the performance the collector and the heat pump individually to be majorly significant.

In addition to the experimental research, many analytical and numerical methods have been proposed for analysing the heat transfer in GHEs. The most widely used analytical tools for analysing heat transfer in GHEs are Kelvin's theory of heat sources and the Laplace transform method (Ingersoll et al. 1954). One of the first was the cylindrical heat source theory proposed by Carslaw and Jaeger (1946). The theory is based on a long isolated pipe surrounded by an infinite solid of constant properties. This model is simple to use and to understand. It was used by many authors. The review of the modelling approaches was performed in (Atam et al. 2016, part 1 and 2). Thermal response factor models represent a group of techniques based on infinite or finite line source theory. The concept of a special functions, as "g-function" as was introduced by Eskilson (1987). A "g-function" is defined as a non-dimensional temperature response factor which relates the borehole wall temperature and the heat rate per borehole length from the ground, through the ground thermal diffusivity. There are many approaches to deriving the g-functions: numerical, analytical and combination of numerical and analytical methods. A detailed description of the analytical models for heat transfer by vertical ground heat exchangers was presented in (Min et al. 2015). The authors explain that analytical methods and models appear to be more useful than the numerical methods for advancing the GHE technology. Currently, the commercial numerical simulation software is widely used for this purpose (Guan et al. 2017). The finite-difference method (Lee 2011), finite volume method (Rees et al. 2013) and finite-element methods (Wołoszyn et al. 2013) are used to describe the heat transfer process in VGHE. However, these methods are not practical for the engineering applications (Min et al. 2015). Although they are well established and commonly applied to the transfer heat analysis, in many problems, mesh generation can be very laborious and constitutes the most expensive and difficult part of numerical simulations (Werner-Juszczuk et al. 2012).

The alternative for the above-mentioned mesh and mesh free methods is the boundary element method (BEM). The BEM is a method for solving partial differential equations by reformulating as boundary integral equations and then solving them. The Boundary Element Method is often presented as the Boundary Integral Method (BIM) or Boundary Integral Equation Method (BIEM). The great advantage of BEM is the possibility of determining the solution at any point of the domain without the necessity of constructing grids in the considered 2D or 3D space. The discretization is performed only over the boundary. The size of equations to be solved is reduced by one. BEM is successfully applied to the steady and unsteady heat conduction problems (Katsikadelis 2002). The application of BEM requires the knowledge of fundamental solution, which is often treated as a disadvantage of the method, but it stabilizes the numerical solution (Ochiai et al. 2006). Poland has a heating-dominated climate. The north-eastern part of Poland has a colder climate compared to the western and southern Poland. As a result, the heating season is longer and the GSHPS performance can become less efficient (Bakirci et al. 2010). This problem is obvious especially in cold regions, where the heat extracted during the heating season is greater than recovered by the soil and the Sun in the summer time (Fan et al. 2008). Furthermore, the heating and cooling systems based on heat-pump are expected to be reliable for 20 years, while the buildings are designed for 50 years (Bakirci et al. 2011).

The borehole forms the temperature funnel and the global ground exchangers area cools down. There is no special method to check and make the ground thermal balance stable. In a climate like Polish, there is little sunshine during the winter season, however, there is an excess of solar radiation in the summer season. Most often, it is not used. The GSHPS performance may become more efficient using solar thermal energy storage system (Yang et al. 2010). The seasonal storage of solar thermal energy coupling with a heat pump has been the subject of many investigations. The seasonal storage of thermal energy was proposed in the U.S. during the 1960s. In late 1970s, researchers in the north European countries also began investigating the seasonal solar thermal energy storage systems (Sanner et al. 2003). The idea of SGSHPS, which combined the use of solar collector and ground heat exchanger (GHE), enabling the excess solar energy collected during the day time to be saved in the ground by the GHE, was firstly put forward by Penrod et al. (1962). Penrod performed a detailed study on the problem of soil as a heat source. The temperature the undisturbed and disturbed soil, was measured and compared. A complete procedure for designing the proposed SGSHPS as introduced by (Rynkowski 2019).

26.2 SYSTEM DESCRIPTION

The paper presents the experimental studies on the performance of the SGSHPS. The experimental platform was installed in Bialystok University of Technology. Although, in Poland there is a temperate climate with a transitional character between sea and land, Bialystok has a continental climate with harsh winters and hot, dry summers. The climatic indicators show a continental character this part of the country, expressed in large amplitudes of annual temperatures, the lowest average annual temperatures, comparable to mountain areas, and the longest periods of snow cover deposition. The solar collectors were coupled with ground vertical heat exchangers through a water tank. The SGSHPS operates as two independent systems. The first comprises the SGSHPS system, as ground thermal storage system with solar collectors as main components, whereas the second consists of GSHPS with VGHE. The exact description of the SGSHPS system with measurement devices is presented by the author in (Rynkowski 2019). During the SGSHPS operation, the GSHPS does not operate in the spring and summer time. The main SGSHPS components: heat meters from solar collectors, buffer container and heat exchanger were shown in Figure 26.1. The solar collectors were coupled with ground vertical ground heat exchangers through a water tank. The types of solar collectors are: flat type about total absorber area 2 × 2.32 m^2 and heat-pipe type about 2 × 3.03 m^2 total absorber area. The efficiency of solar collectors declared by the manufacturer are 86.3% and 81.3%, respectively. By the heat exchanger (Figure 26.1) the heat from buffer container is transferred by VGHE to the ground. The system has three VGHEs and three control probes. Each of them is 100 m long. During the experiment system, only VGHE No. 2 was operating (Figure 26.2). Two other VGHEs, i.e. No. 1 and No. 3, were working as control probes, combined with others control probes – No. 4, 5 and 6.

The location of the VGHEs and the control probes was shown in Figure 26.2. The distances between the heat source of energy, from VGHE No. 2 to the control points that are VGHE no. 1 and no. 3 and control probes no. 4, 5 and 6 are:

- to VGHE no. 1 and no. 3–5 m,
- to control probe no. 4–10 m,
- to control probe no. 5–2.2 m,
- to control probe no. 6–4.4 m (Figure 26.3).

The system was operating from April 4 to November 12, 2018. The quantity of heat transferred to ground during this time was calculated from (26.1):

$$E_{VGHE} = \left[c_p \sum_{i=1}^{N} m_i \left(T_{i,in} - T_{i,out} \right) t_i \right] / \left(3{,}6 \cdot 10^6\right) \quad [\mathrm{kWh}] \qquad (26.1)$$

where c_p is specific heat at constant pressure of the glycol, J/(kgK); m is mass flow rate, kg/s; t is operation time (in the study every five minutes) and **i** is operation period.

The total quantity of heat transferred to ground was equal to 5 785.3 kWh. The average transfer factor from solar radiation to soil in SGSHPS in 2018 was about 42.2%. The variation in temperatures throughout the system operation at the inlet to VGHE was within the limits of 22–28°C. The mean inlet

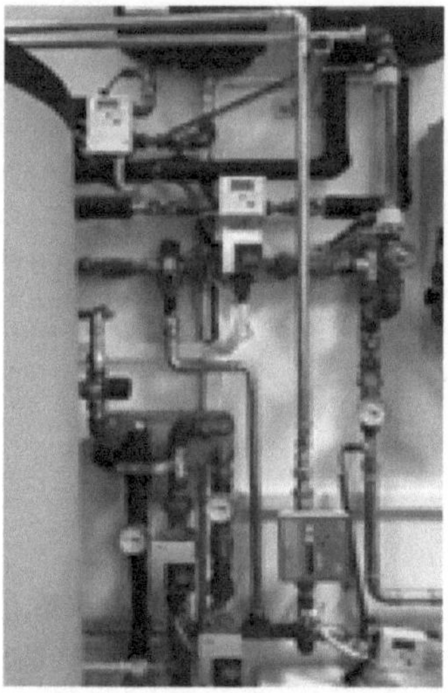

Figure 26.1. The main SGSHPS components: heat meters from solar collectors, buffer container and heat exchanger. exchanger.

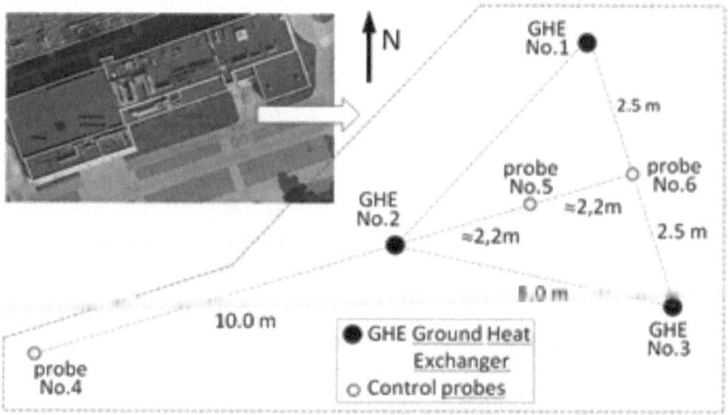

Figure 26.2. Location the VGHEs and control probes at the SGSHP system at Bialystok University ofTechnology.

temperature to VGHE No. 2 during experiment was is about 25°C. The average temperature difference between the supply and return was about 7°C. In the later part of the season, the temperatures were considerably lower, due to the lower solar radiation. During the system stagnation, the inlet and outlet temperature were close. The daily thermal storage was determined by the weather conditions.

26.3 A BRIEF OF BOUNDARY ELEMENT METHOD

The thermal processes, in which the heat conduction is the main mechanism, are described by Fourier-Kirchhoff equation. The unsteady heat conduction in homogeneous solid substance with constant material properties without inner heat sources, is described by the heat conduction equation:

$$\nabla^2 - \frac{1}{\alpha}\frac{\partial}{\partial t} T(x, y, t) = 0 \qquad (26.2)$$

where $\alpha = \lambda/c$ is the thermal diffusivity, in which λ is the thermal conductivity and c is the volumetric specific heat and $\partial/\partial t$ is the local time derivative. The boundary conditions (26.3a, 26.3b) and initial condition (26.3c) are as follows:

$$T(x,y,t) = T_L(x,y,t), \quad (x,y) \in L_q \tag{26.3a}$$

$$q(x,y,t) = -\lambda \frac{\partial T(x,y,t)}{\partial n}, \quad (x,y) \in L_T \tag{26.3b}$$

$$T(x,y,0) = T_0(x,y), \quad (x,y) \in \Lambda \tag{26.3c}$$

The boundary conditions (26.3a) is known as Dirichlet boundary condition (the value of temperature on boundary Lq), (26.3b) – Neumann boundary condition – the value of heat flux on boundary L_T, (26.3c) – the initial condition inside the domain (Λ) at initial time T_0.

Assuming that in a two dimensional area (Λ) bound by the edge line (Figure 26.3), part of the line is described by the boundary condition (26.3a), while the second part of the line is described by the boundary condition (26.3b), an integral equation describing the transient temperature has the general solution of the integral form as follows (Brebbia et al. 1984):

$$\begin{gathered}\chi(\mathbf{p})T(\mathbf{p},t) + \int_{t_0}^{t}\int_{(\Lambda_g)} T(\mathbf{q},\tau)\,F_Q(\mathbf{p},t;\mathbf{q},\tau)dL_{gq}d\tau + \sigma\int_{t_0}^{t}\int_{(\Lambda_f)} q(\mathbf{q},\tau)\,F_T(\mathbf{p},t;\mathbf{q},\tau)dL_{fq}d\tau \\ = \tilde{T}_L(\mathbf{p},t) + \tilde{q}_L(\mathbf{p},t) + \tilde{T}_{\Lambda 0}(\mathbf{p},t_0) + \tilde{q}_\upsilon(\mathbf{p},t)\end{gathered} \tag{26.4}$$

where p and q are source and field points, respectively, within the domain (Λ) or on the boundary (Figure 26.3) with t_0 and t as the analysed time interval. Coefficient χ (p) is related to the local geometry of the boundary at point (p). The fundamental solution of heat conductivity equation (26.2), also called Green function for heat equation, and its normal derivative for two dimensional problems are given by:

$$\begin{aligned}F_T(\mathbf{p},t;\mathbf{q},\tau) &= \frac{1}{4\pi a\,(t-\tau)}\exp\left(\frac{-(r_{pq})^2}{4a\,(t-\tau)}\right)\\ F_Q(\mathbf{p},t;\mathbf{q},\tau) &= \frac{a d_{pq}}{8\pi[a\,(t-\tau)]^2}\exp\left(\frac{-(r_{pq})^2}{4a\,(t-\tau)}\right)\end{aligned} \tag{26.5}$$

and:

$$\begin{aligned}\tilde{T}_L(\mathbf{p},t) &= \int_{t_0}^{t}\int_{(\Lambda_f)} T(\mathbf{q},\tau)F_Q(\mathbf{p},t;\mathbf{q},\tau)dL_{fq}d\tau\\ \tilde{q}_L(\mathbf{p},t) &= \sigma\int_{t_0}^{t}\int_{(\Lambda_g)} q(\mathbf{q},\tau)\,F_T(\mathbf{p},t;\mathbf{q},\tau)dL_{gq}d\tau\end{aligned} \tag{26.6}$$

$$\begin{aligned}\tilde{T}_{\Lambda 0}(\mathbf{p},t_0) &= \iint_{\Lambda} T(\mathbf{v},t_0)\,F_T(\mathbf{p},t;\mathbf{v},t_0)d\Lambda_v\\ \tilde{q}_\upsilon(\mathbf{p},t_0) &= \sigma\int_{t_0}^{t}\iint_{(\Lambda)} q_\upsilon(\mathbf{v})F_T(\mathbf{p},t;\mathbf{v},\tau)d\Lambda_v d\tau\end{aligned} \tag{26.7}$$

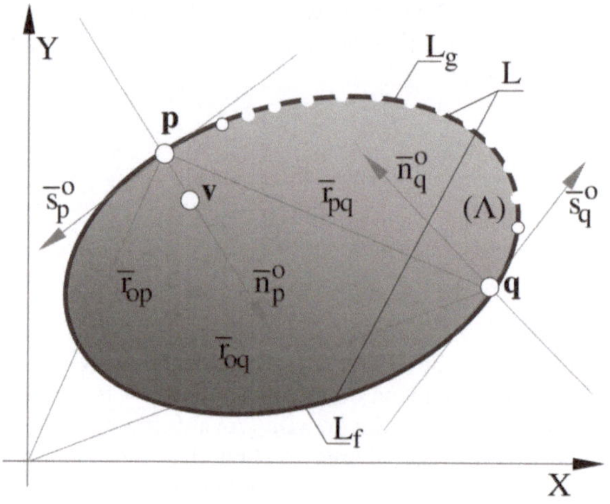

Figure 26.3. Sketch for the two dimensional boundary problem analysis of Boundary Element Method.

Functions $F_T(\mathbf{p}, t; \mathbf{v}, t_0)$ i $F_T(\mathbf{p}, t; \mathbf{v}, \tau)$ are equal to:

$$F_T(\mathbf{p}, t; \mathbf{v}, t_0) = \frac{1}{4\pi a\,(t - t_0)} \exp\left(\frac{-(r_{pv})^2}{4a\,(t - t_0)}\right)$$

$$F_T(\mathbf{p}, t; \mathbf{v}, \tau) = \frac{1}{4\pi a\,(t - \tau)} \exp\left(\frac{-(r_{pv})^2}{4a\,(t - \tau)}\right) \tag{26.8}$$

After determining the unknown values $T(\mathbf{p},t)$ and $q(\mathbf{p},t)$ on parts of the shoreline (L_q) and (L_f), by solving the integral equation (26.4), the temperature $T(\mathbf{u}, t)$ at any point ($\mathbf{u}$) of the flat area (Λ) can be determined from the relationship (26.9):

$$\begin{aligned} T(\mathbf{u}, t) = & \int_{t_0}^{t} \oint_{(L)} T(\mathbf{q}, \tau)\, F_Q(\mathbf{u}, t; \mathbf{q}, \tau) dL_q d\tau + \sigma \int_{t_0}^{t} \oint_{(L)} q(\mathbf{q}, \tau)\, F_T(\mathbf{u}, t; \mathbf{q}, \tau) dL_q d\tau \\ & + \iint_{(\Lambda)} T(\mathbf{v}, t_0) F_T(\mathbf{u}, t; \mathbf{v}, t_0) d\Lambda_v + \sigma \int_{t_0}^{t} \iint_{(\Lambda)} q_v(\mathbf{v}) F_T(\mathbf{u}, t; \mathbf{v}, \tau) d\Lambda_v d\tau \end{aligned} \tag{26.9}$$

The numerical solution of integral equations describing the unsteady the transient heat conduction in two-dimensional space is carried out by replacing the shoreline with an appropriate system of partial lines and integrating the equation on individual elements.

26.4 INFINITE LINE SOURCE ANALYSIS

According to the infinite line source model (Carslaw et al. 1946), an infinite line source with a constant heat rate per unit length q in an infinite medium produces a temperature increase depending on the radial distance and on time as follows:

$$T(r, t) - T_0 = \frac{q}{4\pi\lambda} E_i\left[\frac{r^2}{4\alpha t}\right] = \frac{q}{4\pi\lambda} \int_{\frac{r^2}{4\alpha t}}^{\infty} \frac{e^{-u}}{u} du \tag{26.10}$$

whereT_0 is the undisturbed temperature, λ and α are the medium thermal conductivity and thermal diffusivity, respectively, and E_i is the exponential integral function. For condition (26.11):

$$\frac{\alpha t}{r^2} > 5 \tag{26.11}$$

equation (26.10) becomes:

$$T(r,t) - T_0 \cong \frac{q}{4\pi\lambda}\left[\ln\left(\frac{4\alpha t}{r^2}\right) - \gamma\right] \tag{26.12}$$

Equation (26.12) can be used to evaluate the temperature increase at the borehole wall (Angelotti et al. 2018).

26.5 NUMERICAL AND EXPERIMENTAL DATA ANALYSIS

In this section, The 2D unsteady heat transfer problem was considered to show the accuracy and performance of the present SGSHPS. Firstly, BEM was used to analyse the unsteady heat transfer problem in the vicinity of the VGHE. The author's program has been written in the Fortran programming language to solve the problem. Secondly, the infinite line source model was used.

Starting from the information and considering the reference thermal–physical properties of soil, the average thermal conductivity and thermal diffusivity were estimated as 1.0 W/(mK) and $1.17{\cdot}10^{-6}$ m^2/s for soil, for PEX pipe 0.35 W/(mK) and for bentonite, as filing, 2.0 W/(mK). The bore diameter is 150 mm. The diameter of each pipe is 40 mm. According to the citied equation (26.2, 26.4) the problem was limited to the heat conduction problem. For the initial condition, the ground temperature was assumed on the basis of previous findings, as 8°C. The Dirichlet boundary condition at the edge of the area was set to 8°C. The boundary condition on the internal boundary lines of the supply and return pipes were obtained from the experiment. The weighted average temperatures during the whole period of the system operation, on the inlet and outlet, were assumed as 12.93°C and 9.93°C, respectively.

For the needs of the line source model, the average heat flux reaching the ground was obtained from the quantity of heat transferred to ground (1) and the length VGHE No. 2. The average heat flux reaching the ground was equal 9.4 W/m.

26.5.1 *Boundary Element Method for temperature field*

Figure 26.4 shows the mesh of domains with U-tube configuration used in numerical modeling for the needs of BEM. As can be seen, soil meshing around the shaft pile and pile meshing around the U-tubes were refined in order to increase the precision of results related to the heat transfer through these zones. Four zones were used in the model: two zones as tubes wall, one zone as drilling and the last zone as soil. The data on material properties of fluid, tube and filing were obtained from reality.

As the result of the numerical calculations using BEM, the temperature fields in the vicinity of VGHE after the first day of the SGSHP system operation and at the end of work during the summer season, after 200 days were presented in Figure 26.5(a)–(d). The images clearly show the range of the heat flux influence.

26.5.2 *Line source model*

On the basis of the experiment data, the average heat flux transferred through the ground heat exchanger to the ground was determined.

As previously mentioned, the weighted average value of the heat flux during the SGSHP operation was 9.4 W/m. The heat flux, together with the assumed soil parameters, were adopted as given to solve the equation (26.10). Analogous results to item 5.1 are presented in Figure 26.6.

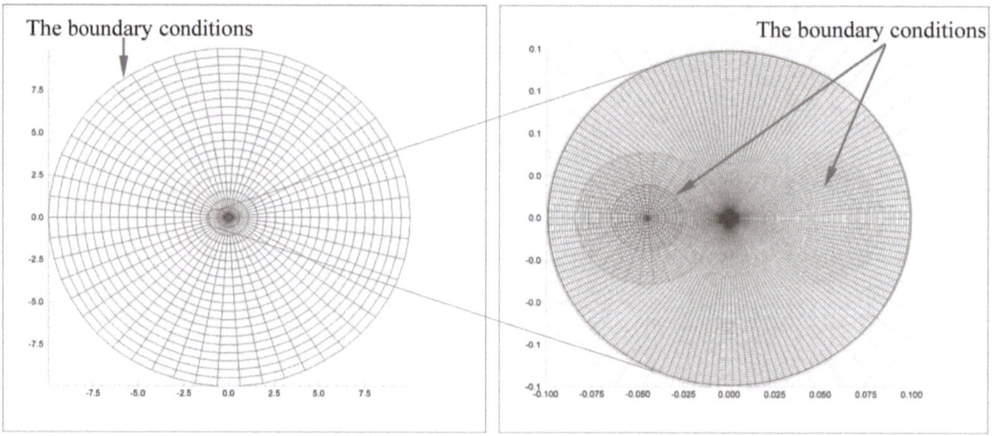

Figure 26.4. The mesh of domains with U-tube configuration used in numerical modeling (linear dimension in m).

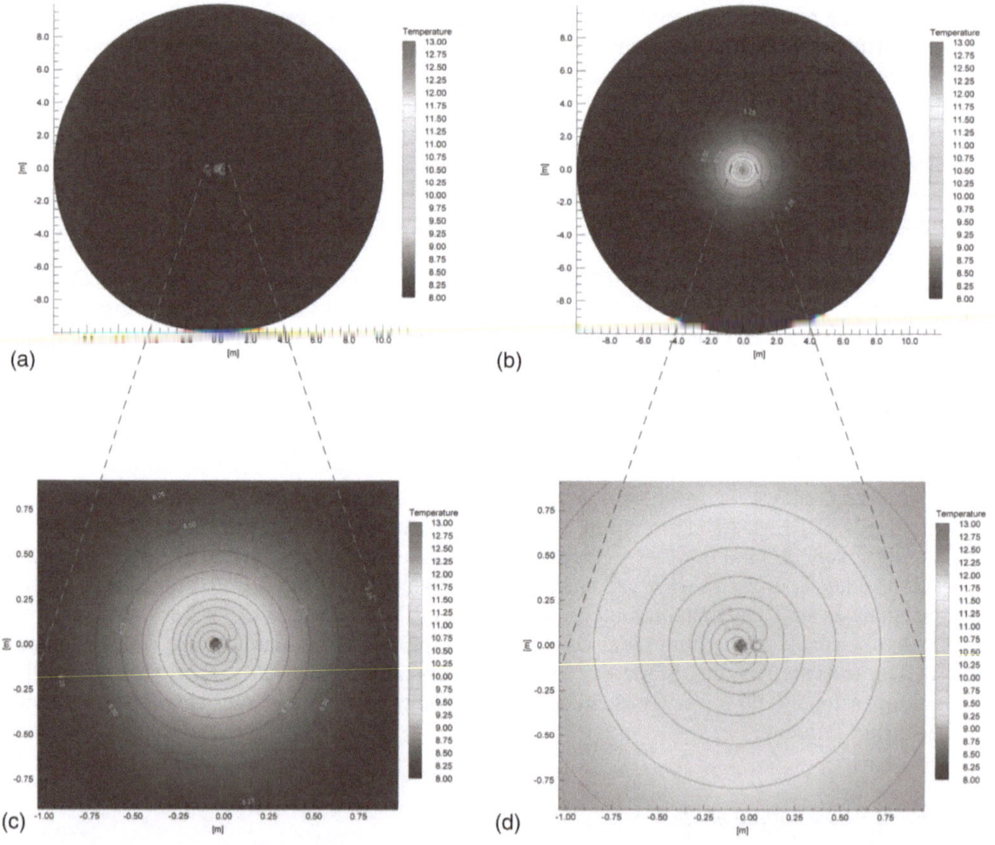

Figure 26.5. (a) The temperature fields in the vicinity of VGHE after 1 day operation, for the diam. of 10 m. (b) The temperature fields in the vicinity of VGHE after 200 day operation, for the diameter of 10 m. (c) The temperature fields in the vicinity of VGHE after 1 day operation, for the diameter of 1 m. (d) The temperature fields in the vicinity of VGHE after 200 day operation, for the diameter of 1 m.

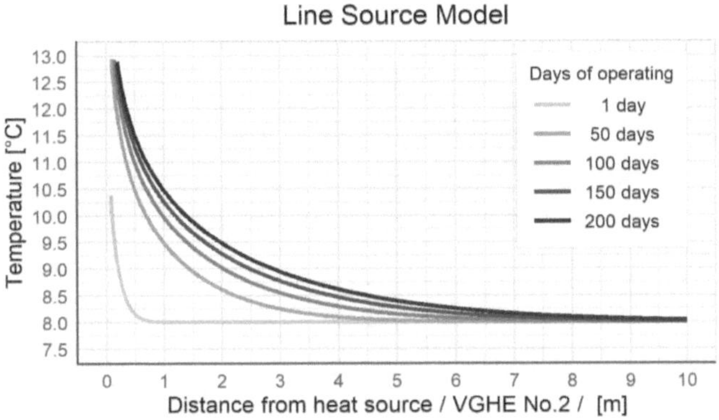

Figure 26.6. The temperature profile in the vicinity of VGHE after selected time operation, | according to line source model.

26.6 COMPARISON BETWEEN NUMERICAL AND EXPERIMENTAL RESULTS

During the experiment, the soil temperatures were measured in 5 minutes steps. The vertical ground heat exchanger No. 2 was a source of energy for ground. VGHE No. 1 and No. 3 and control probes No. 4, No. 5 and No. 6 were control wells for the measurement data. The temperature of the ground after the selected control days of SGSHPS operation, as the result of numerical calculation and the experiment were presented in Figure 26.7a–26.7d. Each graph shows a slight increase in the ground temperature as a function of system operation. The detailed results of BEM give the higher values compared to linear source model. There are discrepancies between the temperatures data obtained from numerical models and experiment. It is clearly noticeable, the temperatures discrepancies mainly concern a point of 2.2 m away. Although the values increase, they differ slightly in both models. It may indicate a lack of knowledge about the soil and the phenomena occurring in it. This result has to be treated with caution and should be validated if possible. What can also be seen is the two identical values of temperature for 5.0 m point away, for the control probes which are 5 m apart. A slight discrepancy is visible after 50 day of operating SGSHPS (Figure 26.7). For longer working times (Figure 26.7b, 26.7c and Figure 26.7d) the temperature values overlap. From an engineering point of view, the values should nevertheless be considered as satisfactory.

26.7 CONCLUSION

An experiment was conducted in order to explore and demonstrate the use of the ground as a heat accumulator in a summer season. In summer, excess heat from solar radiation is directed to the ground. This solves the problems what to do with the excess heat stored in the buffer container and allows for effective regeneration of the soil. The numerical and experiment results with solar assisted ground source heat pump system in the summer season were presented. As mentioned in chapter 2, the average efficiency ratio of heat transferred from solar radiation to soil in sghp system was 42.3%.

The experiment showed that for a relatively small active solar collectors area (flat type 2×2.32 m^2 and heat-pipe 2×3.03 m^2) and only one operating vghe can effectively regenerate soil and accumulate the heat in the soil. The temperature increase, from the initial temperature 8°C, after the summer season was, for 2.2 m up to 8.38°C, for 4.4 m–8.53°C, and for 5 m–8.25°C. At a distance of 10 m, no effect of heat flux was observed. The numerical experiment showed two things. First, if there is no complete ground information, applying the complex method, which is undoubtedly boundary element method compering to line source model is troublesome and the results cannot be considered reliable from a scientific point of view. of course, the analysed example concerned one vghe and a complex problem requires numerical methods which are a great tool. Second, the lower temperature at a distance 2.2 m – 8.44°C comparing to the higher temperature 8.65°C at 4.4 m away, shows that the results are highly

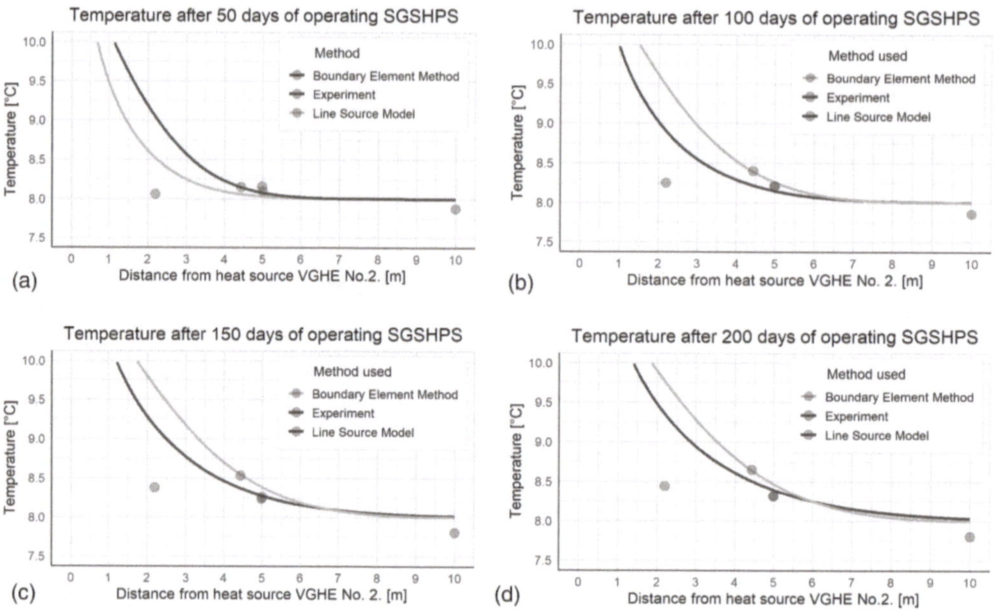

Figure 26.7. (a) The temperature of the ground after 50 days of SGSHPS operation. (b) The temperature of the ground after 100 days of SGSHPS operation. (c) The temperature of the ground after 150 days of SGSHPS operation. (d) The temperature of the ground after 200 days of SGSHPS operation

dependent on the input data, which requires the knowledge of a strongly heterogeneous soil structure. For another measuring point, for 5.0 m, it can be stated that the results of the numerical calculation correspond to the experiment. having long-term simulation the real benefit is soil regeneration and improve coefficient of performance (cop). The rise of temperature is very beneficial to the cop of the gshps. The influence of the regeneration on the heat pump performance will be visible during the heat pump operation in the next heating season.

In a heating dominated climate, like the polish one, the technology for solar energy storage with a heating system based on heat pump puts soil in a state of thermal equilibrum and improves the gshps efficiency in the winter season. The future works of interest include studying the effects of multiple pipes and more complex control mechanisms.

ACKNOWLEDGEMENTS

This work was performed within the framework of Grant No. WZ/WBiIS/9/2019 of Bialystok University of Technology and financed by the Ministry of Science and Higher Education of the Republic of Poland.

REFERENCES

America Society of Heating Refrigerating and Air-Conditioning Engineers (2015) Inc. Geothermal Energy. ASHRAE Handbook - Heating, Ventilating and Air-conditioning Applications. SI Edition.

Angelotti A. et al. (2018) On the applicability of the moving line source theory to thermal response test under groundwater flow: considerations from real case studies. *Geothermal Energy.* https://doi.org/10.1186/s40517-018-0098-z.

Atam E. et al. (2016). Ground-coupled heat pumps: Part 1 – Literature review and research challenges in modeling and optimal control. *Renewable and Sustainable Energy Reviews* 54:1653–1667. https://doi.org/10.1016/j.rser.2015.10.007

Atam E. et al. (2016) Ground-coupled heat pumps: Part 2 – Literature review and research challenges in modeling and optimal control. *Renewable and Sustainable Energy Reviews* 54:1668–1684. https://doi.org/10.1016/j.rser.2015.07.009

Bakirci K. et al. (2010) Evaluation of the performance of a ground-source heat-pump system with series (GHE) ground heat exchanger in the cold climate region. *Energy* 35:3088–3096. https://doi.org/10.1016/j.energy.2010.03.054

Bakirci K. et al. (2011) Energy analysis of a solar-ground source heat pump system with vertical closed-loop for heating applications. *Energy* 36(5):3224–32. https://doi.org/10.1016/j.energy.2011.03.011

Brebbia C.A. et al. (1984) Boundary Element Techniques. Theory and Applications in Engineering. Springer-Verlag.

Carslaw H.S. et al. (1946) Conduction of heat in solids. Oxford : Claremore Press.

Deng N. et al. (2012) Numerical analysis of three direct cooling systems using underground energy storage: A case study of Jinghai County, Tianjin, China. *Energy and Buildings* 47:612–618. https://doi.org/10.1016/j.enbuild.2011.12.038

Emmi G. et al. (2015) An analysis of solar assisted ground source heat pumps in cold climates. *Energy Conversion and Management* 106:660–675. https://doi.org/10.1016/j.enconman.2015.10.016

Eskilson P. (1987). Thermal analysis of heat extraction boreholes [Ph.D.thesis]. University of Lund, Sweden.

Fan R. et al. (2008) Theoretical study on the performance of an integrated ground-source heat pump system in a whole year. Energy 33:1671–9. https://doi.org/10.1016/j.energy.2008.07.017

Fidorow N. et al. (2015) The influence of the ground coupled heat pump's labor on the ground temperature in the boreholes – Study based on experimental data. *Applied Thermal Engineering* 82:237–245. https://doi.org/10.1016/j.applthermaleng.2015.02.035

Guan Y. et al. (2017) 3D dynamic numerical programming and calculation of vertical buried tube heat exchanger performance of ground-source heat pumps under coupled heat transfer inside and outside of tube. *Energy and Buildings* 139:186–196. https://doi.org/10.1016/j.enbuild.2017.01.023

Ingersoll L.R. et al. (1954) Heat conduction with engineering, geological, and other applications. The University of Wisconsin Press: Revised ed. Madison.

Katsikadelis J.T. (2002) Boundary Elements. Theory and Applications. Elsevier Science Ltd.

Lee C.K. (2011) Effects of multiple ground layers on thermal response test analysis and ground-source heat pump simulation. *Apply Energy* 88:4405–4410. https://doi.org/10.1016/j.apenergy.2011.05.023

Min L. et al. (2015) Review of analytical models for heat transfer by vertical ground heat exchangers (GHEs). A perspective of time and space scales. *Applied Energy* 151:178–191. https://doi.org/10.1016/j.apenergy.2015.04.070

Nam Y. et al. (2008) Development of a numerical model to predict heat exchange rates for a ground-source heat pump system. *Energy and Buildings* 40:2133–2140. https://doi.org/10.1016/j.enbuild.2008.06.004

Ochiai Y. et al. (2006) Transient heat conduction analysis by triple-reciprocity boundary element method. *Engineering Analysis with Boundary Elements* 30:194–204. https://doi.org/10.1016/j.enganabound.2005.07.010

Parisch P. et al. (2014). Investigations and model validation of a ground-coupled heat pump for the combination with solar collectors. *Applied Thermal Engineering* 62:375–381. https://doi.org/10.1016/j.applthermaleng.2013.09.016

Penrod E.B. et al. (1962) Design of a flat-plate collector for a solar-earth heat pump. *Solar Energy* 6(1):9–22.

Rad F. et al. (2013). Feasibility of combined solar thermal and ground source heat pump systems in cold climate. Energy Buildings 61:224–232. https://doi.org/10.1016/j.enbuild.2013.02.036

Rees S.J. et al. (2013) A three-dimensional numerical model of borehole heat exchanger heat transfer and fluid flow. *Geothermics* 46:1–13. https://doi.org/10.1016/j.geothermics.2012.10.004

Rynkowski P. (2019) The influence of the solar-ground heat pump system on the ground temperature in the boreholes and in their vicinity – study based on experimental data. E3S Web of Conferences 116, 0006. https://doi.org/10.1051/e3sconf/201911600067.

Sanner B. et al. (2003) Current status of ground source heat pumps and underground thermal energy storage in Europe. *Geothermics* 32(03):579–88. https://doi.org/10.1016/S0375-6505(03)00060-9

Samii B. M. et al. (2016) Solar assisted heat pump systems for low temperature water heating applications: A systematic review. *Renewable and Sustainable Energy Reviews* 55:399–413. https://doi.org/10.1016/j.rser.2015.10.157

Verma V. et al. (2018) Experimental study of solar assisted ground source heat pump system during space heating operation from morning to evening. *Journal of Mechanical Science and Technology* 32(1):391–398. https://doi.org/10.1016/j.enbuild.2017.01.041

von Cube H. L. et al. (1981) Heat Pump Technology. London: Butter Worths.

Wang Q. et al. (2011) Research on integrated solar and geothermal energy engineering design in hot summer and cold winter area. *Procedia Engineering* 21:648–655. https://doi.org/10.1016/j.proeng.2011.11.2061

Werner-Juszczuk A. et al. (2012) Application of boundary element method to solution of transient heat conduction. *Acta mechanica et automatic* vol.6, no.4.

Wołoszyn A. et al. (2013) Modelling of a borehole heat exchanger using a finite element with multiple degrees of freedom. *Geothermics* 47:13–26. https://doi.org/10.1016/j.geothermics.2013.01.002

Yang W. et al. (2010) Current status of ground-source heat pumps in China. *Energy Policy* 38(1):323–32. https://doi.org/10.1016/j.enpol.2009.09.021

Yang W. et al. (2015) Experimental investigations of the performance of a solar-ground source heat pump system operated in heating modes. *Energy and Buildings* 89:97–111. https://doi.org/10.1016/j.enbuild.2014.12.027

CHAPTER 27

Low radiator design temperatures – analysis of the ability to utilise solar heat gains and create thermal comfort based on dynamic simulations

E. Figiel
West Pomeranian University of Technology, Szczecin, Poland

ABSTRACT: The paper analyses how the lowering of the heating medium design temperature in a central heating system affects the possibility of utilising solar heat gains for heating purposes and the thermal comfort in the heated premises. Advanced non-commercial building simulation software, namely Q4HKTM, was used to obtain the results described in the paper. It enabled to simulate the thermal states of a building and its hydraulic central heating installation. The primary aim of this research was to increase the level of knowledge about the impact of radiator design temperatures on the energy consumption. The analysis conducted in the present study indicates that lowering the design temperatures in central heating installations makes it possible to better utilise solar heat gains for heating purposes and create more favourable thermal conditions in order to achieve thermal comfort.

27.1 INTRODUCTION

Using Poland as an example, it can be seen that up until the 1990s, building regulations for the permitted levels of heat loss to the exterior have been steadily lowering. Due to the significant reduction of transmission energy loss in buildings, solar heat gains become an increasingly important element of the thermal balance of heated spaces. Their appropriate utilisation for heating purposes can contribute to a meaningful reduction of energy consumption. The increased energy efficiency of buildings has also enabled the design temperatures of radiators and other heat emitting elements to be lowered. In addition, it is highly favourable to use low temperatures with radiators, when using some renewable energy sources, such as heat pumps and in case of using condensing boilers. This has led to the radiator systems in Poland today usually being designed for the supply/return design temperatures of $\tau_z/\tau_p = 70/50°C$ or 70/55°C in heating systems with radiators and 55/45°C or even lower in the floor or wall heating systems with a heat pump or condensing boiler. In the past, most central heating systems in Poland were designed for a forward/return temperature of 90/70°C. It is of interest to determine the influence of lowering the radiator design temperatures on the possibility of utilising solar heat gains and thermal comfort. The paper presents this assessment also taking into consideration the quality parameters of temperature control in rooms (self-adjusting thermostatic valves). The comparative analysis includes selected parameters of low temperature heating systems ($<75°C$) and high design temperatures 90/70°C, which were in the past defined as commonly used standard-design temperatures. The analysis of how temperatures of the heating medium affect utilisation of solar heat gains was conducted on the basis of a numerical simulation. The Q4HKTM computer program (Figiel 1997) was used to this end. This software was created and carefully validated in Hermann Rietschel Institut, TU Berlin and proven to provide highly accurate data in such system comparison calculations.

This enabled to simulate the thermal states of the heated object in connection with the hydraulic simulation of the central heating system with thermostatic radiator valves. There have been only few studies that investigated the thermal performance of water heating systems by coupling a hydraulic distribution network and thermal system together. This approach can be found in Gamberti et al (2009), Henzle & Floss (2011), Rhee et al. (2009) and Yu (2012). However, this approach assumes that all thermostatic valves operate in an ideal manner. In reality, their performance would be affected by such factors as hysteresis, thermal conductance, sensitivity to differential pressure, valve authority.

DOI 10.1201/9781003171669-27

The paper presents this assessment also taking into consideration the quality of temperature control in rooms.

27.2 METHODS

The adopted method described in Zöllner (1978), Figiel (1997) and Yoo (1994) made it possible to fully reflect the dynamic states occurring in heated buildings and in heating water systems for any configuration of the investigated parameters. The developed program, written in Fortran 95, is able to cover the dynamic thermal characteristics of a building and an entire heating system. A set of heat balance equations has been derived to describe the thermal characteristics of a room. These equations take into account the heat storage of both the room envelope and air as well as radiative and convective heat transfer within the room. This method assumes that the heat flow through the building elements is one dimensional, i.e. effects of room corners or other irregularities are ignored, and air is at uniform temperature throughout a room. The simulated building was divided into several rooms, thermally connected through the shared inside walls. For every wall and for the air in a room, energy balance equations were noted. Differential equations, which were created later after digitisation of space and time variables, were solved using the Crank-Nicolson method (Crank 1996). Part of the program input contains a specification of the building envelope (dimensions and materials), heating system (design temperatures of the heating medium, pipes and TRVs) as well as occupant behaviour represented by a time varied heat output of appliances and people and the setpoint of the room temperature. For the hydraulic simulation, the algorithms for pressure calculation and network structure construction described by Glück (1988) were used. This approach was further modified and extended in order to derive a model for the hydraulic simulation. In this method, the identification of the network occurs automatically. A network consists of several pipes (branches), which are hydraulically connected to each other at junctions (nodes). The construction of the network structure is carried out independently. In the modelling of a heating system, the behaviour of thermostatic valves is a very important component. It couples the occupant's behaviour (choosing the set point of the valve), the building (room temperatures are the "input" of the valve) and the heating system (water flow is the "output" of the valve). All important effects on the thermal and hydraulic behaviour of the thermostatic valves were taken into account (hysteresis, influence of the water temperature on the TRV temperature, time constant and the authority of TRV) but as the simulation results from Figiel (1997) show, only the valve authority had a significant influence on the water distribution. Under some simplifying assumptions, the relationship between the water flow and temperature of the thermostat was idealized, as shown in Figure 27.1 and was mainly introduced to limit mathematical effort of the calculations.

In addition, a separate program for sizing or design calculations of the heating system was created. As input, it receives the design heat load calculated with accordance to the PN-EN 12831 norm as well as the design temperature of the heating medium.

For the selection of the pipe diameter from an internal database the program assumed a maximal pressure drop of 150 Pa/m along the pipe per unit length. The designed central heating system consists of radiators equipped with classic thermostatic valves TRV. When choosing the design of the TRV a valve authority of at least 0.3 was assumed. From the internal database, which contained the flow coefficient *kv* of different TRVs from a manufacturer, an appropriate valve was selected. On the basis of the computed pressure drop, the program selected a pump from the database, which also contained a collection of characteristics of pumps with a maximal feed pressure of 60 kPa.

The numerical calculations were performed for a typical detached house in a ribbon development with the rooms oriented facing north- south. All simulation results shown in this paper were conducted for the three rooms from the southern face, which vary in their usable room space and window dimensions.

The simulated object with good thermal insulation belongs to a class of so-called massive buildings. Every room has one double-glazed window with standard solar radiation transmittance.

A water central pumped heating installation with a double pipe system and riser feed was equipped with single-plate water heating radiators (without ribbing). Each radiator has a thermostatic radiator valve on the inflow. The installation was designed in accordance with the adopted rules, guaranteeing

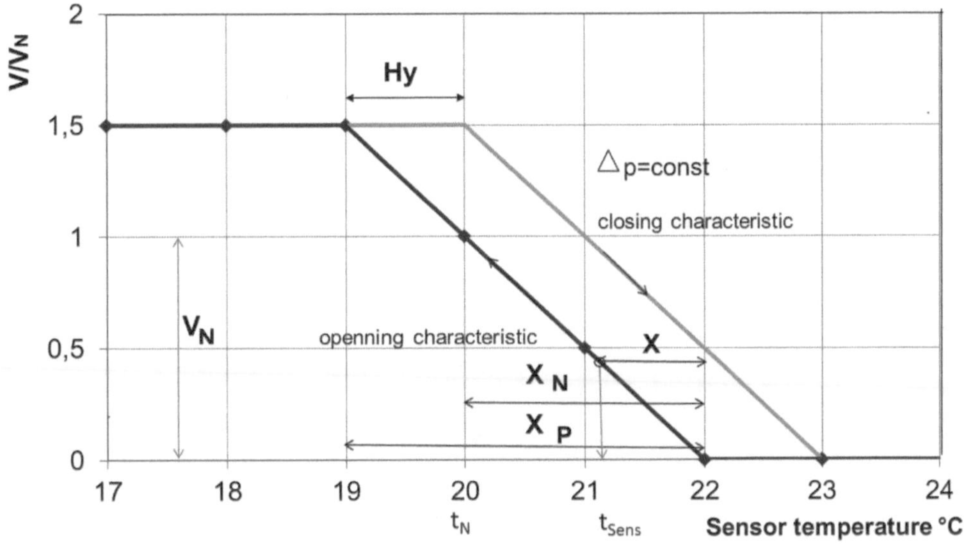

Figure 27.1. Linearized relationship of TRV sensor temperature and valve volume flow V/V_N (N- design value), $X_P = 3K$ (proportional band), $t_N = 20°C$ (set temperature), $X_N = 2K$ – P-Band (valve sizing), X – proportional offset, Δp – pressure drop is constant.

high valve authority. The total heat load of the simulated building was 8.5 kW in accordance to PN-EN 12831. The design heat load was 1880 W, 710 W and 680 W in the analysed rooms numbered 1, 6 and 7, respectively. The simulated central heating system consists of 15 radiators equipped with classic thermostatic valves. The classic thermostats are proportional regulators working without auxiliary energy. They regulate the room temperature by varying the flow volume of the heating water. The setting point of the TRVs is 20°C for rooms and 24°C for bathrooms. For the investigated design supply/return temperature pairs of the heating medium, the surface area of the radiator was changed each time. For example, for the temperature pair 55/45°C compared to the pair 90/70°C the surface area was increased by a factor of 2.46.

The numerical calculations were conducted given the following boundary conditions:

1. All the temperature variants of the heating medium were investigated for the same weather conditions on the basis of weather data for a comparative year (TRY) for the city of Berlin, as found in DWD (1991).The analysis was performed on one selected day from the period of time when flats are centrally heated, when direct sun radiation was substantial and outside temperature was low. The course of outside temperature during the day (4th Feb.) and the intensity of the total radiation on the southern facade of the building are presented in Figures 27.2 and 27.3.
2. Rooms contain no internal heat sources other than the heating elements.
3. Air change rate is constant, and it equals $0.5\,h^{-1}$.
4. The heating system is on non-stop and night mode is not activated.

A reduction in the amount of heat emitted by a radiator compared to the amount of heat gain can be treated as a reliable criterion of the degree of heat gains utilisation. This criterion, denoted as *SG*, has not been described so far in the literature on the subject in question. It can be determined by the following equation:

$$SG = \frac{Q_{GO} - Q_G}{Q_Z} \tag{27.1}$$

The reduction of the amount of heat given off by a radiator depends on the magnitude of heat gains and their profile at a given time. The *SG* values quoted in the paper are therefore closely connected

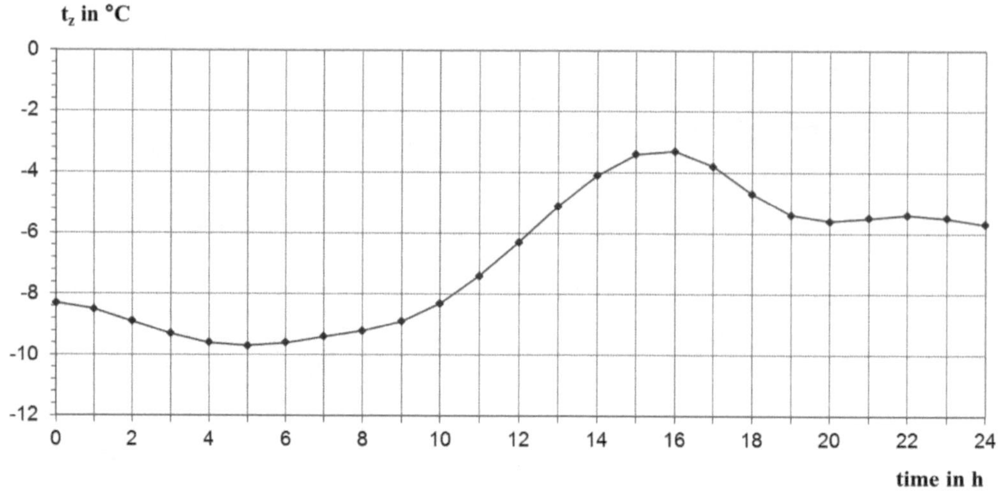

Figure 27.2. Changes of the outside air temperature t_z on 4th Feb. (DWD 1991).

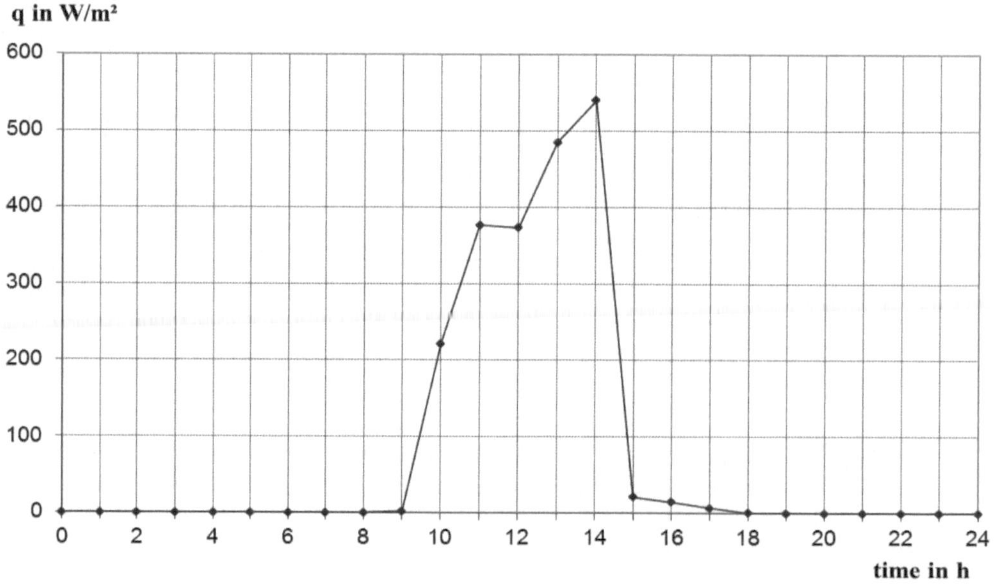

Figure 27.3. Changes of the solar radiation intensity q on the southern facade on 4th Feb. (DWD 1991).

with the assumed boundary conditions and are applicable only in one specific, investigated case. Only on the basis of marked tendencies can generalizations be formed.

27.3 RESULTS AND DISCUSSION

27.3.1 *Design temperatures of radiators – impact on the ability to utilise heat gains*

Two additional values were used for the assessment of the extent the design system temperatures of a heating medium affect the thermal efficiency of a heating element:

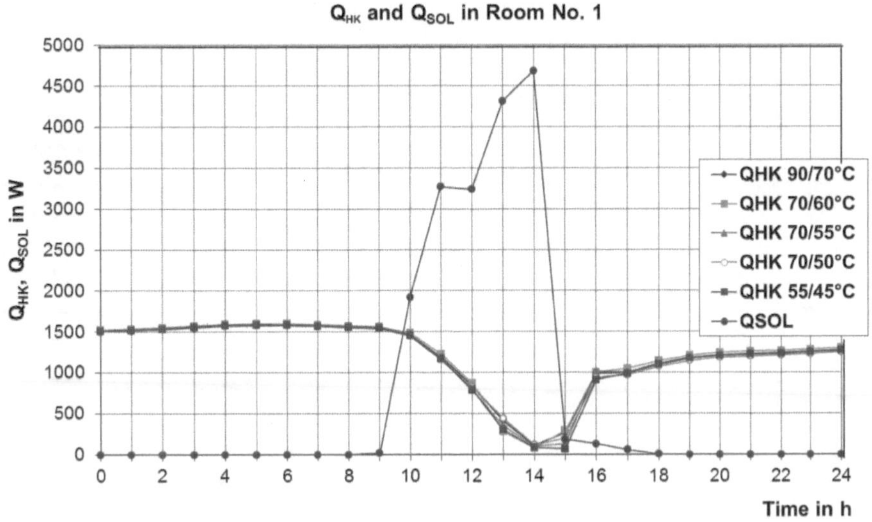

Figure 27.4. The course of heat output from the radiator Q_{HK} and solar heat load Q_{SOL} in room no.1 for different design supply/return temperature pairs on 4th Feb. (DWD 1991).

Φ_N, which can be determined by the following equation:

$$\Phi_N = \frac{\tau_z - \tau_p}{\tau_Z - \tau_i} \tag{27.2}$$

and Δt_N – logarithmic mean temperature difference between heating medium and ambient air

$$\Delta t_N = \frac{\tau_z - \tau_p}{ln \frac{\tau_z - \tau_i}{\tau_p - t_i}} \tag{27.3}$$

The value Φ_N connected with heat transfer, is according to Bach (1997) one of the most important parameters which determines the thermal energy requirements.

The calculated values of Φ_N and Δt_N for the investigated design supply/return temperature pairs of the heating medium at the air temperature in the room $t_i = 20°C$ amount to:

90/70°C $\Phi_N = 0.286 \Delta t_N = 59.8°C$
55/45°C $\Phi_N = 0.286 \Delta t_N = 29.7°C$
70/60°C $\Phi_N = 0.200 \Delta t_N = 44.8°C$
70/55°C $\Phi_N = 0.300 \Delta t_N = 42.1°C$
70/50°C $\Phi_N = 0.400 \Delta t_N = 39.2°C$

All the variants of the heating medium temperatures were investigated under the same weather conditions on the basis of weather reports for a comparative year for the city of Berlin (DWD 1991).

The impact of the examined design temperatures on the course of the heat output from the radiator Q_{HK} for the rooms no.1 and 7 on the day Feb. 4 (DWD 1991) as well as the course of solar heat load Q_{SOL} in these rooms are depicted in Figures 27.4-27.5. In room no. 6 the course of Q_{SOL} is analogous to that in room no. 7. The course of Q_{HK} is also close to that depicted in Figure 27.5.

The values of *SG* coefficients obtained in a numerical simulation for various parameters of the heating medium are presented in Figure 27.6. These values illustrate the degree to which heat gains are utilised depending on the solar radiation intensity in the selected three rooms of southern exposition on a day with a significant share of direct sunlight and low outside air temperature (4th Feb.). The caption of the figure contains both pairs of system temperatures τ_z/τ_p, Δt_N and Δ_N values.

A dependence of the coefficient *SG* on the value of mean temperature difference Δt_N can be observed in Figure 27.6. The mean temperature difference between heating medium and ambient air

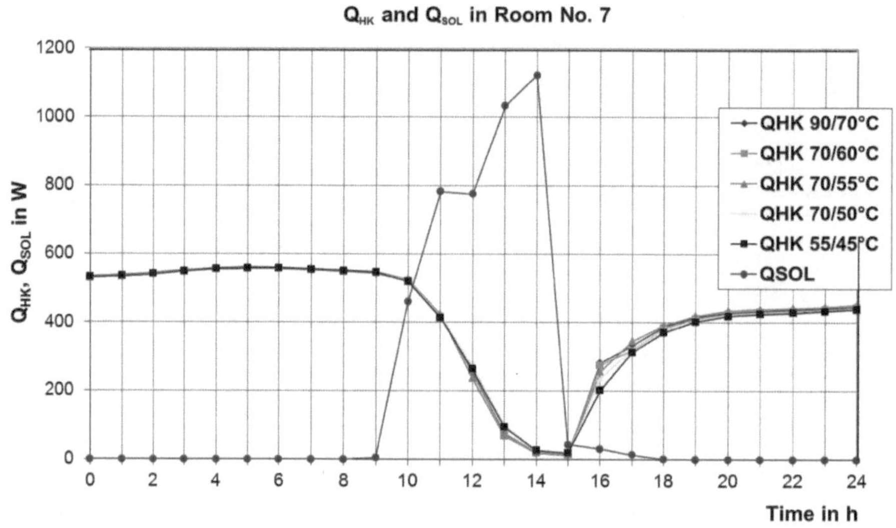

Figure 27.5. The course of heat output from the radiator Q_{HK} and solar heat load Q_{SOL} in room no.7 for different design supply/return temperature pairs on 4th Feb. (DWD 1991).

is the main driving force for the heat exchange. The degree of heat gain utilisation in the premises increases when Δt_N decreases. The influence of Φ_N is of secondary importance. Only when pairs of temperatures for the same value of the mean temperature difference Δt_N are compared, the benefits of choosing a lower value of the coefficient Φ_N can be seen. In this case, lower energy consumption occurs at a higher difference between supply and return temperatures $\Delta\tau = \tau_z - \tau_p$. This tendency is illustrated in Figure 27.7. This agrees with the findings presented by Sarbu & Sebarchievici (2015).

The dependence of the coefficient *SG* on the mean temperature difference between the heating medium and the ambient air can be caused by a higher "self-regulating effect" of heating installations with low values of Δt_N – i.e. the so-called low temperature installations. The "self-regulating effect" should be understood as the ability of a system to adapt to the changing heat requirements without individual room control (for example thermostatic radiator valves TRV).

The differences between the values of the *SG* coefficients obtained as a result of the simulation, for the thermally best variant, i.e. when the temperatures were 55/45°C and for the worst investigated parameter – 90/70°C, were between 5 and 7 percentage points. In the case of the heating medium temperature reaching 90/70°C, the energy consumption increases by 3 to 5% in comparison with the parameters 55/45°C.

An influence of low parameters on the installations containing only central weather compensation, meaning that the water temperature is adjusted proportionately to the outdoor temperature (no individual room control), was also analysed. The obtained *SG* coefficients are presented in Figure 27.8.

The finding confirms the fact stressed by Zöllner (1979) that low temperature heating installation has better "thermal self-stability".

The differences between the values of the *SG* coefficients for temperature pairs 55/45°C and 90/70°C are between 4 and 8 percentage points.

Figure 27.9 shows that for low parameters the room temperature increases, the amount of heat given off by the radiators decreases to a larger degree than that at higher parameters. For example, at 2pm the heat output from the radiator for the parameters 55/45°C is about 12% lower than for 90/70°C. This is also why the low temperature installations are easier to control centrally, than the systems with higher temperatures of the heating medium.

On the other hand, if the central weather compensation control at the heat source is not conducted properly, it must be taken into account that every deviation of supply temperature from the desired

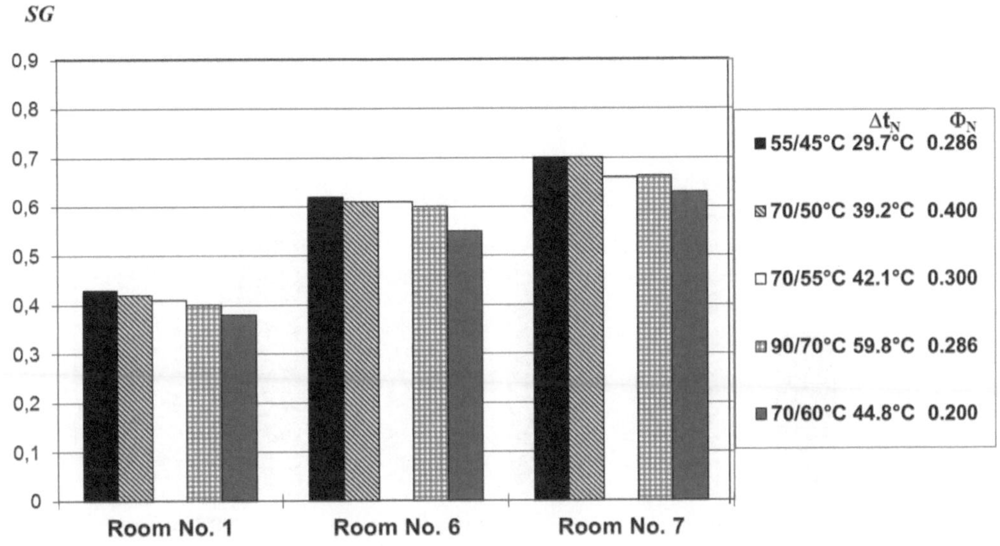

Figure 27.6. *SG* coefficients for various τ_z/τ_p. Individual room control (thermostatic radiator valves) plus central weather compensation at the heat source.

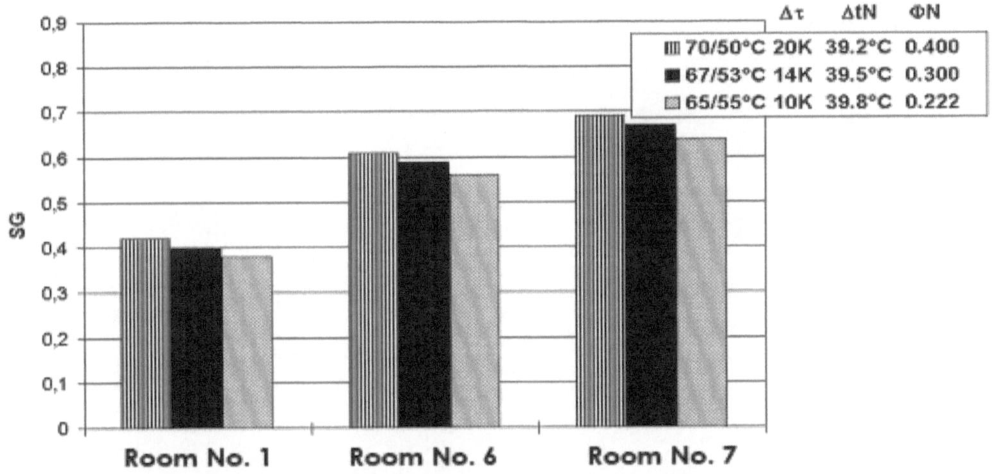

Figure 27.7. *SG* coefficients for τ_z/τ_p with similar values of mean temperature difference Δt_N. Individual room control (thermostatic radiator valves) plus central weather compensation at the heat source.

value, in the installations with low value of mean temperature difference Δt_N, will result in a larger extent in the amount of energy being given off to the heated premises.

27.3.2 *Design temperatures of radiators – impact on the ability to create thermal comfort*

The function of a modern heating installation is not only to make sure that the required indoor temperature is reached in the premises, but also to provide the required amount of thermal energy and the conditions necessary for thermal comfort. In spite of the constantly increasing thermal protection requirements, the thermal insulation of exterior wall barriers, especially windows, is not good enough so that a negative influence of low temperatures of these areas on the perceived thermal comfort could be neglected. In apartment buildings, the temperature of the inside window panes, given the current

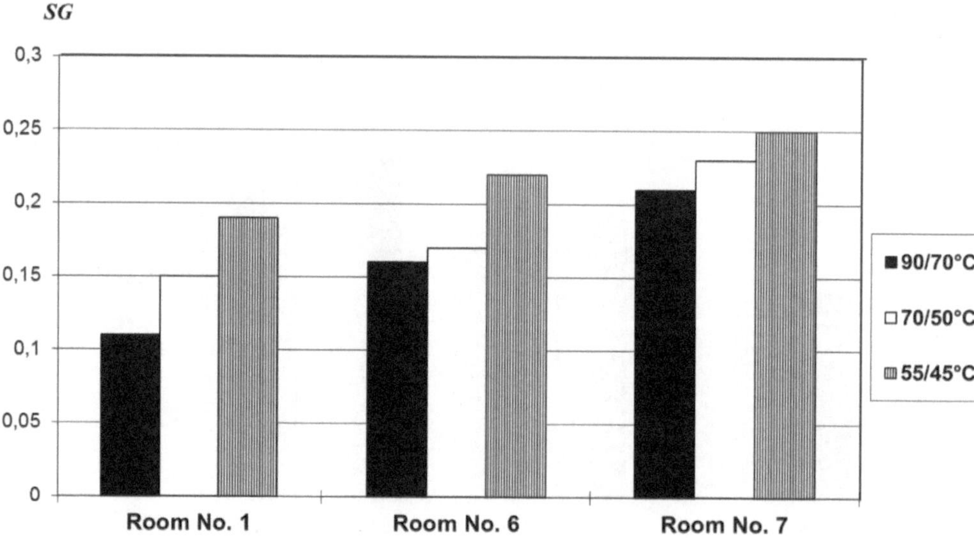

Figure 27.8. *SG* coefficients for various temperature variants of heating medium. Central weather compensation at the heat source without individual room control (TRV).

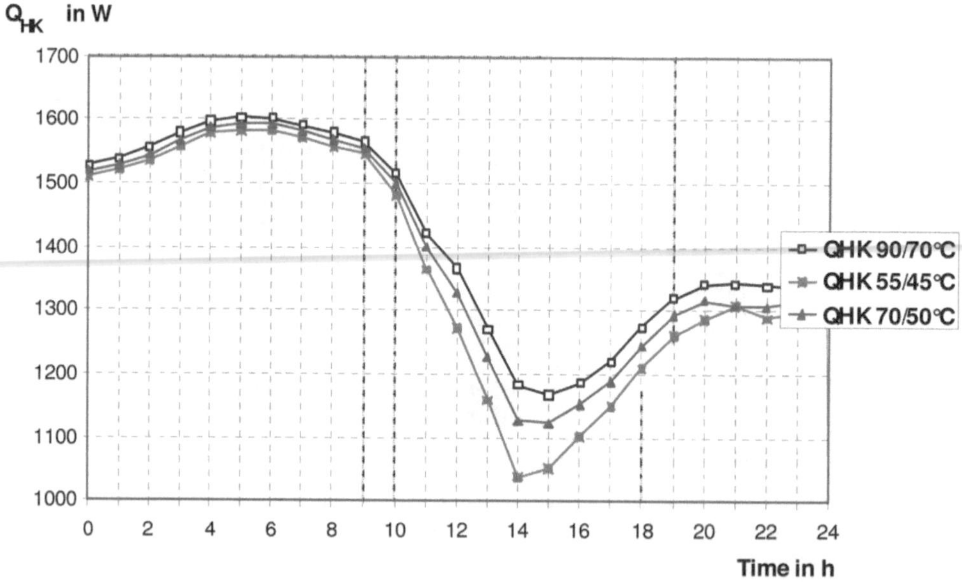

Figure 27.9. The course of heat output from the radiator Q_{HK} in room no.1. Central weather compensation at the heat source without individual room control (TRV).

thermal insulation requirements of wall barriers, is usually lower by several degrees than the air in the investigated premises. The negative influence exerted by cold radiation of the windows panes can only be compensated by an appropriately selected heating element. Numerous studies (Gendelis et.al 2015, Hesaraki & Holmberg 2013, Sarbu & Sebarchievici 2015) for different radiator types emphasise that using a low temperature heating system increases not only energy efficiency but also thermal comfort.

As a result of numerical simulation, the courses of the mean radiant temperature were obtained. The courses, which are to be seen in Figures 27.10 and 27.11, present the mean radiant temperature

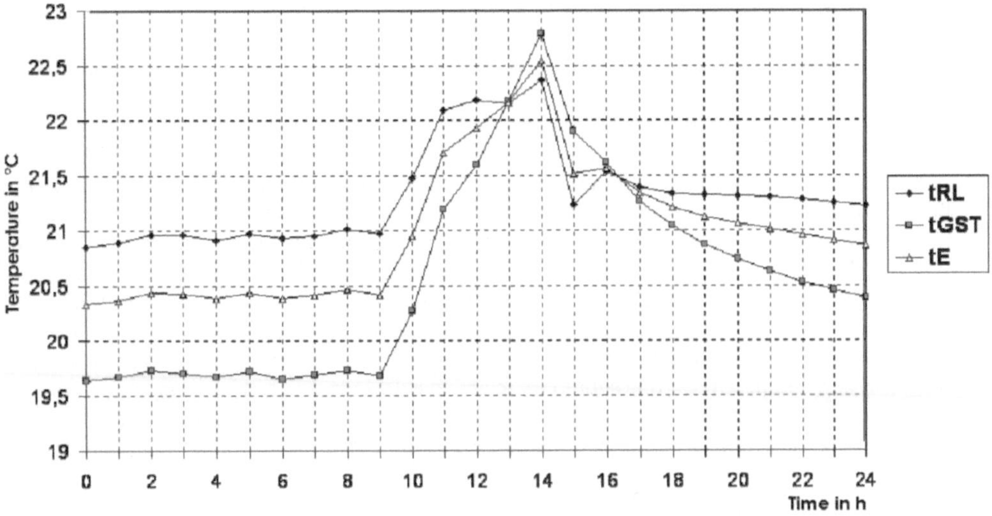

Figure 27.10. The course of air temperature t_{RL}, operative temperature t_E, mean radiant temperature t_{GST} for 90/70°C.

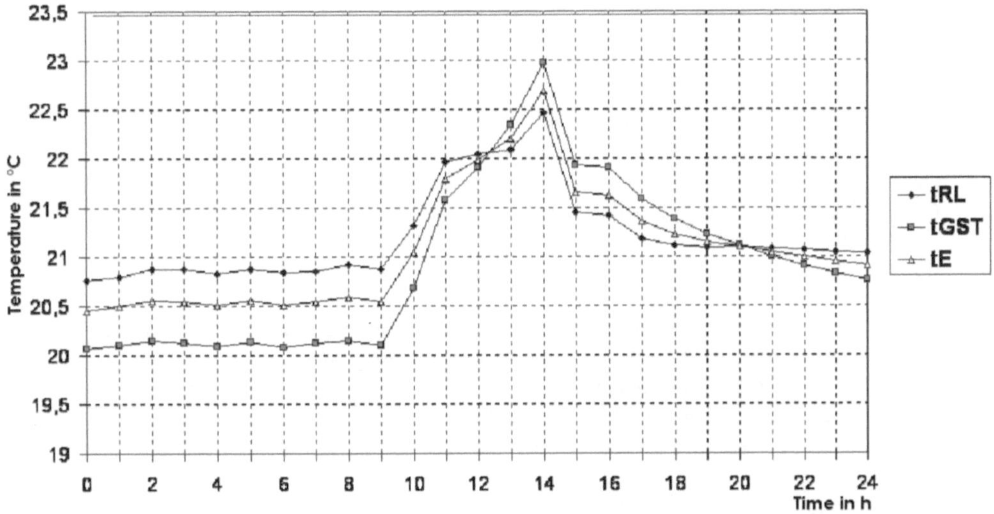

Figure 27.11. The course of air temperature t_{RL}, operative temperature t_E, mean radiant temperature t_{GST} for 55/45°C.

of wall barriers t_{GST}, indoor air temperature t_{RL}, and operative temperature t_E in the heated premises, which illustrate the possibilities of compensating the cold wall barriers radiation for two pairs of temperatures, i.e. 55/45°C and 90/70°C.

The figures show that the difference between the temperatures t_{GST} and T_{RL} is smaller for low temperature installations than that for the installations with 90/70°C parameters. Consequently, lower design temperatures of a heating medium make it possible to obtain better conditions of thermal comfort in the heated premises.

27.4 CONCLUSIONS

The analysis on the parameters of a heating medium conducted in the present study indicates that lowering the design temperatures in central heating installations below the value of the supply temperature of 75°C makes it possible to better utilise solar heat gains for heating purposes and create more favourable thermal conditions in order to achieve thermal comfort.

The utilisation of solar heat gains increases, when the mean temperature difference between the heating medium and the ambient air Δt_N decreases. A dependence between the utilisation of solar heat gains and the value of Φ_N coefficient is to be perceived only, when comparing the temperatures with similar Δt_N values. Installations with supply temperature $\tau_z < 75°C$ due to their high "thermal self-stability" allow reducing the energy consumption as well, when only central weather compensation at the heat source is conducted. Owing to the benefits resulting from saving thermal energy, the fact that the heating surface increases could be considered of secondary importance. While selecting design system temperatures, one should try to achieve the lowest possible mean temperature difference between a heating medium and the ambient air (Δt_N). However, it should also be taken into consideration that all possible errors, either in selecting a heating surface or in not maintaining the required temperature of the heating medium in the installation, have stronger effect on the room temperature for lower values of Δt_N.

NOMENCLATURE

t_i	design air temperature in the room, °C
t_z	outside air temperature, °C
t_E	operative temperature in the room, °C
t_{RL}	air temperature in the room, °C
t_{GST}	mean radiant temperature, °C
q	solar radiation intensity, W/m^2
Q_{G0}	heat given off by radiator with no heat gains, kWh
Q_G	heat given off by radiator, when heat gains have influence on the room conditions, kWh
Q_{HK}	heat output from the radiator, W
Q_{SOL}	solar heat load in the room, W
Q_Z	heat gains, kWh
SG	coefficient, -
Δt_N	mean temperature difference between the heating medium and ambient air, °C
$\Delta\tau$	difference between supply and return design temperatures, °C
τ_z, τ_p	design supply and return temperature of the radiator, °C
Φ_N	coefficient connected with heat transfer, -

REFERENCES

Bach, H. 1997. Mit der Dämmung der Gebäude steigt die Anforderung an die Nutzenübergabe. *HLH* 48: 32–37.

Crank, I. & Nicolson, P. 1996. A practical method for numerical evaluation of solutions of partial differential equations of the heat conduction-type. *Advances in Computational Mathematics* 6: 207–226.

Deutscher Wetterdienst (DWD) 1991, *Testreferenzjahre (TRY)- Meteorologische Grundlagen für technische Simulationvon Heiz- und Raumlufttechnischen Anlagen.* Karlsruhe: BINE- Profi- Info- Service.

Figiel, E. 1997. *Regelgüte der Raumtemperatur und Fremdenergienutzung in beheizten Gebäuden bei Anwendung thermostatischer Heizkörperventile. Rechnerische Gebäude- und Anlagensimulation, PhD thesis, TU Berlin.* Berlin.

Figiel, E. 2014. Modelling approach for hydraulic heating system in building application. In AGH-UST (ed.), *XXI Fluid Mechanics Conference; Proc. intern. conf. Krakow. 15–18 June 2014*. Krakow. Retrieved: http://iopscience.iop.org/1742-6596/530/1

Gamberti, M. et. al. 2009. Simulink Simulator for Building Hydraulic Heating Systems Using the Newton-Raphson Algorithm. *Energy and Buildings* 41: 848–55.

Gendelis, S. et. al. 2015. Experimental research of thermal comfort conditions in small test buildings with different types of heating. *Energy Procedia* 78: 2929 – 2934.

Glück, B. 1988. *Bausteine der Heizungstechnik. Hydrodynamische und Gasdynamische Rohrströmung.* Berlin: VEB Verlag für Bauwesen.

Henze, G. & Floss, A. 2011. Evaluation of Temperature Degradation in Hydraulic Flow Networks. *Energy and Buildings* 43: 1820–28.

Hesaraki, A. & Holmberg, S. 2013. Energy performance of low temperature heating systems in five new-built Swedish dwellings: A case study using simulations and on-site measurements. *Building and Environment* 64: 85–93.

Rhee, K. et. al. 2009. Simulation study on hydraulic balancing to improve individual room control for radiant floor heating system. *Building Services Engineering Research and Technology* 31:57–73.

Sarbu, I. & Sebarchievici, C. 2015. A study of the performances of low-temperature heating systems. *Energy Efficiency* 8: 609–627

Zöllner, G. 1978. *Jahresenergieverbrauch von HD-Induktionsklimaanlagen. Rechnersche Prozeßsimulation und Verbrauchsmessung, PhD thesis, TU Berlin*. Berlin.

Zöllner, G. 1979. Niedertemperaturheiztechnik, *HLH* 30: 3–22.

Yoo, S. H. 1994. *Raumklimatische und energetische Beurteilung baukonstruktiver Maβnahmen im Hinblick auf klimagerechtes Bauen mit Solarenergienutzung, PhD thesis, TU Berlin*. Berlin.

Yu, Y. et. al. 2012. Development and evaluation of a simplified modelling approach for hydraulic systems; Proc. *International High Performance Buildings Conference at Purdue*. 16–19 July 2012. West Lafayette.

CHAPTER 28

Variability of average consumption of water per resident in a multi-apartment residential buildings

W. Szaflik
Faculty of Civil and Environmental Engineering, West Pomeranian University of Technology, Szczecin, Poland

ABSTRACT: The article presents the method of determining the probability of the average daily water consumption based on the number of residents in multi-apartment buildings. It has been assumed that the average daily water consumption by a single resident of a building is a random variable with a normal distribution of the expected value $\overline{q}_j$ and standard deviation $S_{\overline{q}_j}$. Using those assumptions the distribution parameters for total (hot and cold) average daily water consumption per resident were determined. For the purpose of the study, the results of measurements of the average annual water consumption in 2019 from sixty-five residential buildings located in Szczecin were obtained. Knowing those parameters it is possible to determine the value of water consumption per resident in regard to its probability. The average water consumption per resident was calculated for probability levels of 90%, 95% and 99% in relation to the number of residents. The obtained results were compared on a graph.

28.1 INTRODUCTION

Each person consumes water from the water supply system in a different way. The amount of water consumed per day by each resident is thus random. Water consumption by a single resident is influenced by many factors (Ghavidelfar et al. 2016, Englart et al. 2019, Stec 2019), which can be divided into two groups. The first one is related to the characteristics of the building's water supply system, while the second depends on the characteristics of residents. To the first group we can include:

- existence of hot water installation in the building,
- climate changes and climate zone in which the building is located (House-Peters et al. 2010),
- draw-off points in apartments,
- the method of water consumption account,
- pressure at draw-off points,
- type of installation and appliances (Behling 1992, Grafton et al. 2011, Englart et al. 2019),

The factors related to the residents include:

- the number of residents in particular apartment,
- habits and activities of each individual resident related to the water consumption,
- type of day (working day, weekend, holiday),
- demographic structure of the residents,
- social structures of the residents,
- the time of staying in the apartment (work and activities of residents),
- the standard of living,
- hygienic habits of the residents and their health (Kalbusch et al. 2020),
- behaviour of residents related to the time of the day or year, and meteorological conditions,
- attractiveness of TV shows.

An important factor that influences water consumption is the price of water (de Souza et al. 2017; Grafton et al. 2019, 2011). The average water consumption changes over the years (Ferreira et al.

DOI 10.1201/9781003171669-28

2020; Ghavidelfar et al. 2016; Jaszewska et al. 2020; Kalbusch et al. 2020; Kowalski et al. 2019; Stec 2019).

The above mentioned factors randomly influence the average daily water consumption per resident in a longer period. It can be assumed that the average water consumption by each resident is a random variable.

Results of measurements for residential building with varying number of residents has shown that the average water consumption per resident in buildings with a lower number of residents has significantly higher variability than in building with higher number of residents (Szaflik 2020).

The literature on water consumption in residential buildings and its modelling is rich. Shamim et al. (Shamim et al. 2020) have performed an analysis of more than 20 studies from recent years concerning the water consumption behaviours in society. De Souza et al. (de Souza et al. 2017) conducted a study on estimating the water consumption in multi-apartment residential buildings which includes measurements from 30 buildings of different characteristics. An extensive study on the water consumption in residential dwellings was performed in the 1990s in Great Britain. Edwards et al. (Edwards et al. 1995) presented the results from data acquired from the initial year.

Ferreira et al. (Ferreira et al. 2020) proposed a stochastic simulation for modelling water demand in residential buildings. (Mui et al. 2007) proposed the application of the Monte-Carlo method to simulate consumption. The model presented in the study has been verified empirically by measuring water consumption per resident in 60 residential units. Different study (Szaflik 2020) presents the method of determining the average daily water consumption per resident with a specific probability in multi-family residential buildings. However, the study does not include the influence of number of residence on the water consumption probability.

Study performed by Matsiyevska et al. (Matsiyevska et al. 2020) presents an analysis of water consumption in residential building based on data from Poland, Czechia, Romania, Slovakia and Ukraine. Study conducted by Longlong et al. (Longlong et al. 2019) shows a method of prognosing the daily water consumption by the residents using BP neural network and correlation analysis. Compared to neural network model without correlation analysis the errors of prognosed water consumption were lower. The method can be used for prognosing average annual water consumption per resident.

The review of literature did not show any studies that take into account the influence of number of residents and randomness of water consumption on the average water consumption per resident. The article presents a probabilistic method that allows to determine, based on measurements, distribution parameters for variability of average daily water consumption per resident in relation to the number of residents.

28.2 MODEL DESCRIPTION

Daily water consumption per resident in a building depends on many factors, the most important of which are listed in the introduction. The model assumes that the average daily consumption per resident is a random variable with a normal distribution of the expected value $\overline{q}_j$ and standard deviation $S_{\overline{q}_j}$. On this basis, the average daily unit water consumption was calculated with a certain probability in relation to the number of inhabitants of the building. The density of this distribution is determined by (28.1) and the distribution by equation (28.2) (Beniamin et al. 1970):

$$f(q_j) = \frac{1}{\sqrt{2\pi}} \exp\left(-\frac{1}{2}\frac{(q_j - \overline{q}_j)^2}{S_{\overline{q}_j}^2}\right) \tag{28.1}$$

$$F(q_j) = \frac{1}{\sqrt{2\pi}} \int_{-\infty}^{q_j} \exp\left(-\frac{1}{2}\frac{(q_j - \overline{q}_j)^2}{S_{\overline{q}_j}^2}\right) dq_j \tag{28.2}$$

For every normal distribution the sum of variables with a normal distribution also has a normal distribution (Beniamin et al. 1970).

By knowing the general distribution parameters based on the rule described above, it is possible to determine the parameters of the distribution of the sum of **m** values of individual water consumptions, using following relationships:

$$\overline{q}_m = m\,\overline{q}_j \tag{28.3}$$

$$S^2_{\overline{q}_m} = m\,S^2_{\overline{q}_j} \tag{28.4}$$

$$S_{\overline{q}_m} = \sqrt{m}S_{\overline{q}_m} \tag{28.4a}$$

Based on formulas below the distribution parameters for the variability of average water consumption per building with **m** residents was determined:

$$\overline{q}_j\,(m) = \frac{mq_j}{m} = \overline{q}_j \tag{28.5}$$

$$S_{\overline{q}_j}\,(m) = \frac{\sqrt{m}S_{\overline{q}_m}}{m} = \frac{S_{\overline{q}_j}}{\sqrt{m}} \tag{28.6}$$

In case of this simplified model, the key part, which is described further, is the methodology of calculating standard deviation for a single result of average water consumption per resident based on the measurements acquired from those residential buildings.

The models used typically are based on a constant value of the average water consumption per resident. The advantage of proposed model is that it allows to determine the value of water consumption based on the probability and number of residents in a building. This allows to perform additional risk assessments related to the height of consumption.

28.3 CHARACTERISTICS BUILDINGS AND MEASUREMENTS

For the purpose of the study the measurements of water consumptions were performed in multi-apartment residential buildings located in Szczecin, West Pomerania, Poland. For chosen location the average annual temperature based on data from 1931 to 2018 is 12.0°C, while the annual precipitation is 542 mm (Warunki 2020). The buildings are supplied with water by the Water and Sewerage Service, while hot water is prepared in the heat centers. The analysis was carried out on the data provided by one of the Housing Cooperatives in Szczecin. Measurements of water consumption and the number of residents living in studied buildings were performed in 2019. The measurements from water meters were taken on 31st December 2018, 30th April 2019, 31st August 2019 and 31st December 2019. The water meters had proper verification marks. For the purpose of this study 65 residential buildings (3778 apartments) were analyzed, in which the number of residents was 5874.

Figure 28.1 presents the histograms characterizing studied buildings: number of apartments, number of residents and the average number of residents per apartments.

To draw-off points in apartments we can include baths or showers, washbasins and washing machines (mostly automatic) installed in bathrooms and kitchen sinks, and dishwashers in kitchens. The installation was equipped with residential water meters that measure the water consumption.

The average number of apartments in buildings was 58.12 (apartments/building), the average number of residents in the buildings was 90.37 (residents/building), The average population was rather low, resulting in 1.56 people per apartment in 2019.

The number of residents was determined by the Housing Cooperative on the basis of self-reports by apartments' owners sent to the administration. The number varied between 23 and 249 people per building. The actual number of people living in a particular building was practically impossible to determine.

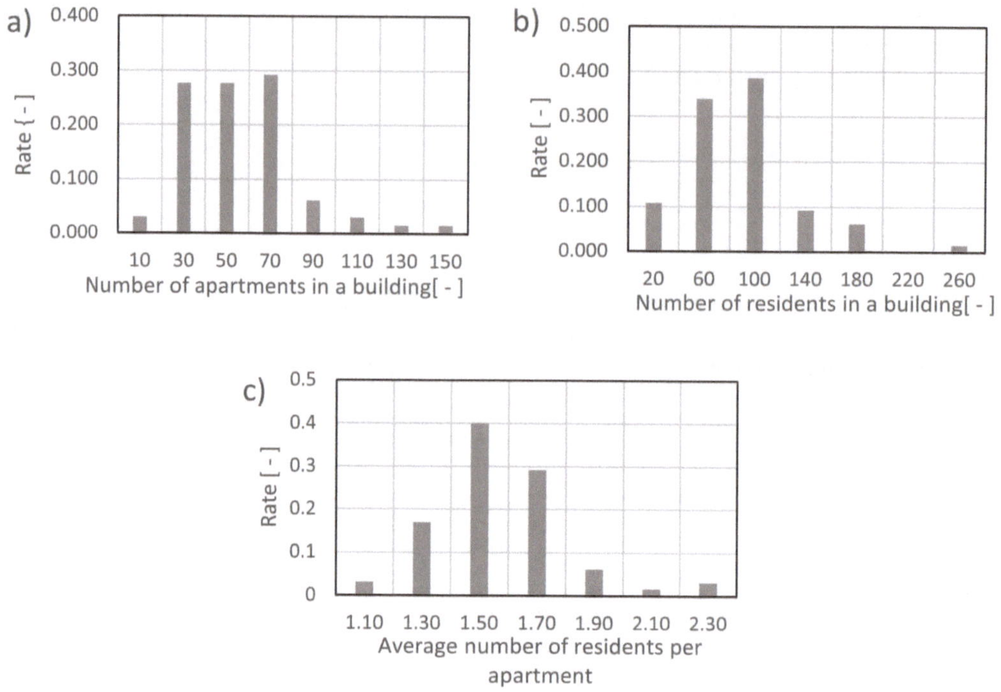

Figure 28.1. Building characteristics a) number of apartments in a building, b) number of residents in a building, c) average number of residents per apartments.

28.4 CALCULATION METHODOLOGY

The values of water consumption in buildings were determined on the basis of the values indicated by the water meters. Using the number of residents in the analyzed buildings given by the Housing Cooperative, the average water consumption per day by one person for each building "***k***" was calculated:

$$\overline{q}_{j_k} = \frac{Q_k}{365 \times m_k} \tag{28.7}$$

The weighted average value of daily water consumption per resident for all buildings was calculated using the formula:

$$\overline{q}_j = \frac{\sum_{k=1}^{n} m_k \times \overline{q}_{j_k}}{\sum_{k=1}^{n} m_k} \tag{28.8}$$

The average water consumption per resident was determined on the basis of a very large sample (the sample size in the examined buildings was 5874 people). It was assumed that the calculated weighted average value should be closely correlated to the expected value. Due to the large number of buildings, it was assumed that the difference between the average value determined for a single building and the average value for all buildings corresponds to the mean square error of the average water consumption for particular building $\overline{S}_{\overline{q}_{j_k}}$:

$$\overline{S}_{\overline{q}_{j_k}} = \overline{q}_j - \overline{q}_{j_k} \tag{28.9}$$

Table 28.1. Average value of annual water consumption per resident, standard deviation and the value of the trend line slope

Lp.	Average value of water consumption per resident	Standard deviation	Trend line slope
	[dm^3/(resident·day)]	[dm^3/(resident·day)]	[–]
1	110.22	12.02	−0.0217

For the building "***k***" the mean error is calculated:

$$\bar{S}_{\bar{q}_{jk}} = \frac{S_{\bar{q}_{jk}}}{\sqrt{m_k}} = \sqrt{\frac{\sum_{i=1}^{m_k}\left(q_{j_i} - \bar{q}_{m_k}\right)^2}{m_k\left(m_k - 1\right)}} \tag{28.10}$$

By transforming the (28.10) it was possible to determine the sum of squares of deviations for water consumption per residents of the building "***k***":

$$\sum_{i=1}^{m_k}\left(q_{j_i} - \bar{q}_{m_k}\right)^2 = m_k\left(m_k - 1\right)\bar{S}^2_{\bar{q}_{jk}} \tag{28.11}$$

It was possible then to determine the sum of the squares of deviations of the water consumption for the residents of each building. Using the standard deviation, for all buildings, a relationship (28.13) was obtained for the average value of the standard deviation of the water consumption per resident:

$$S_{\bar{q}_j} = \sqrt{\frac{\sum_{k=1}^{k=n} m_k\left(m_k - 1\right)\bar{S}^2_{\bar{q}_{jk}}}{\left(\sum_{k=1}^{k=n} m_k\right) - 1}} \tag{28.12}$$

Knowing the mean value of the standard deviation of the average water consumption per capita $S_{\bar{q}_j}$ allows to determine from the presented relationship (28.7) the mean value of the standard deviation for a population of any size $S_{\bar{q}_j}(m)$. The formula also allows to calculate the average annual water consumption per resident of any probability for this population.

To calculate the average water consumption per resident at certain probability depending on the number of residents it was possible to use the relationship for standardized normal distribution:

$$q_j(m,p) = \bar{q}_j + z(p)S_{\bar{q}_j}(m) \tag{28.13}$$

28.5 RESEARCH RESULTS

Table 28.1 shows the average value of daily water consumption per resident, standard deviation and the value of the trend line slope in relation to number of residents.

Figure 28.2 presents the results of water consumption per resident and their rate of occurrence.

The distribution of the water consumption levels are strictly correlated to the normal distribution curve. The most frequent values of water consumption levels are the closest to the average (110.22 [dm^3/(resident·day)]).

The values of the water consumption per resident for each studied building, based on the measurements, are presented in Figure 28.3.

It can be noticed that the impact of the number of people living in the building on the value of the average water consumption per resident is practically insignificant. The trend line's slope for measured water consumptions has a rather small, negative value. This indicates that the more people are living in a building, the lower the theoretical value of water consumption per resident.

Using presented in chapter 4 method the value and standard deviation of average daily water consumption per resident in studied sample was determined. Results were presented in Table 28.2.

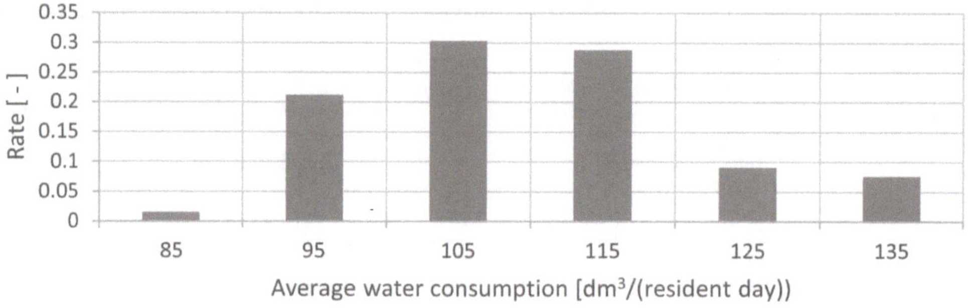

Figure 28.2. Histogram of average water consumption per resident.

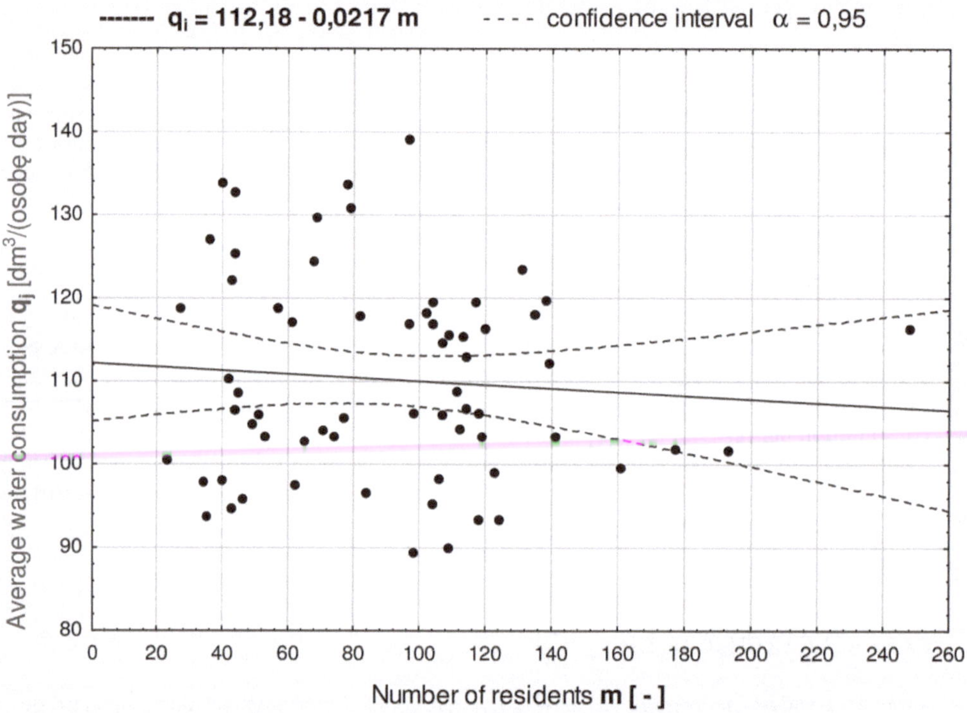

Figure 28.3. Value of the average water consumption per resident in correlation to the number of residents.

The average values determined on the basis of measurements as well as for probability of 90%, 95% and 99% in relation to the number of residents are presented in Figure 28.4.

Based on the data presented in Figure 28.4 it can be noticed, that as expected, the higher the probability the higher the average water consumption per resident. With the increase of number of residents the average water consumption per resident at assumed probability of 90%, 95% and 99% monotonically decreases. This means that for building with lower number of residents water consumption above the average will occur more of than in building with higher number of residents.

Table 28.2. Weighted average of water consumption per resident and standards deviation for the sample size.

Lp.	Weighted average of water consumption per resident	Standard deviation
	[dm^3/(resident·day)]	[dm^3/(resident·day)]
1	109.77	43.28

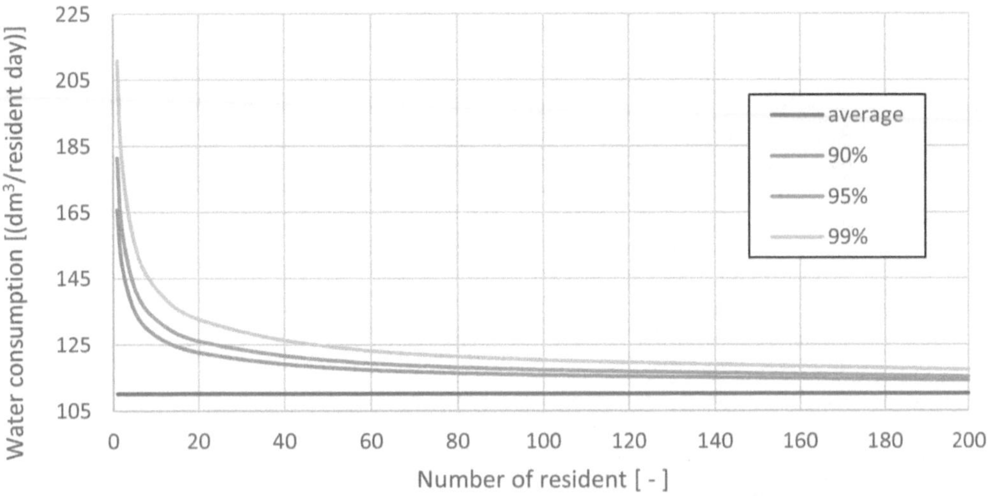

Figure 28.4. Water consumption [dm^3/(resident·day)], average value and probability (90%, 95%, 99%) in relation to number of residents.

28.6 CONCLUSIONS

The article presents a probabilistic model of average daily water consumption per resident. It was assumed that the water consumption is a random variable with a normal distribution of the expected value $\overline{q}_j$ and standard deviation $S_{\overline{q}_j}$. Those assumptions allowed to propose a methodology for determination of the standard deviation $S_{\overline{q}_j}$ based on the measurements of water consumption from multiple residential buildings. Knowing the parameters of the distribution allows to calculate the water consumption per resident with an assumed probability.

The models used typically are based on a constant value of the average water consumption per resident. The advantage of proposed model is that it allows to determine the value of water consumption based on the probability and number of residents in a building. This allows to further analyze the probability of certain levels of water consumption.

Proposed probabilistic model the average water consumption includes variance in water consumption. The model allows to determine the average water consumption with certain probability in relation to number of inhabitants.

The article present calculations of the distribution parameters based on the measurements of the average annual water consumption in 2019 by residents of sixty-five residential buildings (5874 residents) located in Szczecin, Poland were used to validate the model.

The study allowed to draw following conclusions:

- Weighted average water consumption was 109.77[dm^3/(resident·day)] for studied buildings
- Standard deviation for the average water consumption was 43.28 [dm^3/(residents·day)] in this study.

Average water consumption [dm^3/(resident·day)] almost completely does not depend on the number of residents. The trend line slope has a negative value, which implies that with the increase of the number of residents the average consumption slowly decreases.

The higher the number of resident in a building the lower the average water consumptions for particular probability level.

For a set number of residents the average water consumption with higher probability is higher than the average water consumption determined for lower probability.

NOTATIONS

f – probability density function,
F – distributor,
i – resident variable "***i***",
k – building variable "***k***",
m – number of residents,
n – number of buildings,
Q_i – annual water consumption per building [m^3/day],
q_j – average daily water consumption per resident, [dm^3/(resident/day)],
$\overline{q}_j$ – expected value of the average daily water consumption per resident, [dm^3/(resident/day)],
$S_{\overline{q}_j}$ – standard deviation of average daily water consumption per resident, [dm^3/(resident/day)],
$\overline{S}_{\overline{q}_{jk}}$ – mean error of the average result for a single building "***k***",
$z(p)$ – values determined from standard normal distribution tables for the probability p.

REFERENCES

Behling P. J., Bartilucci N. J.: Potential Impact of Water-Efficient Plumbing Fixtures on Office Water Consumption. Journal – American Water Works Association. Vol. 84, 01 October 1992.

Beniamin J. R., Cornell C. A.: Probability, Statistics and Decisionfor Civil Engineers. Mc Graw-Hill, 1970.

Ferreira T. D. V. G., Goncalves O. M. Stochastic simulation model of water demand in residential buildings. Building Services Engineering Research and Technology Volume 41, 2020, Pages 544–560.

Ghavidelfar S., Shamseldin A. Y., Melville B. W.: A Multi-Scale Analysis of Low-Rise Apartment Water Demand through Integration of Water Consumption, Land Use, and Demographic Data. Journal of the American Water Resources Association. Vol. 52, 15 July, 2016.

Grafton R. Q., Ward M. B., Hang T., Kompas T.: Determinants of residential water consumption: Evidence and analysis from a 10-country household survey. Water Resources Research. Volume 47, Issue 8, August 2011.

Edwards A., Martin L.: A Methodology for Surveying Domestic Water Consumption. Water and Environment Journal Volume 9, Issue 5, 1995.

House-Peters L., Pratt B., Chang H.: Effects of Urban Spatial Structure, Sociodemographics, and Climate on Residential Water Consumption in Hillsboro, Oregon1. Journal of the American Water Resources Associaton. Vol 46, 07 June, 2010.

Englart S., Jedlikowski A.: He influence of different water efficiency ratings of taps and mixers on energy and water consumption in buildings. *SN Applied Sciences*. 1(6); Springer International Publishing, 2019.

Jaszewska M., Szaflik W.: Domestic hot and cold water consumption in households in Szczecin between 2006 – 2019. INSTAL 4/2020, p. 22–25.

Kalbusch A., Henning E., Brikalski M. P., de Luca F. V. (Konrath A. C.: Impact of coronavirus (COVID-19) spread-prevention actions on urban water consumption. Resources Conservation and Recycling. Volume: 163, 2020.

Kowalski D., Kowalska B., Skwarek M., Czuryło D.: Water demand changes in selected European countries in the year 1990–2016. INSTAL 6/2019 str. 42–44.

Longlong L., Wenzhu L., Xin L.: Prediction of Daily Water Consumption of Residential Communities Based on Correlation Analysis and BP Neural Network. Published under licence by IOP Publishing Ltd. IOP Conference Series: Earth and Environmental Science, Volume 371, Issue 3.

Matsiyevska O., Kapalo P.,Vrana J., Jacob C.: Analysis of the Water Consumption in the Apartment House – Case Study. International Scientific Conference EcoComfort and Current Issues of Civil Engineering EcoComfort 2020: Proceedings of EcoComfort 2020 pp 294–302.

Mui K. W., Wong L. T., Law L. Y.: Domestic water consumption benchmark development for Hong Kong. Building Serv. Eng. Res. Technol. 28,4 (2007) pp. 329–335.

Shamim M. A., Cheema F. A., Omer M.: The Study of Water Consumption Behavior Under the Societal, Industrial and Environmental Dynamics: A Confirmatory Analysis from the Metropolitan City of Karachi. South Asian Journal of Management Sciences Vol: 14(2)/2020: 159–186.

de Souza C. Kalbusch A. Estimation of water consumption in multifamily residential buildings. Acta Scientiarum Technology. vol 39/2017 str. 161.168.

Stec A.: Demand for Water in the Building. Sustainable Water Management in Buildings. Water Science and Technology Library, volume 90, 2019, pp 21–32.

Szaflik W.: Water consumption in residential multifamily buildings. INSTAL 10/2020. p. 1822.

Szczecin – climatic conditions. http://hikersbay.com/climate-conditions/poland/szczecin/warunki-klimatyczne-w-szczecinie.html?lang=pl#weather-temperature-months z dnia 16.06.2020

Author index